Robotics, Automation, and Control in Industrial and Service Settings

Zongwei Luo
South University of Science and Technology of China, China

A volume in the Advances in Civil and Industrial Engineering (ACIE) Book Series

ENGINEERING SCIENCE REFERENCE
An Imprint of IGI Global

Managing Director:	Lindsay Johnston
Managing Editor:	Austin DeMarco
Director of Intellectual Property & Contracts:	Jan Travers
Acquisitions Editor:	Kayla Wolfe
Production Editor:	Christina Henning
Development Editor:	Austin DeMarco
Typesetter:	Michael Brehm
Cover Design:	Jason Mull

Published in the United States of America by
Engineering Science Reference (an imprint of IGI Global)
701 E. Chocolate Avenue
Hershey PA, USA 17033
Tel: 717-533-8845
Fax: 717-533-8661
E-mail: cust@igi-global.com
Web site: http://www.igi-global.com

Library of Congress Cataloging-in-Publication Data

Robotics, automation, and control in industrial and service settings / Zongwei Luo, editor.
pages cm
Includes bibliographical references and index.
Summary: "This book meets the challenges presented by the rise of ubiquitous computing by providing a detailed discussion of best practices and future developments in the field of automation and robotics"-- Provided by publisher.
ISBN 978-1-4666-8693-9 (hardcover) -- ISBN 978-1-4666-8694-6 (ebook) 1. Automation. 2. Robotics. 3. Service industries--Information technology. I. Luo, Zongwei, 1971-
T59.5.R573 2015
629.8--dc23

2015015068

This book is published in the IGI Global book series Advances in Civil and Industrial Engineering (ACIE) (ISSN: 2326-6139; eISSN: 2326-6155)

British Cataloguing in Publication Data
A Cataloguing in Publication record for this book is available from the British Library.

For electronic access to this publication, please contact: eresources@igi-global.com.

Advances in Civil and Industrial Engineering (ACIE) Book Series

ISSN: 2326-6139
EISSN: 2326-6155

Mission

Private and public sector infrastructures begin to age, or require change in the face of developing technologies, the fields of civil and industrial engineering have become increasingly important as a method to mitigate and manage these changes. As governments and the public at large begin to grapple with climate change and growing populations, civil engineering has become more interdisciplinary and the need for publications that discuss the rapid changes and advancements in the field have become more in-demand. Additionally, private corporations and companies are facing similar changes and challenges, with the pressure for new and innovative methods being placed on those involved in industrial engineering.

The **Advances in Civil and Industrial Engineering (ACIE) Book Series** aims to present research and methodology that will provide solutions and discussions to meet such needs. The latest methodologies, applications, tools, and analysis will be published through the books included in **ACIE** in order to keep the available research in civil and industrial engineering as current and timely as possible.

Coverage

- Ergonomics
- Production Planning and Control
- Structural Engineering
- Quality Engineering
- Construction Engineering
- Engineering Economics
- Urban Engineering
- Optimization Techniques
- Hydraulic Engineering
- Transportation Engineering

IGI Global is currently accepting manuscripts for publication within this series. To submit a proposal for a volume in this series, please contact our Acquisition Editors at Acquisitions@igi-global.com or visit: http://www.igi-global.com/publish/.

The Advances in Civil and Industrial Engineering (ACIE) Book Series (ISSN 2326-6139) is published by IGI Global, 701 E. Chocolate Avenue, Hershey, PA 17033-1240, USA, www.igi-global.com. This series is composed of titles available for purchase individually; each title is edited to be contextually exclusive from any other title within the series. For pricing and ordering information please visit http://www.igi-global.com/book-series/advances-civil-industrial-engineering/73673. Postmaster: Send all address changes to above address.

Titles in this Series

For a list of additional titles in this series, please visit: www.igi-global.com

Contemporary Ethical Issues in Engineering
Satya Sundar Sethy (Indian Institute of Technology Madras, India)
Engineering Science Reference • copyright 2015 • 343pp • H/C (ISBN: 9781466681309) • US $215.00 (our price)

Emerging Issues, Challenges, and Opportunities in Urban E-Planning
Carlos Nunes Silva (University of Lisbon, Portugal)
Engineering Science Reference • copyright 2015 • 380pp • H/C (ISBN: 9781466681507) • US $205.00 (our price)

Technology and Practice in Geotechnical Engineering
Joseph B. Adeyeri (Federal University Oye-Ekiti, Nigeria)
Engineering Science Reference • copyright 2015 • 836pp • H/C (ISBN: 9781466665057) • US $225.00 (our price)

Fracture and Damage Mechanics for Structural Engineering of Frames State-of-the-Art Industrial Applications
Julio Flórez-López (University of Los Andes, Venezuela) María Eugenia Marante (Lisandro Alvarado University, Venezuela) and Ricardo Picón (Lisandro Alvarado University, Venezuela)
Engineering Science Reference • copyright 2015 • 602pp • H/C (ISBN: 9781466663794) • US $225.00 (our price)

Computer-Mediated Briefing for Architects
Alexander Koutamanis (Delft University of Technology, The Netherlands)
Engineering Science Reference • copyright 2014 • 321pp • H/C (ISBN: 9781466646476) • US $180.00 (our price)

Technologies for Urban and Spatial Planning Virtual Cities and Territories
Nuno Norte Pinto (The University of Manchester, UK) José António Tenedório (Universidade NOVA de Lisboa, Portugal) António Pais Antunes (University of Coimbra, Portugal) and Josep Roca Cladera (Technical University of Catalonia, BarcelonaTech, Spain)
Information Science Reference • copyright 2014 • 349pp • H/C (ISBN: 9781466643499) • US $200.00 (our price)

Formal Methods in Manufacturing Systems Recent Advances
Zhiwu Li (Xidian University, People's Republic of China) and Abdulrahman M. Al-Ahmari (King Saud University, Saudi Arabia)
Engineering Science Reference • copyright 2013 • 531pp • H/C (ISBN: 9781466640344) • US $195.00 (our price)

Production and Manufacturing System Management Coordination Approaches and Multi-Site Planning
Paolo Renna (University of Basilicata, Italy)
Engineering Science Reference • copyright 2013 • 377pp • H/C (ISBN: 9781466620988) • US $195.00 (our price)

IGI GLOBAL
DISSEMINATOR OF KNOWLEDGE
www.igi-global.com

701 E. Chocolate Ave., Hershey, PA 17033
Order online at www.igi-global.com or call 717-533-8845 x100
To place a standing order for titles released in this series, contact: cust@igi-global.com
Mon-Fri 8:00 am - 5:00 pm (est) or fax 24 hours a day 717-533-8661

Table of Contents

Detailed Table of Contents

The number of robots operating in human environments is increasing every day. In order to operate in such environments, the robot must be able to navigate, interact with human, pick and place different objects. This chapter presents a mobile humanoid robot that is able to localize itself, navigate to the target location, and generates the arm motion based on the specific task. The robot utilizes the Laser Range Finder, camera and compass sensor for localization and navigation. In addition, the robot generates the arm motion satisfying multiple motion criteria, simultaneously. This chapter evolves neural controllers that generate the humanoid robot arm motion in dynamic environment optimizing three different objective functions: minimum time, distance and acceleration. In a single ran of Multi-Objective Genetic Algorithm, multiple neural controllers are generate and the same neural controller can be employed to generate the robot motion for a wide range of initial and goal positions.

The interest in developing cooperative systems has increased due to the advantages they offer. Such systems can perform tasks that a single robot would be impossible to achieve. In this chapter, a summary of the cooperative robots's study, a classification of the type of grips, and path planning is presented. In addition, the properties and characteristics of the dynamic model, and the effects of torque and friction in contact tasks are shown. General considerations that should be made to analyze a cooperative system are introduced, and finally, the principle of orthogonalization, which separates the position and the force using a projection matrix which allows us to develop a control-observer scheme, is presented.

This chapter presents a new sensor-based path planner, which gives a fast local or global motion plan capable to incorporate new obstacles data. Within the first step, the safest areas in the environment are extracted by means of a Voronoi Diagram. Within the second step, the fast marching method is applied to the Voronoi extracted areas so as to get the trail. This strategy combines map-based and sensor-based designing operations to supply a reliable motion plan, whereas it operates at the frequency of the sensor. The most interesting characteristics are high speed and reliability, as the map dimensions are reduced to a virtually one-dimensional map and this map represents the safest areas within the environment.

Research on the area of sculptured surface machining optimization is currently directed towards the implementation of artificial intelligence techniques. This chapter aims at presenting a novel approach of optimizing machining strategies applied to manufacture complex part geometries. Towards this direction a new genetic-evolutionary algorithm based on the virus theory of evolution is developed as a hosted module to a commercial and widely known CAM system. The new genetic algorithm automatically evaluates pairs of candidate solutions among machining parameters for roughing and finishing operations so as to optimize their values for obtaining optimum machining programs for sculptured parts in terms of productivity and quality. This is achieved by introducing new directions of manipulating manufacturing software tools through programming and customization. The environment was tested for its efficiency and has been proven capable of providing applicable results for the machining of sculptured surfaces.

In a world in continuous evolution, the different industrial actors need to be reactive to remain competitive and to conquer new market trends. To achieve this, they are constrained to improve their way of industrial management, both at the strategic level, to adapt to technological advances and follow market trends. In this chapter, we introduce a new simulation method that makes it easy to understand the results of a given simulation. This is of crucial importance because the design stage of a manufacturing system usually implies not specialist actors. The objective of the chapter is to present the main advantages of using the virtual reality (VR) to the manufacturing processes simulation. To this end, a state of the art will compose the first part of the chapter. In the second part, we address the issue of the contribution of the VR to the industrial simulation.

Section 3
Industrial Applications of Robotics, Automation and Control

In this chapter, a systematic and practical design process and methodology is presented and applied to design a new high- technology product: a litter-collecting robot. Although considerable theoretical and practical models have been developed in product design and development, there are still limited effective models on the practical design process on a detailed level. This chapter elaborates on recent relevant research in the design methodology field and try to improve the details of product design process and apply it to a litter-collecting robot design. The detailed and practical approach demonstrated on the design of a high- tech product in this paper, can be applied effectively to the design process of industrial products.

We have surveyed polypeptides with the optimal conformations of nests which are the common anion-binding motifs comprising 8% of the amino acids which are characterized by a structural depression or a hole. Using automated bioinformatics algorithm, novel ring structure of the nest has been found. Using automated algorithm, models of polypeptides were made in-silico (computationally) and oxygen atoms are inserted along the extension of the NH groups. These sophisticated algorithms allow insertion of atoms along the NH group at the correct distance which causes extension of the group thus forming hydrogen bond. Optimal conformations of these structures are found from these customized models. This study chapter provides a demonstration of an important discovery of optimum conformations of RL and LR nests by the use of sophisticated bioinformatics automation pipeline and a unique application of automation and control in bioinformatics.

This chapter presents an approach to the strategic role of information and information technology in the shop floor control in footwear industry sector, pointing and tracking through the various stages of the production process. Discusses the importance of industries perform monitoring of production

Preface

INTRODUCTION

Robotics, as one of the representative intelligent technologies, is expected with significant demand not only to help address the labor cost rising and labor management problems, but also to provide human-like services and capabilities. Fast development in artificial intelligence and natural language processing has made robotics systems to interact with human with more friendly and natural interfaces. These user friendly and natural interfaces are especially valuable for robotics systems as expected pervasive adoption of service robotics in health-care and other human intensively-present environment would lead to humanoid robotics that are demanding more on human like interactions.

Traditional industrial robotics has less weight on Human Machine Interaction (HMI) as automation, precision, and dexterity is what is necessary. The recent industry trend of mass customization throughout whole manufacturing chains has led to attention of human intervention and relationship management in deriving agile and flexible manufacturing assembly lines. Meanwhile, robotics focused on service delivery will have to demand more on HMI performance. Typical interfaces include device interfaces like keyboards, joysticks, mouses, and touch screens, providing basic means for humans to interact with a robot. Natural language interface has become a very attractive means to allow humans to interact with a robot, thanks to fast advances of voice recognition and natural language processing technologies. Virtual reality enabled interactive technologies and motion capturing interface like MS Kinect are another means and have been becoming popular.

In enabling human like interactions, Internet of Things and Big Data Computing (IoT/Big Data) present promising ways for developing devices that sensing human and environment, and develop analytic algorithms and systems to identify and discover human motion and human affection. Cognitive service systems with human reasoning capabilities are one way towards this direction of development. While research in cognitive systems generally focuses on cognitive phenomena such as perception, attention, anticipation, planning, learning, and reasoning, it is more attractive to develop cognitive service systems based on IoT and Big Data Computing to explore new paradigms, methodologies and algorithms to develop intelligent service systems and applications.

ABOUT THIS BOOK

Vision for robotics, automation and control in industrial and service settings has been driven by fast advances in information technology (e.g. RFID, sensor, Internet of Things and Cloud). A smarter world vision supported by ubiquitous interconnection and intelligence has generated considerable interest and demand for the next generation of robotics, automation and control technologies and their applications towards enabling smart manufacturing and human centric services.

This book would provide a forum of innovative findings in advanced robotics, automation and control research and development. It aims to promote an international knowledge exchange community involving a multidisciplinary participation from researchers, practitioners, and academics with insight addressing issues in real life problems towards smarter manufacturing and human centric services. By disseminating latest developments in robotics, automation, control innovation and transformation upon current and/or emerging technology opportunities and market imperatives, this book covers both theoretical perspectives and practical approaches for smart manufacturing and human-centric service research and development.

The target audience would include multidisciplinary participants from society, industry, academia, and government. The book would be suitable as a good reference book for college students and professors.

CONTENT OF THIS BOOK

Nowadays, there's an increasing number of robots in various environments. Intelligent robots have risen in agriculture, industry and service business. It involves fields such as food manufacturing, material processing and intelligent navigation. This book mainly focuses on robotics, automation and control, and their applications in industrial and service areas. It consists of 11 chapters classified in four sections, i.e. 1) introduction on robotics and the basic theories, algorithms and designing process, 2) introduction on automation and control, including basic information technologies and methodologies, 3) the industrial applications of robotics, automation and control, such as in bioinformatics, human-computer interaction, CNC machining optimization, etc., and 4) the service applications of robotics, automation and control, including advertising and risk management, etc. .

Section I on Introduction of Robotics Includes the Following Chapters

- **Chapter 1:** Assistive Intelligent Humanoid Robot in Human Environment, Zulkifli Mohamed, Genci Capi
- **Chapter 2:** Cooperative robots, Pablo Sánchez-Sánchez and Marco Antonio Arteaga-Pérez
- **Chapter 3:** Mobile Robot Path Planning using Voronoi Diagram and Fast Marching, S. Garrido and L. Moreno

Chapter 1 introduces a mobile humanoid robot that is able to localize itself, navigate to the target location, and generates the arm motion based on the specific task. The robot utilizes the Laser Range Finder (LRF), camera and compass sensor for localization and navigation. In addition, the robot generates the arm motion satisfying multiple motion criteria, simultaneously. This chapter evolves neural controllers that generate the humanoid robot arm motion in dynamic environment optimizing three different objective functions: minimum time, minimum distance and minimum acceleration. An advantage of proposed method is that in a single ran of Multi-Objective Genetic Algorithm, multiple neural controllers are generated. The same neural controller can be employed to generate the robot motion for a wide range of initial and goal positions.

Chapter 2 introduces the basic knowledge of cooperative robots. Cooperative robots are increasingly contributing to achieve greater flexibility optimization and application. Due to control techniques have enabled robotics implemented in different applications. Gripping objects of a set of robots require minimal integration architecture sensors, due to various factors such as the cost involved in getting encoders, the space required for integrating a manipulator, more transducers and the number of inputs and outputs can be contained in the data processing cards. A robot is a re-programmable, multi-functional manipulator designed to handle materials, tools or specialized devices through programmed movements. Movements include interaction with objects and the environment. A cooperative system consists of multiple robot manipulators which aimed to hold an object. Therefore, the position of the end-effector of each robot is limited geometrically. These constraints modeling the object and cause a reduction in degrees of freedom. This is because the end-effector of each robot must maintain contact with the object. Consequently, it cannot be moved in all directions. The degrees of freedom lost becomes in force contact. Therefore, they should be included in the dynamics of each robot to form the cooperative system.

Chapter 3 presents a new sensor-based Path Planner, which gives a fast local or global motion plan capable to incorporate the new obstacle data. For navigation in complicated environments, a robot must reach a compromise between the requirementl for having efficient and optimized trajectories and also the need for reacting to sudden events. This paper presents a new sensor-based Path Planner, which gives a fast local or global motion plan capable to incorporate the new obstacle data. Within the first step the safest areas in the environment are extracted by means of a Voronoi diagram. Within the second step the fast marching methodology is applied to the Voronoi extracted areas so as to get the trail. the strategy combines map-based and sensor-based designing operations to supply a reliable motion plan, whereas it operates at the frequency of the sensor. The most interesting characteristics are speed and reliability, as the map dimensions are reduced to a virtually one-dimensional map and this map represents the safest areas within the environment for moving the robot. Additionally, the Voronoi diagram is calculated in open areas, and with all reasonably shaped obstacles, that permits to use the planned trajectory methodology in advanced environments wherever different strategies of planning based on Voronoi don't work.

Section II on Introduction of Automation and Control Includes the Following Chapters

- **Chapter 4:** Interactive and Collaborative Virus-Evolutionary CNC Machining Optimization Environment, N.A. Fountas, N.M. Vaxevanidis, C.I. Stergiou, R. Benhadj-Djilali
- **Chapter 5:** Simulation of Manufacturing Processes via Virtual Reality, Mohamed-Amine Abidi, Barbara Lyonnet, Pierre Chevaillier, Rosario Toscano, Patrick Baert

Chapter 4 focuses on a novel approach of optimizing machining strategies applied to manufacture complex part geometries. A new genetic-evolutionary algorithm based on the virus theory of evolution is developed and new directions of manipulating manufacturing software tools are introduced. Research on the area of sculptured surface machining optimization is currently directed towards the implementation of artificial intelligence techniques. This chapter aims at presenting a novel approach of optimizing machining strategies applied to manufacture complex part geometries. Towards this direction a new genetic-evolutionary algorithm based on the virus theory of evolution is developed as a hosted module to a commercial and widely known CAM system. The new genetic algorithm automatically evaluates pairs of candidate solutions among machining parameters for roughing and finishing operations so as to optimize their values for obtaining optimum machining programs for sculptured parts in terms of productivity and quality. This is achieved by introducing new directions of manipulating manufacturing software tools through programming and customization. The environment was tested for its efficiency and has been proven capable of providing applicable results for the machining of sculptured surfaces.

Chapter 5 introduces a new simulation system that makes it easy to understand the results of a given simulation. We deal with the main advantages of using the virtual reality (VR) to the manufacturing processes simulation and end up with a proposal of solution which allows to integrate VR with the simulation of production flows through a software architecture. In a world in continuous evolution, the different industrial actors need to be reactive to remain competitive and to conquer new market trends. To achieve this, they are constrained to improve their way of industrial management, both at the strategic level, to adapt to technological advances and follow market trends. At the strategic level, this leads manufacturers to update and adapt their ways of production management, improve the performance of manufacturing processes and reduce production deadlines to deal with the arrival of new products and certainly new competitors. In this chapter, we introduce a new simulation system that makes it easy to understand the results of a given simulation. This is of crucial importance because the design stage of a manufacturing system usually implies important actors that are not necessarily specialist of the mathematical concepts implied in the discrete event processes simulation. The objective of the chapter is to introduce the main advantages of using the virtual reality (VR) to the manufacturing processes simulation. To this end, a survey of the simulation of discrete event systems, also the main simulation tools and the different research works in VR that treats issues related to the industry will compose the greater part of this chapter. And we ends up with a proposal of solution which allows to integrate VR with the simulation of production flows through a software architecture.

Section III on Industrial Applications of Robotics, Automation and Control Includes the Following Chapters

- **Chapter 6:** Conceptual Process for Designing High-Technology Products: Case Study of a Litter-Collecting Robot, Arsalan Safari
- **Chapter 7:** Investigation of Optimum Conformations and Structure Analysis of RL and LR Nests using Ramachandran Plot, Sumukh Deshpande, Saikat Kumar Basu and Pooja Purohit
- **Chapter 8:** Strategic Role of Information and Information Technology in Shop Floor Control in Footwear Industry Sector, Sergio Ricardo Mazini
- **Chapter 9:** An Intuitive Teleoperation of Industrial Robots: Approach Manipulators by Using Visual Tracking Over A Distributed System, Andrea Bisson, Stefano Michieletto, Valentina Ferrari, Fabrizio Romanelli, Emanuele Menegatti

In Chapter 6, the authors study the design of a new high-technology product: a litter-collecting robot. The process includes problem definition and analysis, customer survey, market evaluation, requirement engineering, and product characteristics. Finally this approach we studied can be applied effectively to the design process of industrial products. In this study, a systematic and practical design process and methodology is applied to design a new high- technology product: a litter-collecting robot. Although considerable research has been conducted in product design and development, there is limited documented research on the practical design process on a detailed level. The design process discussed in this paper includes problem definition and analysis, customer survey, market evaluation, requirement engineering, and product characteristics. These steps are followed by product design specifications, critical factors determination, and design solution generation. At the end, the system identification matrix, main system, subsystems, outputs, inputs, and product architecture schematic of the litter-collecting robot are designed and developed. The detailed and practical approach demonstrated on the design of a high- tech product in this paper, can be applied effectively to the design process of industrial products.

In Chapter 7, the study provides a demonstration of an important discovery of optimum conformations of RL and LR nests by the use of sophisticated bioinformatics automation pipeline and a unique application of automation and control in bioinformatics. They have surveyed polypeptides with the optimal conformations of nests which are the common anion-binding motifs comprising 8% of the amino acids which are characterized by a structural formation in which the main chain NH groups of three successive residues bind an anionic atom or group forming a depression or a hole. Using automated bioinformatics algorithm, novel ring structure of the nest has been found. The anion-binding site is characterized by alternating residues of αR or αL main-chain dihedral angles. Using automated algorithm, models of polypeptides were made in-silico (computationally) and oxygen atoms as if hydrogen-bonded are inserted along the extension of the NH groups. These sophisticated algorithms allow insertion of atoms along the NH group at the correct distance which causes extension of the group thus forming hydrogen bond. Optimal conformations of these structures are found from these customized models when one oxygen atom bridges two NH groups by forming an extra hydrogen bond with the next but one residue. This chapter provides a demonstration of an important discovery of optimum conformations of RL and LR nests by the use of sophisticated bioinformatics automation pipeline and a unique application of automation and control in bioinformatics.

Chapter 8 presents an approach to the strategic role of information and information technology in the shop floor control in footwear industry sector, pointing and tracking through the various stages of the production process. It discusses the importance of industries perform monitoring of production processes, with the goal of identifying information needs, actions and solutions that will contribute to the improvement and efficiency of the production process. The chapter also discusses the contribution of information technology to the information systems of companies, through the resources and solutions available today, such as Enterprise Resource Planning - ERP, Manufacturing Resource Planning - MRP and Shop Floor Control - SFC. The research method is the case study conducted in firm located in an industrial Brazilian footwear. This study examines the use of a solution called GradeSFC tracking and pointing of the production process.

Teleoperation of manipulator robots with RGB-D sensors is now mainly done using inverse kinematics techniques. In Chapter 9, instead, authors described an intuitive way to teleoperate an industrial manipulator robot using a vision sensor, in order to control directly the manipulator joints by retargeting specific human motion. In this way the human operator has the full control of robot movements with practically no training, because of the intuitivity of this teleoperation method. The remapping into the robot joints is done by computing angles between vectors built from positions of human joints, tracked by the used vision sensor. The system developed for this work uses a Comau manipulator robot and a Microsoft Kinect as vision sensor for the hardware part, and the Robot Operating System (ROS) framework for the software part in order to fulfill the teleoperation task. The system obtained is very modular which allows to change either the tracking sensor or the robot model with some small changes. Finally, the developed teleoperation system has been successfully tested on two real Comau robots, revealing to be fast and strongly reliable.

Section IV on Service Applications of Robotics, Automation and Control Includes the Following Chapters

- **Chapter 10:** A Gamification Mechanism for Advertising in Mobile Cloud, Zongwei Luo, Qixing Zhuang, Tao Jiang, Yang Liu, Feng Yi
- **Chapter 11:** Robotic Transformation and Its Business Applications in Food Industry, Anas Mathath and Yudi Fernando

In Chapter 10, authors propose a gamification mechanism for supporting advertising in mobile devices. Gamification for advertising uses game thinking and mechanism in non-game contexts to engage users in developing advertising content and delivering it into mobile devices. To support this gamification advertising mechanism, we develop a cloud based service platform for media integration and distribution, supporting flexible interactions and collaboration among media content providers, advertisers, and developers. Media content providers supply the advertising resources to the cloud. Advertisers provide requirements to customize the media contents for advertising. Developers will offer systems and tools to assemble advertising resources and integrate them with game content. Contribution of this chapter is it introduces game theory and mechanism design into gamification for advertising which is demonstrated as feasible and just in time. And the gamifiation for advertising is the first in the literature ever discussed as presented in the context of mechanism design. A layering solution with introduction of an advertising layer for developing gamified applications for mobile devices is also the first ever in the literature as we know.

Chapter 11 discusses the robotic transformation and its business applications in food industry. The role of robots is becoming substantial for industrial applications and business competitiveness. The robot transformation in food industry has increased business productivity, reduced cost and enhanced customer experiences. The usage scale of robots has an increasing trend globally when industries modernize and increase the production capacities with ability in handling complex tasks. In food industry robots should fulfill the basic requirements like hygiene and ease of programming. Robots serve mainly in production systems for material handling and packaging operations. Although robotics provides a better interface to raise productivity, small scale food companies are often reluctant to invest in robotics and infrastructure. High initial investment and maintenance costs are the obstacles. They also need to spent additional costs to employed skilled employees for its programming. Besides that the challenges of robot transformation are being faced by the large scale companies as well. There are two points raised in the discussion, would the robot technology which has been developed only capable owned by large scale food companies and the experiences gained in the restaurant which serves by robots can replace the human touch. At the end of this chapter, some solutions are given to shed light on the application of robot in food industry and deepen critical analysis for researchers, technocrats and business practitioners.

Zongwei Luo
South University of Science and Technology of China, China

Acknowledgment

The editor would like to express his gratitude to a number of people who have contributed to the completion of this book in various ways and to thank them all for their assistance and encouragement.

First, we wish to thank all of the authors for their excellent contributions to this book. All of you also served as reviewers for manuscripts written by other authors. Thank you all for your contributions and your constructive reviews.

Second, we wish to thank all Editorial Advisory Board (EAB) members. Many of you have delivered responsive and valuable reviews.

In addition, we are thankful for the support of the startup fund of SUSTC (Y01236115/Y01236215).

Finally, we wish to thank the staff of IGI Global for their help and guidance!

Section 1
Introduction of Robotics

Chapter 1
Assistive Intelligent Humanoid Robot in Human Environment

Zulkifli Mohamed
Universiti Teknologi MARA, Malaysia

Genci Capi
University of Toyama, Japan

ABSTRACT

The number of robots operating in human environments is increasing every day. In order to operate in such environments, the robot must be able to navigate, interact with human, pick and place different objects. This chapter presents a mobile humanoid robot that is able to localize itself, navigate to the target location, and generates the arm motion based on the specific task. The robot utilizes the Laser Range Finder, camera and compass sensor for localization and navigation. In addition, the robot generates the arm motion satisfying multiple motion criteria, simultaneously. This chapter evolves neural controllers that generate the humanoid robot arm motion in dynamic environment optimizing three different objective functions: minimum time, distance and acceleration. In a single ran of Multi-Objective Genetic Algorithm, multiple neural controllers are generate and the same neural controller can be employed to generate the robot motion for a wide range of initial and goal positions.

INTRODUCTION

The need for assistive robotics is expected to grow due to the large number of elderly people and shortage of qualified staff, especially in developed countries. The assistive robots are divided mainly in two categories: 1) Rehabilitation robots and 2) Assistive social robots. The second one is further divided in 2 subcategories: a) Service robots and b) Companion robots.

This chapter focuses on service robots which are mainly used to assist elderly or unable people in everyday life activities. The authors present a humanoid robot mobile platform that is able to self localizes, navigate to the target location, and perform picking and placing tasks using both arms. In performing domestics task such as cleaning, picking and carrying food or household item from one room to another, a mobile humanoid robot are required to have high mobility, and ability to manipulate object in opti-

DOI: 10.4018/978-1-4666-8693-9.ch001

mized manner. There are numbers of constraints and consideration need to be made in order to perform these tasks such as unknown environment, obstacles avoidance, object recognition and manipulation, the execution time, safety features and understanding human instruction.

In order to perform a navigation task such as getting a bottle of juice from the kitchen or guiding humans to the desired location, numbers of considerations need to be made such as understanding the command, determining the environment, obstacles avoidance, object recognition and manipulation and human safety. Chen & Kemp (2011; 2010) has proposed a direct contact between human and robot to guide the human from one location to another, for example leading a child by the hand or assisting nurses in hospitals. A direct physical interface is introduced to enables a user to influence the robot behavior by making contact to the robot body. Testing has been done in a real hospital with volunteered nurses to test the performance of the robot. In their work, a combination of MEKA Robotic arms, an Omni directional Segway and a linear actuator by Festo are used as complete robotic system. The robot wrist is equipped with six axis force sensor to sense the input from human, while holding their hand. A nurse robot, PEARL has been presented in (Pollack et al., 2002) for assisting the elderly. Two main purpose of PEARL is to remind people about routine activities such eating, drinking, bathing and taking medicine and the other function is to guide them through their environment.

In (Stuckler et al., 2009), a personal robot for helping household chores has been presented. The robot called "Dynamaid" adapted four schemes for navigation namely Fast Simultaneously Localization and Mapping (FastSLAM), localization, path planning and safe local navigation. Holz et al. (Holz, Nieuwenhuisen, & Droeschel, 2013), has proposed global-to-local control strategy to navigate the developed robot, Cosero from the transport box to the processing place for bin picking. Adaptive Monte Carlo Localization is used to estimate the robot's pose in a given grid map using LRF. A* search is applied to find the shortest obstacles free path from the estimate position to the target location. For the arm manipulation, LBKPIECE is utilized in order to maximized the performance and reduce the execution time.

During everyday task performance, humans move their arms in different ways satisfying different constraints. For example to move a cup of coffee, human arm need to move such as to minimize the acceleration in order not to spill the coffee. For more complicated tasks such as drawing a straight line or pushing a non-rigid object, the kinematics constraints such as velocity and acceleration are required to be decreased resulting in a longer execution time. Such scenarios inspired researchers to adapt the similar approach in generating robot hand motion for specific tasks.

Many different methods and approaches have been proposed in the past decades on humanoid robot arm motion generation. A minimum time trajectories robot arm motion has been proposed in (Sahar & Hollerbach, 1986). In their work, a general solution is proposed to solve minimum time trajectory path which involves joint space tessellation, a dynamic time scaling and a graph search. The full dynamics of the arm movement and actuator constraints are incorporated. With these features, the arm motion while avoiding obstacle can be easily generated. Flash & Hogans (1985) had proposed minimum hand jerk criteria where the position vector of the hand is defined with respect to the Cartesian coordinate system. Differentiating the position three times will define the jerk of the hand. The arm motion from its initial to the goal position is generated by minimizing the time integral of the square magnitude of jerk. Rosenbaum et al. (1995) has proposed a similar approach of generating arm motion by minimum angle jerk. The coordinate movement of the arm and trunk using optimization criteria defined in the joint space.

Minimum torque change criterion has been introduced by Uno et al. (1989), where control objects are the joint links plan in an intrinsic dynamic-mechanical space. The hand trajectory properties are reproduced based on the arm dynamics, posture, external forces and motion duration. An improve version of minimum torque change has been proposed in (Uno, Y. Kawato, M. Suzuki, 1989). In minimum commanded torque change, the incorrect values of the inertia and viscosity as in minimum torque change have been improved. The same approach has been proposed by (Nakano et al., 1999) using representation of motor commands controlling the muscles. Kawato et al. (1990) utilized minimum torque change to produce a multi joint arm motion while avoiding obstacles and passing through points. Wada et al. (2001) has compared the performance of all four optimal theories namely minimum hand jerk, minimum angle jerk, minimum torque change and minimum commanded torque change. In his study, the minimum commanded torque change show the optimum results and has the closest trajectories to human.

In recent years, there are numbers of arm motion generation technique, Vahrenkamp et al. (2008) suggested Rapid-Exploring Random Trees (RRTs) which can adapt the number of active degree of freedom used in robot motion thus improving the performance and quality of the trajectories. The numbers of degree of freedom used are optimized using RRT. In an eight degree of freedom robot, RRT determines the optimum number of joints needed to complete the task. An improved version of RRT, Rapidly Exploring Dense Tree (RDT) has been proposed in (Vahrenkamp, Kaiser, Asfour, & Dillmann, 2011), with addition of automatic adjustment collision detection system. Ang proposed a minimum time motion planning of robot arm using Pareto based multi-objective Bees Algorithm for a SCARA robot. Four different operators are used to optimize the cubic splines trajectories thus minimizing the travelling time of the robot which are discrete recombination, intermediate recombination, line recombination and path redistribution and relaxation.

Pires et al. (2007), proposed a multi objective motion generation for two and three dof planar robot manipulator optimizing two and five objective criteria. The five chosen objectives are; minimum joint traveling distance, minimum joint velocity, minimum Cartesian distance, minimum Cartesian velocity and minimum energy. The robot manipulator trajectory is minimized via GA adopting direct kinematics. In other works by Ramabalan et al. (2008), two multi-objective evolutionary algorithms (MOEA); elitist non-dominated sorting genetic algorithm (NSGA-II) and multi-objective differential evolution (MODE) are proposed to generate the motion of two robots, Cartesian and PUMA 560. Both robot end effectors are required to do pick and place operation in the workspace avoiding three obstacles. The two objective functions selected to be optimized are the travelling time and consumed energy. In order to select the Pareto optimal front, normalized weighting objective function and fuzzy membership function are used. Liu et al. (2011) proposed an improved version non-dominated differential evolution (NSDE) to generate the two and three degree of freedom (dof) planar redundant manipulator motion. Multiple objectives namely singularity avoidance, obstacles avoidance and joint limit avoidance are chosen to be optimized. Rehman et al. (2010) have proposed MOGA for generating parallel kinematics machine motion while considering three objectives optimization. The optimal path of the three dof parallel kinematics machine is generated optimizing the minimum electric energy used by the actuators, maximum torque and minimizing the shaking force.

In this chapter, the authors present application of intelligent algorithms for humanoid robot navigation and arm motion generation. The robot utilizes the sensors data such as LRF, camera and compass to navigate in the environment. In addition, a neural network based arm motion generation method is proposed. Multi-objective evolutionary algorithm is utilized to generate optimal neural controllers for the robot arm motion to perform simple domestic task. The performance of evolved neural controllers is tested on the mobile humanoid robot developed in our lab.

Mobile Humanoid Robot

In this section, the detail descriptions of the developed mobile humanoid robot are discussed in terms of its kinematics analysis, mechatronics design and software configuration. This robot is originally designed to assist elderly or handicap people in everyday life chores. The design of the mobile humanoid robot considers the safety features, ability to navigate throughout the environment, obstacles avoidance, object recognition and the ability to perform object manipulation tasks (Mohamed & Capi, 2012). The robot system is divided into two main parts, the upper body and the mobile platform as shown in Figure 1. The upper body is designed mainly for object manipulation, while the mobile platform is utilized for the robot navigation.

Upper Body

The upper body of the mobile humanoid robot consists of the head, a pair of arms and space for all the electronics parts and components. The head part has two USB cameras, which are actuated by two servo motors. All six DC motors utilized to actuate the arm joints are aligned in series and kept inside the arm and main body for higher stability. A pair of harmonic drives is attached to the shoulders for smooth motion without gear backlashes. One laser range finder (LRF1) is utilized to detect the obstacles and object to be manipulated. Two grippers are attached to both arms for simple manipulation and they are actuated by servo motors. Aluminium plates and rods are chosen for most of the robot parts due to its light weight and high strength properties. Initially, plastic material is chosen for the robot arms cover, but in the final design, combinations of two L-shape plate are chosen. These covers not only secure the electronics components inside the arm structures but also prevent the arm from twisting while in motion or holding an object. Figure 2 shows the arm design assembly and the real robot arm for right and left hand, respectively.

The key specifications the mobile humanoid robot is shown below:

- Arm length – 54 cm
- Total height – 134 cm
- Total length – 67 cm
- Robot platform width – 52 cm
- Upper body weight – 14 kg
- Mobile platform weight – 16 kg
- Maximum moving speed 1 m/s

Kinematics Analysis

The robot hand positioning is very important in order to perform a task with high accuracy and stability. Proper kinematics analysis is required to determine the current and goal position of the robot hand. The direct and inverse kinematics analysis of the robot arm is discussed in this section.

Figure 1. Mobile humanoid robot system

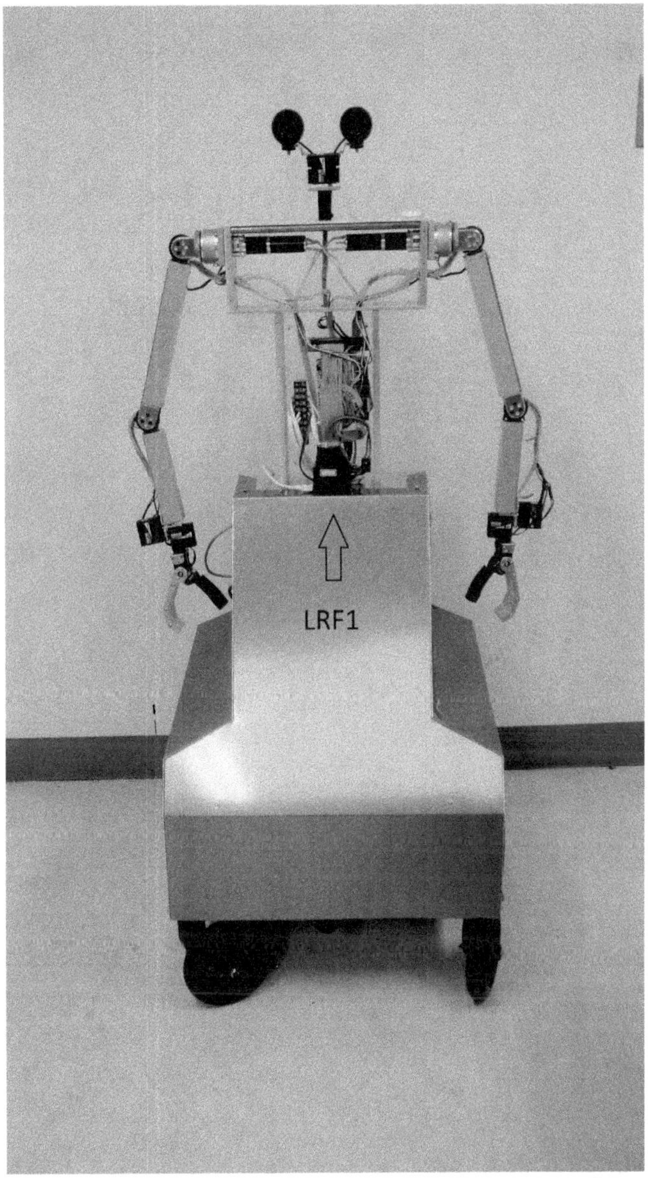

Denavit-Hartenberg (DH) Parameters

There are many methods to determine the direct kinematics of a robot arm. The most well-known method is the Denavit-Hartenberg (DH) (Asada & Slotline, 1986; Jazar, 2007; Sciavicco & Siciliano, 2001; Spong, Lewis, & Abdallah, 1993). In DH analysis, the homogeneous transformation matrix is calculated, which specifies the position and orientation of the robot hand with respect to the base as shown in Figure 3. The D-H parameters for the robot hand assigned frames are given in Table 1. The total transformation equation is as follows:

Figure 2. Left and right hand design of the mobile humanoid robot

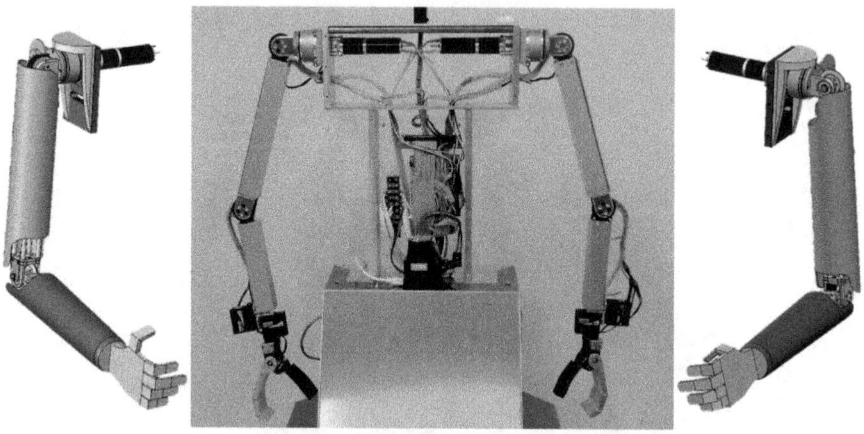

Figure 3. Coordinate frame of the robot upper body

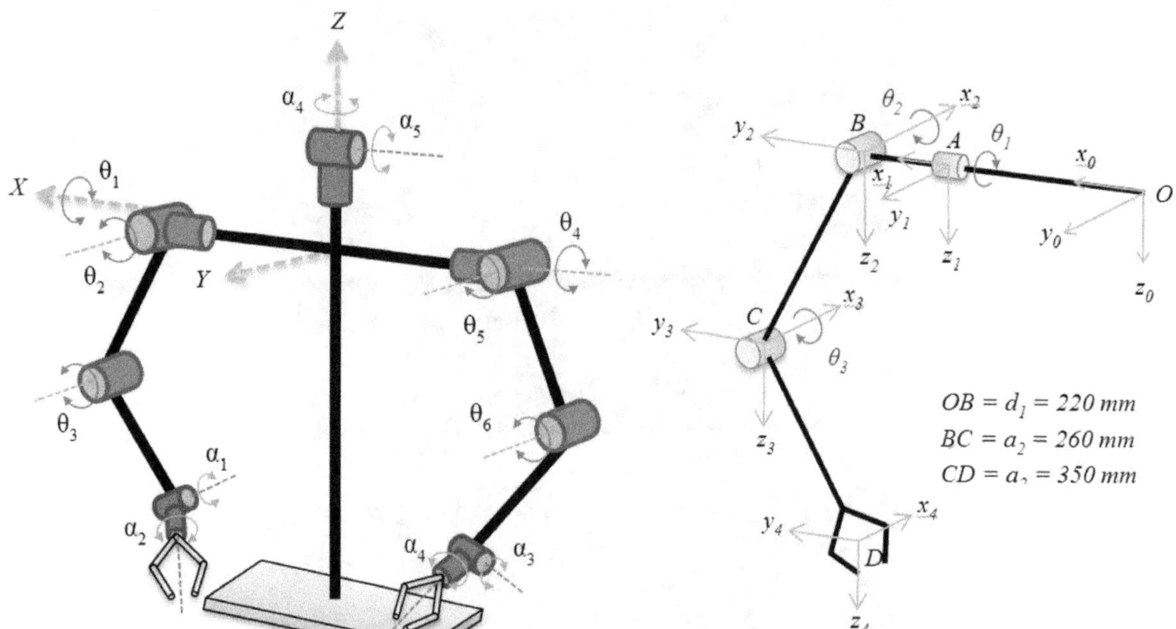

$$T_4^0 = A_1 A_2 A_3 A_3 = \begin{bmatrix} C_1 C_{23} & -C_1 S_{23} & S_1 & C_1\left(a_3 C_{23} + a_2 C_2\right) \\ S_1 C_{23} & -S_1 S_{23} & -C_1 & S_1\left(a_3 C_{23} + a_2 C_2\right) \\ S_{23} & C_{23} & 0 & a_3 S_{23} + a_2 S_2 + d_1 \\ 0 & 0 & 0 & 1 \end{bmatrix} \quad (1)$$

where C_1 is $cos\theta_1$, C_2 is $cos\theta_2$, S_1 is $sin\theta_1$, S_2 is $sin\theta_2$, C_{23} is $cos(\theta_2+\theta_3)$ and S_{23} is $sin(\theta_2+\theta_3)$.

Table 1. D-H Parameters

Joint, i	a_i	α_i	d_i	θ_i
OA	0	90^0	d_1	θ_1
B	a_1	0	0	θ_2
C	a_2	0	0	θ_3
D	0	0	0	Gripper

$$T_4^0 = A_1A_2A_3A_4 = \begin{bmatrix} n_x & o_x & a_x & d_x \\ n_y & o_y & a_y & d_y \\ n_z & o_z & a_z & d_z \\ 0 & 0 & 0 & 1 \end{bmatrix}$$

The first three columns represent the orientation of the end effector whereas the last column represents the position of the robot hand (Figure 3) as follows:

$$\begin{aligned} P_x &= d_1 + a_2 \sin\theta_2 + a_3 \sin\left(\theta_2 + \theta_3\right) \\ P_y &= \sin\theta_1 \left[a_2 \cos 0_2 + a_3 \cos\left(\theta_2 + \theta_3\right)\right] \\ P_z &= \cos\theta_1 \left[a_2 \cos\theta_2 + a_3 \cos\left(\theta_2 + \theta_3\right)\right] \end{aligned} \tag{2}$$

Inverse Kinematics

The direct kinematics analysis in the previous section established a functional relationship between the joint angle and the robot hand position and orientation. In inverse kinematics analysis, the joint angles are determined based on the given position of the robot arm. Inverse kinematics is very important while the robot has to move its hand to a specific location. In this research, a geometric approach is utilized to determine the inverse kinematics of the robot arm. Referring to Figure 4, the three joint angles of the robot arm are determined using equations (3), (4) and (5) for shoulder, upper arm and lower tight arm, respectively.

$$\theta_1 = \tan^{-1}\left(\frac{P_y}{P_z}\right) \tag{3}$$

$$\theta_3 = \cos^{-1}\left(\frac{\left(P_d^2 + \left(P_x - d_1\right)^2 - \left(a_2^2 + a_3^2\right)\right)}{2a_2a_3}\right) \tag{4}$$

Figure 4. Inverse kinematics analysis of the robot hand

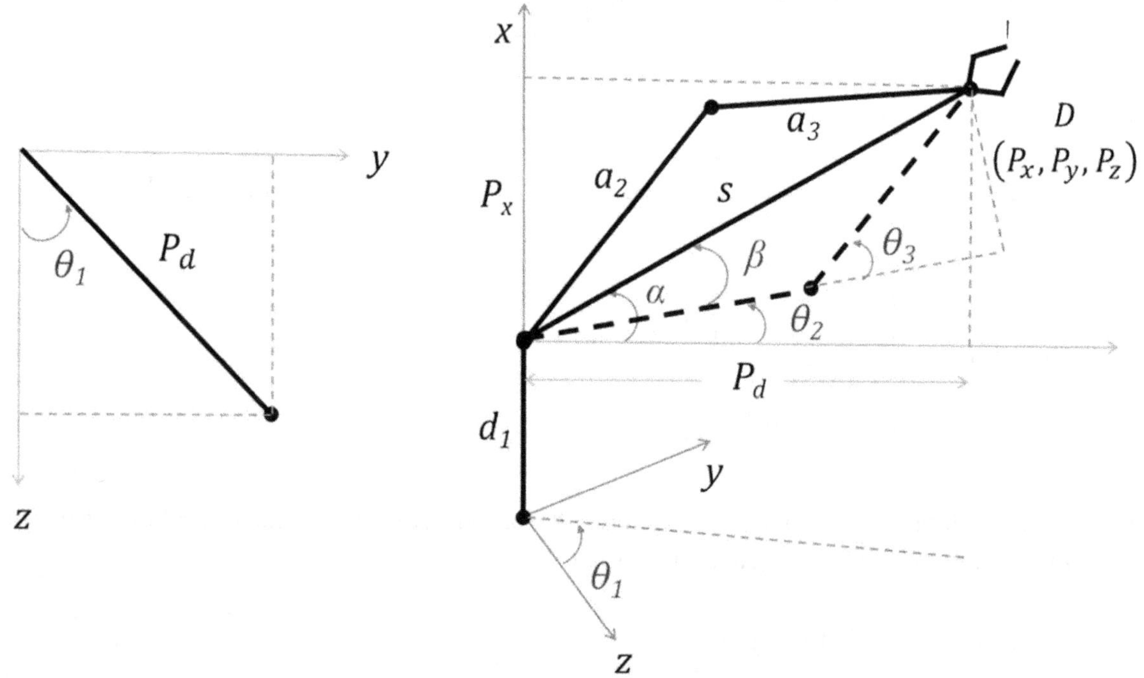

$$\theta_2 = \tan^{-1}\left(\frac{(P_x - d_1)}{P_d}\right) \pm \cos^{-1}\left(\frac{a_2 + a_3\cos\theta_3}{s}\right) \quad (5)$$

These equations are utilized in robot simulator and on the real robot. The same approach is done for the left hand with a minimum modification of the coordinate system. The detail explanation and derivation of the robot hand's inverse kinematics are discussed in (Mohamed & Capi, 2012).

Mobile Platform

The mobile platform is powered by two AC motors with 24V battery, a controller and base for the upper body. The battery, PC, AC motors, LRF2 and the upper body is placed on this platform, as shown in Figure 5. The mobile platform has a maximum speed of 1 ms^{-1} and it can be easily controlled using MATLAB based PC. LRF2 is utilized for robot navigation which can scan from 2 cm to 560 cm in distance with 240^0 range. The accuracy of the measured distance is within 3% at 100 ms/scan and the scanning resolution is 0.360. The LRF has weights 160 g and it is powered by 5V voltage supply.

EVOLUTION OF NEURAL CONTROLLERS

Arm Motion as an Optimization Problem

In performing a domestic task such as picking and placing a bottle, removing the thrash, pushing a box or avoiding obstacles, the humanoid robot arm trajectory and speed must be carefully selected in order to

Figure 5. Mobile platform

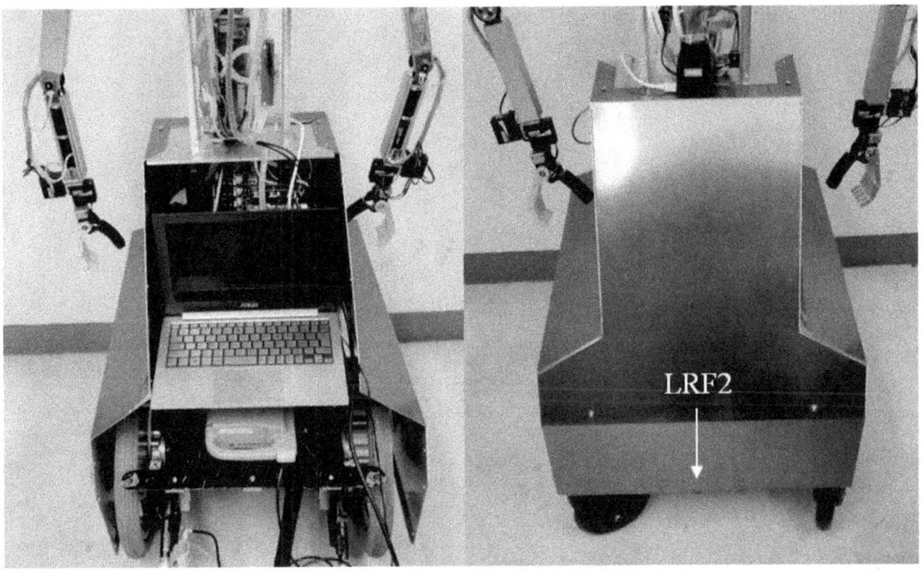

complete the task successfully. Therefore, in each stage of task performance, the main problem is what trajectory and how the moving speed must change connecting the robot hand and goal positions. The humanoid robot has to move the hand (object) from the initial to the goal position, which is connected with an infinite number of trajectories and motion velocities. The main problem is how to determine the joint trajectories in order to reach the goal position in minimum possible time or distance.

In order to find one or more feasible solutions related to one or more objective, optimization is needed to make a system more effective and functional. In this work, the authors proposed an optimal arm motion generation for the developed mobile humanoid robot arms via multi-objective genetic algorithm (MOGA) (Mohamed, Kitani, & Capi, 2013a, 2013b).

Neural Networks

A feed forward neural network (FFNN) is utilized for the arm motion generation as in Figure 6. FFNN receives three inputs: the difference between the robot hand and goal positions coordinate in x, y and z axis. The inverse kinematics, based on potentiometer readings, is utilized to determine the current position of the robot hand. In simulated environment, the goal position is pre-determined while in real situations, it is generated based on the LRF1 data and image processing. The output units directly control the three DC motors, used to move the shoulder, upper arm and lower arm. The output units use a sigmoid activation function (equation (6)) with the range from 0 to 1 where 0 to 0.5 is for one the motor moving direction and 0.5 to 1 for the opposite direction.

$$y = \frac{1}{1 + e^{-x}} \tag{6}$$

Figure 6. FFNN for robot arm

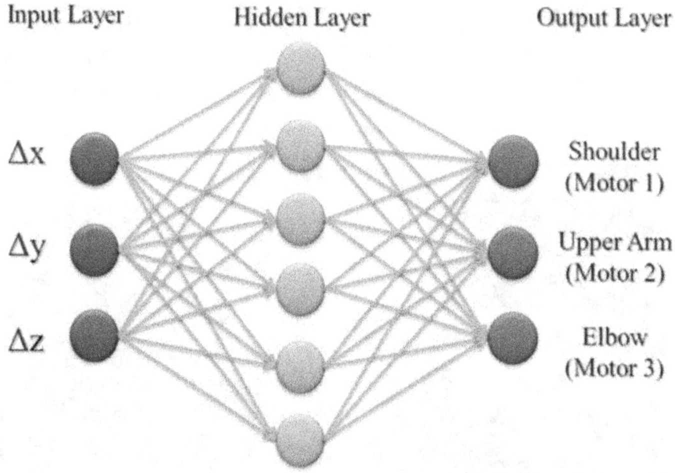

Multi-Objective GA

Multiple Objectives Evolutionary Algorithms have been proved to be well suited for optimization problems, especially when problems are too complex to solve using deterministic techniques such as Jacobian method. The main advantage is that MOEA converge in a set of optimal solutions or the Pareto optimal front (Belding, 1995; Cantu-paz, 2000; Dias & de Vasconcelos, 2002).

NSGA was employed to evolve the neural controller where the weight connections are encoded as real numbers. Dias & de Vasconcelos (2002), has compared the NSGAII with four other MOEAs using two test problems. The comparison show that NSGAII performed better than the others, where the Pareto optimal solutions can be successfully generated. Details explanations on NSGAII are discussed in (Capi, 2007).

Objective Functions

Before the optimal neural controllers can be generated, one important aspect is determining the required objective functions. In this work, three different objective functions are considered: minimum execution time (MT), minimum distance (MD) and minimum acceleration (MA) (Mohamed, Kitani, et al., 2013b; Mohamed, Mano, Kitani, & Capi, 2013).

Minimum Execution Time (MT)

Humans move their arms in many different ways depending on the task. For a simple motion of picking a light object such as picking a pen, or moving freely as waving to someone, humans tend to move in the shortest time. Based on this situation, the first chosen criterion is the minimum execution time. This objective function will minimize the time the robot moves the hand from the initial to the goal location. In our system, the sampling time to process the sensors data and send the motor command is 0.03 second. Therefore, the objective function is to minimize the number of steps, n_{step} for the robot to reach the goal position. The first objective function for the arm motion is as follows:

$$f_1 = n_{step} \tag{7}$$

Minimum Distance (MD)

Another important characteristic in human arm motion is the shortest distance. Normally human move their arms in this manner especially for a specific task, such as drawing a straight line, arranging books and pushing an object. The trajectory connecting the initial and goal positions usually it is the shortest one and this is the reason minimum distance is selected to be one of the objective functions. The minimum distance objective function is as follows:

$$f_2 = \left| \sum rt_i - sd \right| \tag{8}$$

where $\sum rt_i$ is the summation of robot hand moving distance in each time step and *sd* is the shortest distance of the robot hand from its initial position to the goal.

Minimum Acceleration (MA)

For a more complicated task such as moving a cup of tea or a non-rigid object, human will move the arm in a constant velocity and slower speed. A gradually increasing velocity in the beginning and gradually decreasing velocity toward the goal position are required to have a smooth and stable motion. This motion characteristic is chosen as one of the objective function for the robot hand motion. The total acceleration of the robot hand is minimized to have a constant velocity. Two penalty functions are also introduced, one is for the robot to have a gradually deceleration before reaching the goal position and second, minimizing the number of velocity change for a smooth motion throughout the trajectorie. Therefore, the minimum acceleration objective function is as follows:

$$f_3 = \sum a_{hand} + \left(v_{hand\_end} * w \right) + (nvc * w) \tag{9}$$

where Σa_{hand} is the summation of robot hand acceleration in each time step, v_{hand_end} is the robot hand velocity when it approaches the goal position, w is the weight function and *nvc* is number of velocity changes. The number of velocity changes is very important in order to minimize the rapid changes of the robot hand velocity in each time step. The weight function (w) is used to adjust the priority between Σa_{hand}, v_{hand_end} and *ncv*. In the first motion generation, the value of w is set to be 1, and once the value of each term is known, w can be determined. In our implementation the value of w is 60.

Robot Navigation

In the assistive tasks, the robot has to move to the goal location. In this chapter, the authors consider that the environment where the robot will operate is known. However, the environment is considered dynamic with moving and stationary obstacles. The evolved neural controllers for 3 different environments are shown in Figure 7. The robot has to navigate from the elevator into the lab room. The robot is required to navigate in a minimum distance traveled between the starting and the goal position.

A FFNN is utilized for the robot navigation with 16 inputs from LRF2, 6 hidden units and a single output for robot steering (Figure 8). Single objective genetic algorithm (SOGA) is utilized to optimize the connection weight of the FFNN.

Fitness Functions

Environment 1

In the first environment, the robot is required to move from the elevator (point 1) to the middle of the hallway (point 2) in a shortest distance while avoiding obstacles, as in Figure 7. In the simulation, the obstacles position and size are randomly selected. The fitness function utilized is as follow:

$$f_5 = \sqrt{(x_2 - x_1)^2 + (y_2 - y_1)^2} + \frac{sensor_{side}}{n_{step}} \tag{10}$$

where x_1 and y_1 is the position of the robot at point 1, x2 and y2 is the position of the robot at point 2, $sensor_{side}$ is the differences between the left and the right reading of LRF2 and n_{step} is the number of step. This fitness function will minimized the distance between the two points and make sure the robot stay in the middle of the hallway.

Figure 7. Robot navigation in simulation environments

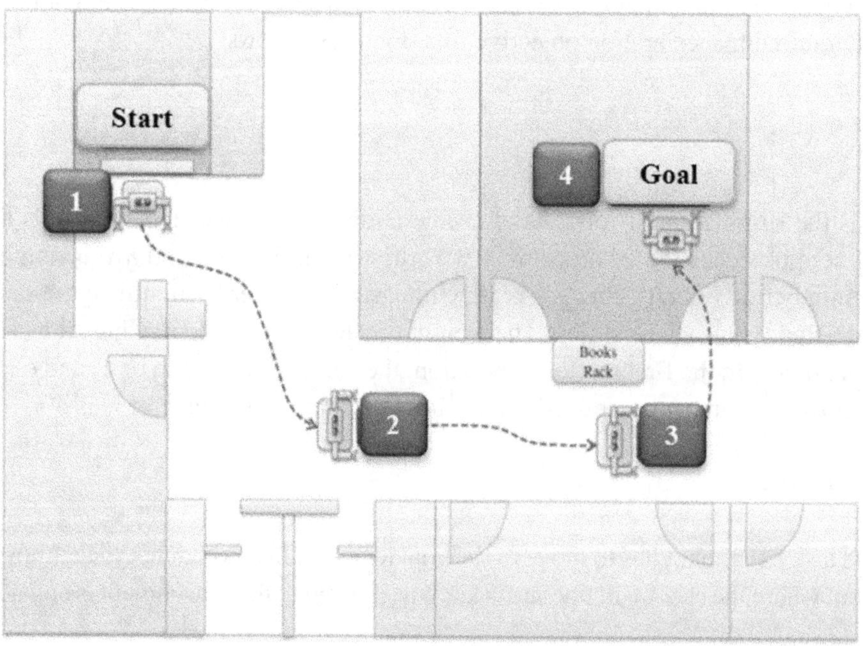

Figure 8. FFNN for mobile platform

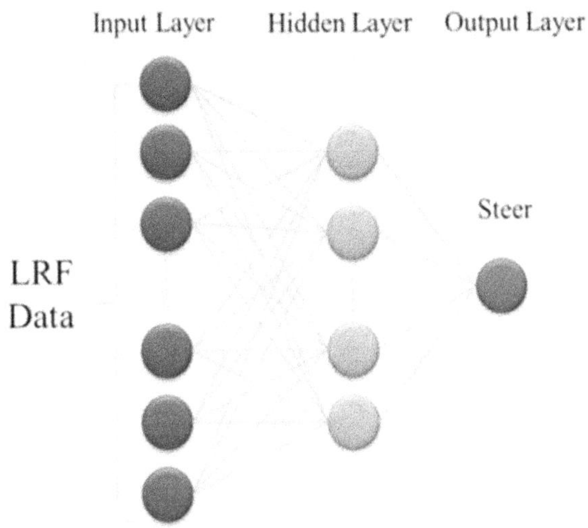

Environment 2

In environment 2, the robot has to maneuver from the starting point of the hallway (point 2) to the entrance of the goal position (point 3), as in Figure 7. The shortest distance between two points in the horizontal x direction is optimized and the sensor data is used for obstacles avoidance and guiding the robot to be in the center line of the hallway. The fitness function for environment 2 is as follows.

$$f_6 = x_3 - x_2 + \frac{sensor_{side}}{n_{step}} \quad (11)$$

where x_2 and y_2 is the position of the robot at point 2, x_3 and y_3 is the position of the robot at point 3 (the goal position entrance).

Environment 3

The robot will determine the desired entrance door to the goal position based on the landmark placed on top of the door. This is similar as in a hospital or a care center where each room will have the room number or its own specific landmarks. In our implementation, a colour landmark is utilized and it will be detected using the robot camera. The size of the colour landmark is 180 mm x 320 mm and it is place 2 m from the floor. The fitness function for the neural controllers is shown below:

$$f_7 = \sqrt{(x_4 - x_3)^2 + (y_4 - y_3)^2} \quad (12)$$

where x_3 and y_3 is the position of the robot at point 3, x_4 and y_4 is the position of the robot at point 4 (the goal position). This function is optimizing the distance between point 3 and 4 while avoiding the door and table inside the room. The door size is 900 mm and it is slightly narrow for the robot to enter with 520 mm width.

RESULTS

The performances of generated neural controllers optimizing multi objective functions are tested on three different tasks, two for the robot arm motion and one for the robot navigation. In task 1, the robot arm is required to perform a simple motion of placing a bottle on the table, as shown in Figure 9(a) and in task 2 a reaching and holding motion is chosen (Figure 9(b)). In task 3, the mobile humanoid robot is required to navigate in the lab environment from a starting point to a goal location inside the room (Figure 7).

Task 1: Placing Motion

Simulation Results

The Pareto front optimizing the three criteria minimum time (MT), minimum distance (MD) and minimum acceleration (MA), are shown in Figure 10(a). Twenty four individuals in Pareto set have been generated with maximum 80 iteration of the MOGA.

Three optimal neural controllers (NN1, NN2 and NN3) are chosen to be tested in simulated and real robot. The selection of these individuals is to compare the performance of the optimal neural controller NN2 which optimizes all 3 objective functions. While NN1 optimizes 2 objective functions (MT and MD) and NN3 optimizes only the MA objective function.

Figure 10 (b) clearly shows that NN1 solution optimizes two objective functions, minimum time (f_1) and minimum distance (f_2). However, moving with a minimum time will make the robot hand moves faster and have a higher total acceleration (f_3) which is 30% more than NN2 and NN3 solutions (Figure 10(d)).

Figure 9. (a) Task 1: Placing a bottle on the table (b) Task 2: Reaching and holding motion

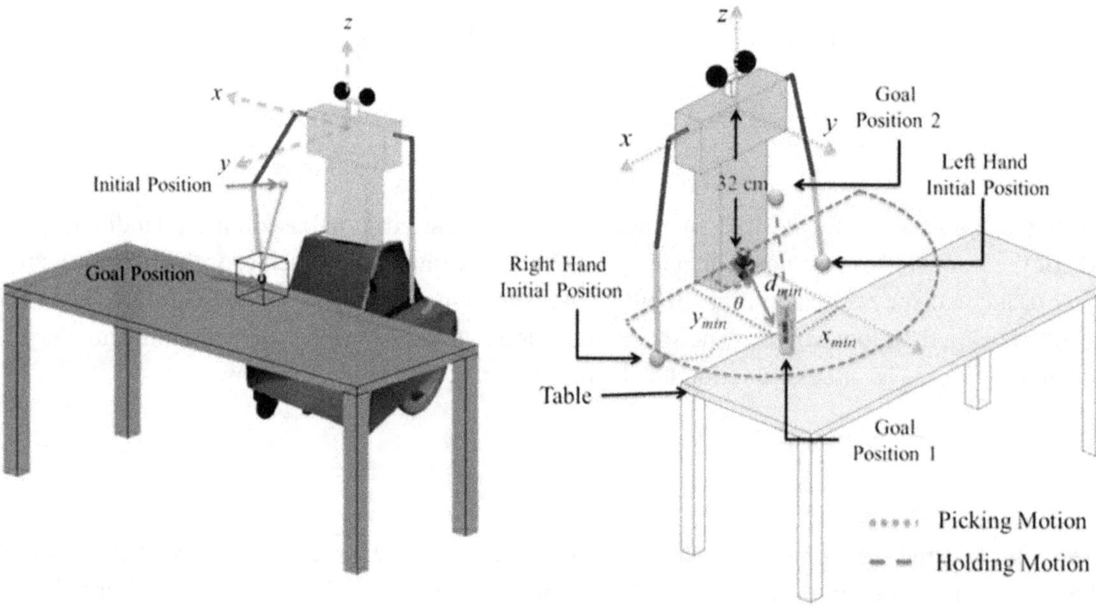

Figure 10. (a) Pareto front of MT-MD-MA objective functions optimization (b) f_2*–*f_1 *view (c)* f_3*–*f_1 *view (d)* f_3*–*f_2 *view*

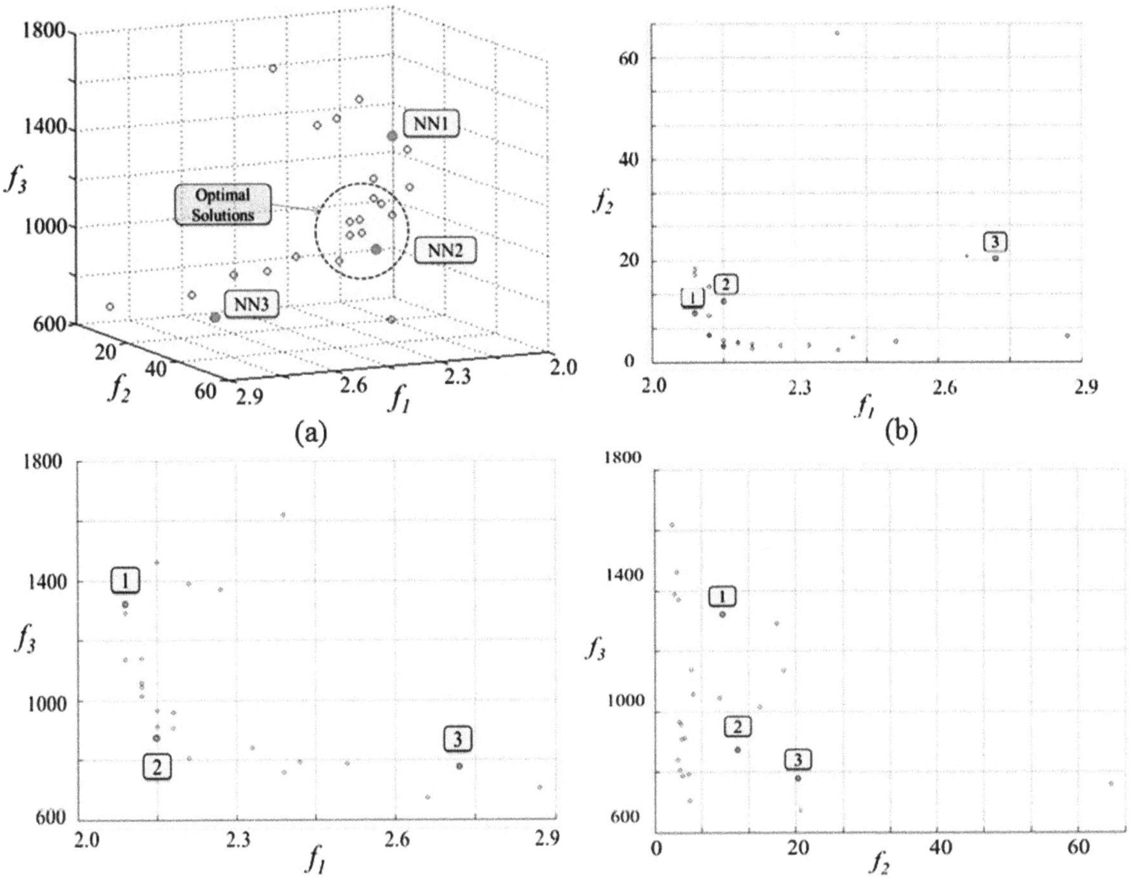

Figure 10 (c) shows NN3 solution which is similar to a single objective optimization, minimizing the total acceleration (f_3) of the robot hand motion. The time required to reach the goal is 20% longer than that of NN2 and NN1. If all three criteria has the same priority in generating the robot hand motion, NN2 performs the best.

Figure 11 shows the simulated motion generated using NN1, NN2 and NN3 solutions. The trajectories of NN1 and NN2 are closer to the shortest distance compared to NN3 trajectory.

Experimental Results

The performance of the generated neural controllers is further tested on the real robot. The video capture of the experiment with the humanoid mobile robot for NN1, NN2 and NN3 are shown in Figure 12(a), Figure 12(b) and Figure 12(c) respectively. The generated motion satisfying all three objective functions can be clearly differentiate and visualized.

Figure 11. Robot hand motion applying NN1, NN2 and NN3 neural controllers

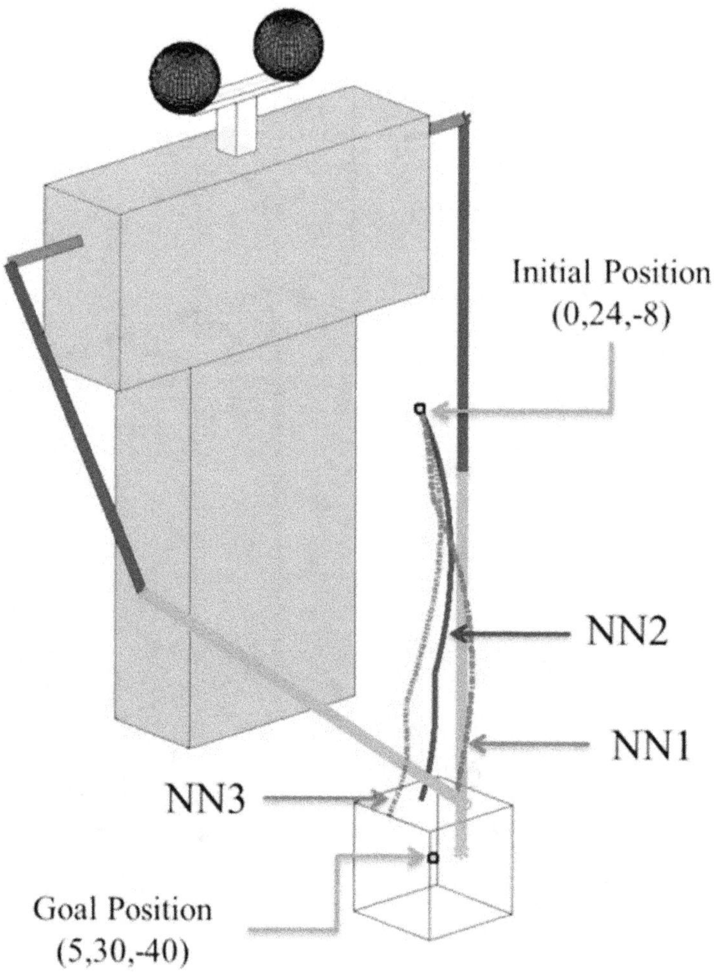

Figure 12. Video capture of robot hand motion applying (a) NN1(b) NN2 (c) NN3 neural controllers

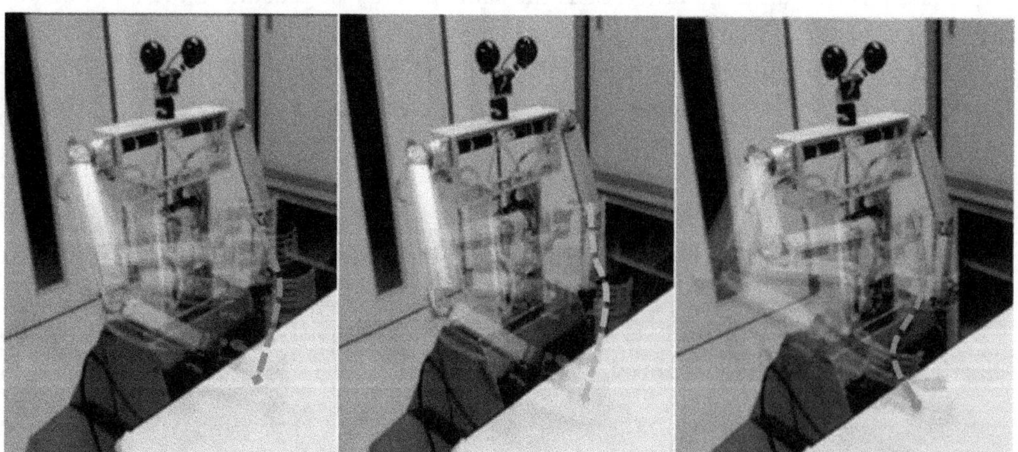

Figure 13. (a) and Figure 13(b) show comparison between the joint angular displacement for shoulder (θ_1), upper arm (θ_2) and lower arm (θ_3) of simulation and experimental results, respectively. The neural controllers show good performance in both environments. However, the results show that it is difficult to maintain a straight line trajectory.

Figure 13. Joint angular displacement comparison of the robot arm for (a) Simulation (b) Experiment

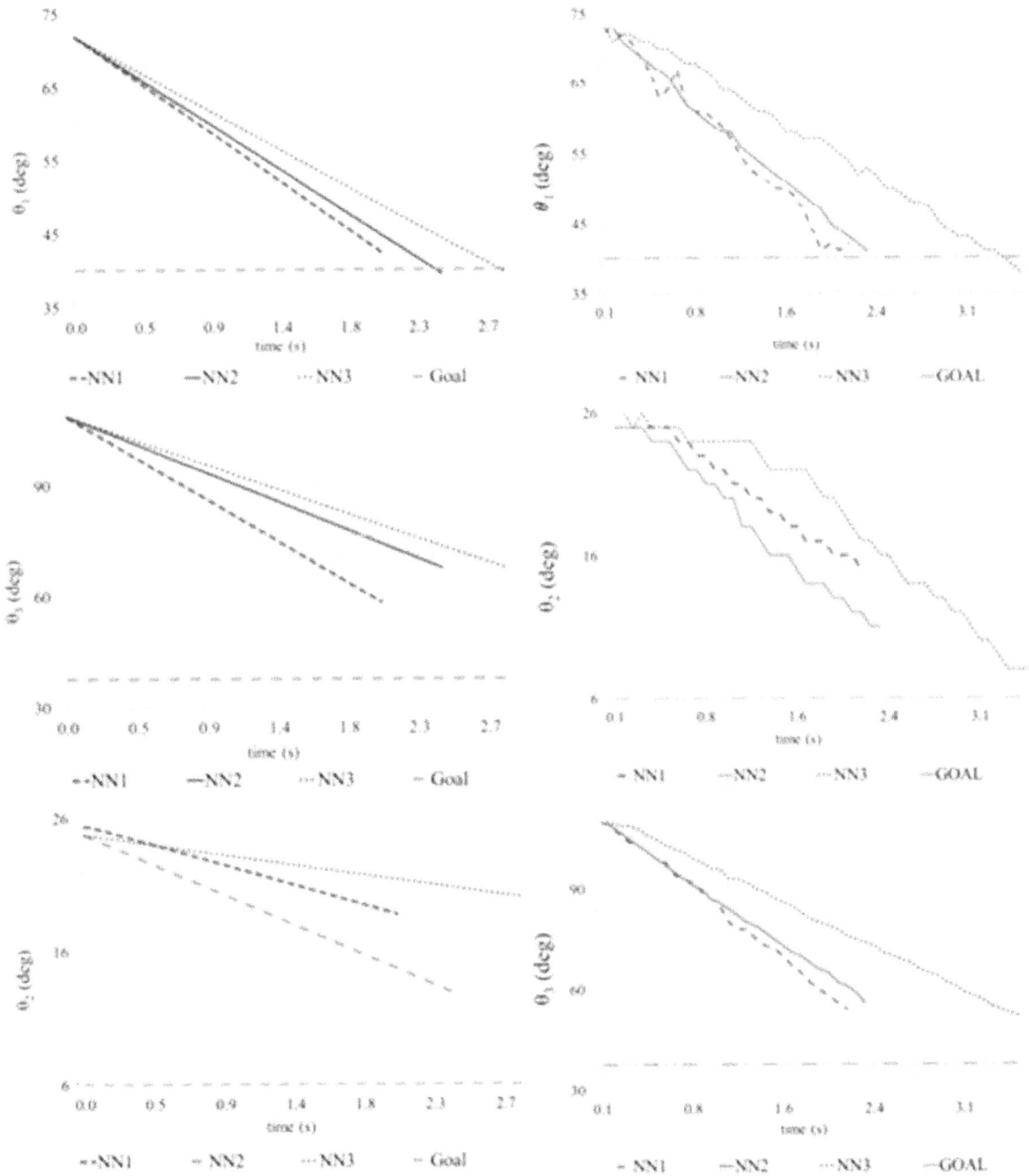

Task 2: Reaching and Holding Motion

In this task, LRF1 is utilized to determine the position of the spray can on the table in x and y directions. LRF1 is placed 32 cm from the reference position of the robot, as shown in Figure14. Once the position of the spray can is determined, the robot has to pick the spray can (goal position 1) while avoiding hitting the table (obstacle). Then move the arm to a holding position (goal position 2), as shown in Figure 9(b). The position of the spray can on the table is randomly chosen in the simulation, but experimentally it is determined using LRF1. The robot will choose left or right hand to perform the task based on the position of the can on the table.

Figure 14shows the object position determination on a standard table height by the robot utilizing LRF1. The minimum distance of the object on the table can be detected by using following equation;

$$\begin{aligned} y_{\min} &= d_{\min} \sin\theta \\ x_{\min} &= d_{\min} \cos\theta \end{aligned} \tag{13}$$

where $d_{\min}$ is the shortest distance determined using LRF1. The function of this LRF is not only for determining the object to be manipulated on the table but also for detecting the obstacle in the workspace area with the assistance of the camera.

Simulation Results

The Pareto front a single run MOGA of 80th generation optimizing all three criteria for the right (Figure 15) and left hand (Figure 16), picking and holding motions show a clear trade-off among objective functions. The Pareto front has 11 and 17 neural controllers for the right hand picking and holding motions,

Figure 14. Object detection via LRF1

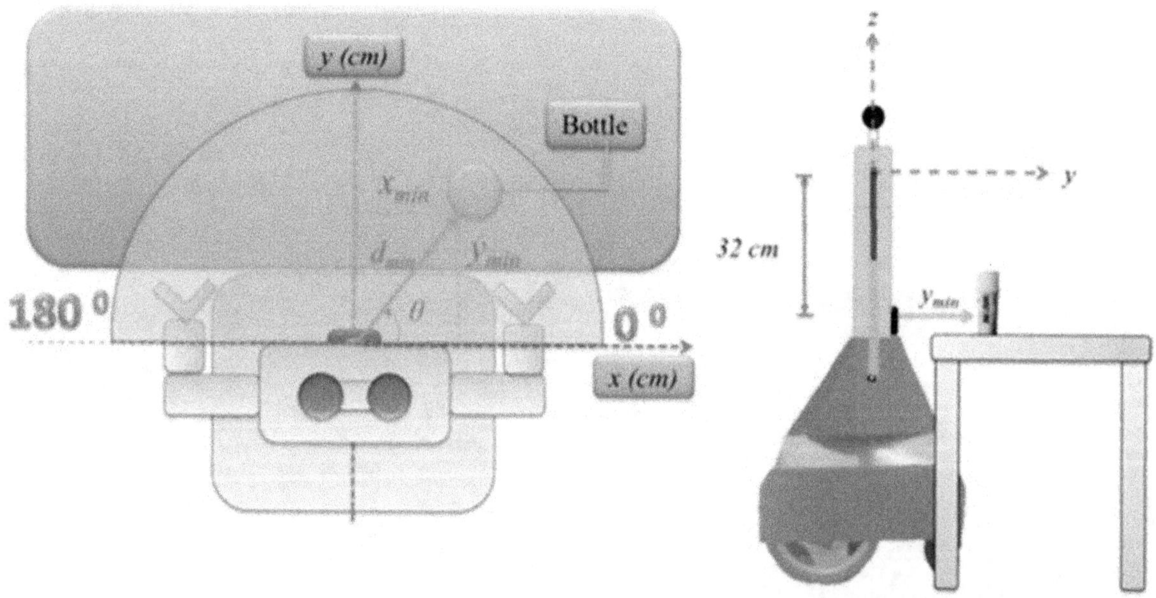

respectively. While the Pareto front of the left hand has 12 and 18 neural controllers. The MOGA did not converge in the same number of the Pareto front solutions. The motion to reach the target object is different from the motion to the holding position because the robot has to avoid the table (Mohamed, Mano, et al., 2013).

Based on the generated Pareto front, two sets of the best optimized robot arm motions have been selected. One neural controller is for picking the spray can, NC1R (Figure 15(a)) and another one for holding it up, NC2R (Figure 15(b)) using the right hand. Another set of neural controllers is for the left hand motion (NC1L and NC2L) are shown in Figure 15(c) and Figure 15(d), respectively.

Random goal positions 1 and 2 are chosen to test the performance of the evolved neural controllers. The robot has to move the hand from its initial position to grasp the spray can place on the table (goal position 1). Then move the spray can to goal position 2. The robot will use its right or left arm based on the spray can location relative to his body. For example, if the spray is on the right hand side, the robot will choose its right hand. Table 2 shows the summary of simulation environment set-ups.

Figure 15. Selected neural controllers for (a) right hand picking (b) right hand holding (c) left hand picking (b) left hand holding

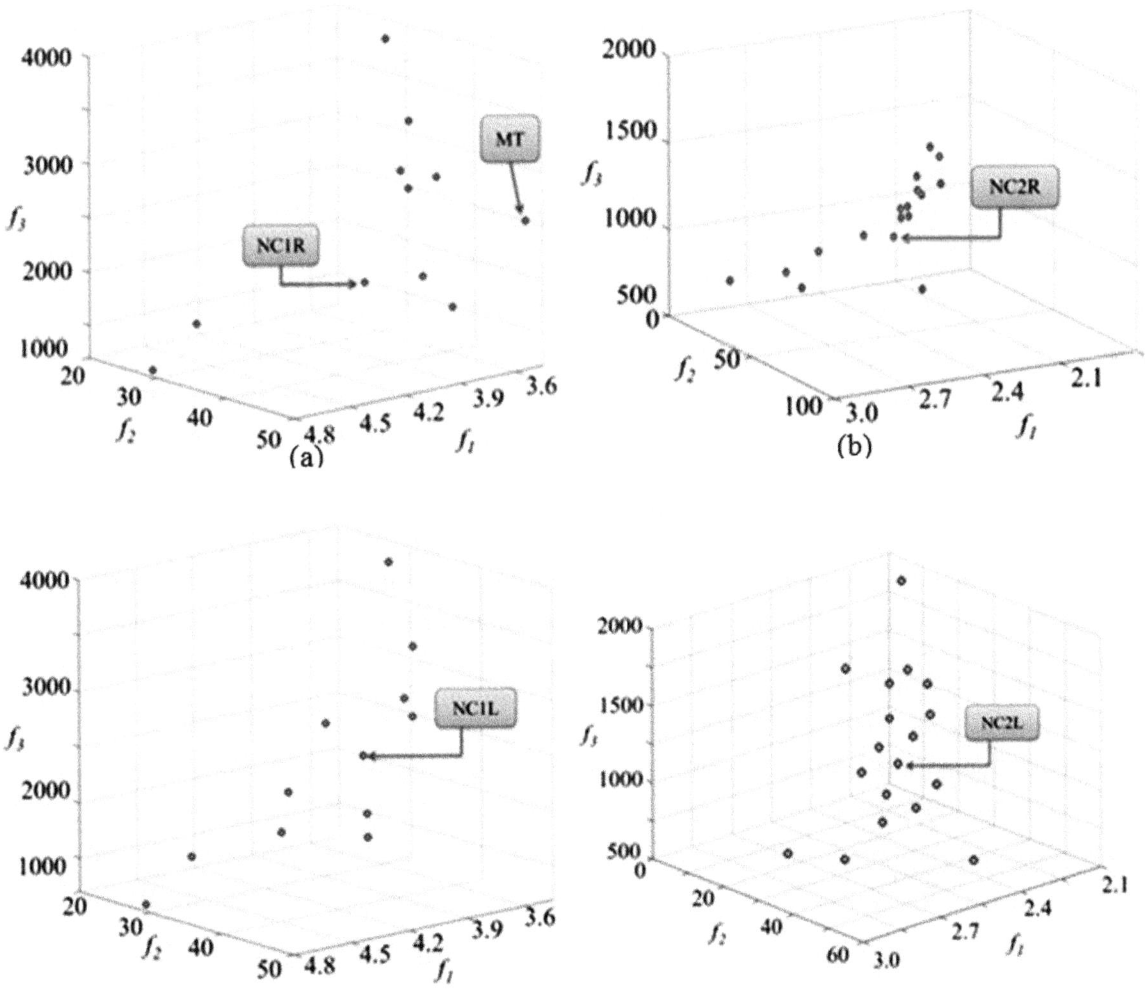

Figure 16. Simulation results for random goal position 1 and 2 for (a) position 1 (b) position 2 (c) position 3

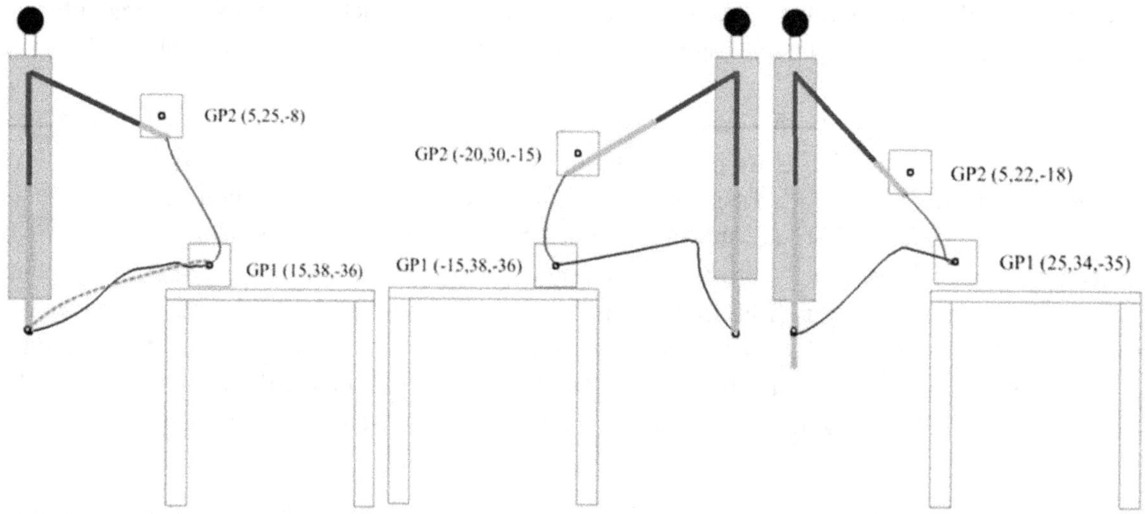

Table 2. Simulation parameters

Experimental Setup	Initial Position (Fixed)			Goal Position 1 (Random)			Goal Position 2 (Random)		
	x_{init}	y_{init}	z_{init}	x_{g1}	y_{g1}	z_{g1}	x_{g2}	y_{g2}	z_{g2}
Position 1	30	0	-48	15	38	-36	5	25	-8
Position 2	-30	0	-48	-15	38	-36	-20	30	-15
Position 3	30	0	-48	25	34	-35	5	22	-18

Three different positions of the spray can on the table have been selected. In the first simulation, the spray can is randomly placed near to the center of the table, as shown in Figure 16(a). The robot hand moves in a constant velocity toward the goal position 1. The velocity is reduced when it gets near to the goal position 1. Once the robot hand reaches the goal position 1, the current position of the robot hand is determined and it is set to be the initial position of the next motion to goal position 2.

In the next set-up, the position of spray can is on the left side of the robot, hence the left hand is chosen by the robot to perform the task (Figure 16(b)). Although the motion characteristic of the left hand compare to the right hand is not similar, the task is completed successfully. This is due to the differences in the motor data obtained from both arms.

In the third simulation, the position of the spray can is slightly further to the right, as shown in Figure 16(c). The kinematics constraints for the robot hand to reach the spray can in this motion is reduced compare to the first motion, thus, by utilizing the same neural controller, the robot hand manages to perform the task successfully.

Experimental Results

The performance of optimized motion generated is tested on the mobile humanoid robot. In the first experiment (position1), the spray can is placed on the same position used in the simulation and the right hand is chosen for this task (Figure 17(a)). The robot performance is similar with the simulated one, where the speed is reduced before the robot reaches the goal position 1 and 2. In difference from simulation, in the real hardware it is hard for the robot to follow a straight line. The reason is that, kinematically the robot arms have some difficulties if the spray can is placed in the middle of the workspace and the communication between the robot and the PC while the robot moves need to be further improved. The experiment is repeated 20 times to verify the performance of the proposed method. Results show that out of 20 times, the robot hand reached the goal position in 90% of the trials.

Figure 17. (b) shows the robot left hand motion for the second experiment (position 2). The spray can is placed on the left side of the table. Similar problem occur in this experiment, the robot hand has some difficulties maintaining the straight line motion. Out of 20 trials, 90% the robot left hand successfully reaches the goal position.

In the third experiment, the spray can is placed on the further right of the table as shown in Figure 17(c) (position 3). The performance of the robot hand is better in this experiment, where the robot motion is smoother following a straight line. The kinematics of the robot arm for this motion is less restricted

Figure 17. Experimental results for random goal positions 1 and 2 (a) Task 1 (b) Task 2 (c) Task 3

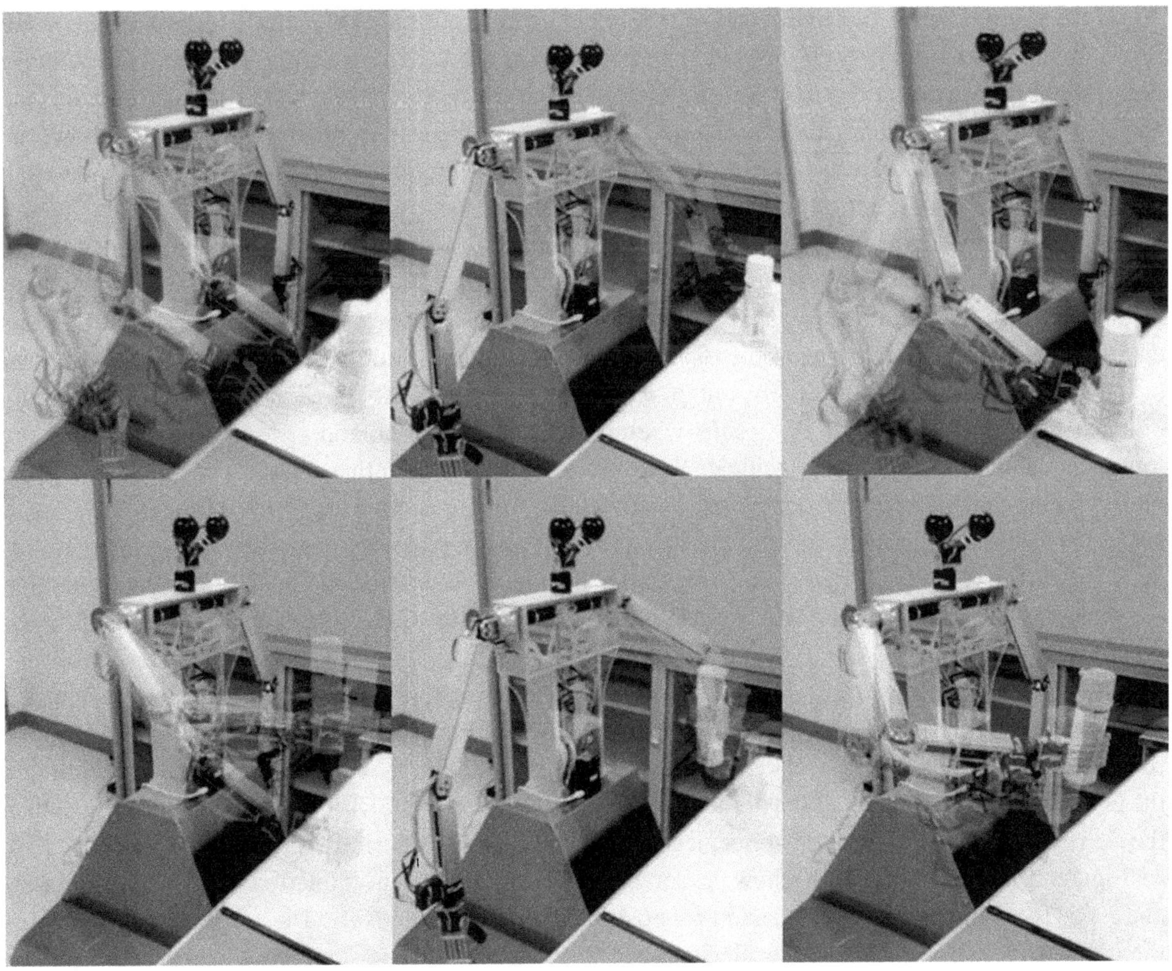

compared to the other two cases. The successful rate of the robot hand reaching the goal position is 95%. The reason that the robot did not reach the goal in 5% of the trials is related with some error in the sensory data.

Robot Navigation

The scanning angle of LRF2 is set to be from 10^0 to 170^0 with obstacle detection within 50 cm to 100 cm. If the obstacles are detected on the right area the mobile platform will steer to the left and vice versa. The obstacles detection via LRF2 can be determined using equations (13).

The performance of neural controllers generated in the simulation is tested on the real robot. The robot moves from the initial to the goal location by switching to the appropriate neural controller. In environment 1, there are some differences between simulation and real robot navigation. This is due to some glass doors which are not detected by the LRF2. Simulated result shows shorter and smoother trajectory from the starting point 1 to point 2 comparing to the real robot motion. The total distance of the real robot is slightly higher but manages to reach the target location successfully. Out of 20 trials that have been carried out, 95% the robot manage to reach at the center of the hallway. Figure 18(a) shows the video capture of the real robot motion in environment 1.

As the robot reaches the environment 2, the robot switches to the next neural controller. The result show good performance in both simulation and on the real robot (Figure 18(b)). The robot has to move in a straight hallway avoiding a book rack before it reach the third environment. The real robot has the ability to move in a similar distance and trajectory as in simulated environment. Figure 18(b) shows the robot motion along the hallway while avoiding the book rack. Out of 20 trials that have been carried out, 95% the robot manage to reach point 3 successfully.

In the third environment the robot has some difficulties to execute the task. The robot needs to enter the room through a 95cm door and stop near the table for manipulation task. Figure 18(c) shows the best result and the real robot move in a slightly in the inner section compares to the simulation result. In this environment, the successful rate of the robot reaching the goal position is reduced. The successful rate of the robot entering the room is 80% out of 20 trials.

Next, this chapter presents the results where the robot navigation and arm motion generation are integrated. The robot has first to navigate to reach the table and then the robot has to pick a spray can placed on the table. The neural controllers generated in previous section are utilized in this experiment. LRF1 is used to determine the position of the spray can on the table and once the spray can position is determined, the robot will decide the best neural controller to for executing the task. The determination of which arm to manipulate the task is depending on the position of the spray can on the table as in the previous experiment.

The simulation results of the optimized arm motion utilizing the same neural controllers as in task 2 and show a good arm motion trajectory. The video capture of the experiment with the humanoid mobile robot is shown in Figure 19(a) and Figure 19(b). Figure 19(a) shows the robot reaching the table utilizing the LRF2 and camera. The position of the spray can is determined by LRF1. The spray can is detected on the right side of the robot, thus the robot picks and hold the spray can using the right hand (Figure 19(b)).

Figure 20 (a) and Figure 20(b) show the comparison results between simulation and experiment in x, y and z axis for goal position 1 (GP1) and goal position 2 (GP2), respectively. These comparisons verified the performance of the neural controllers applied to the real robot and show good performance. Some deterioration in motion can be seen in x and z directions as the robot hand try to maintain the straight

Figure 18. Simulation and experimental results for (a) environment 1(b) environment 2 (c) environment 3

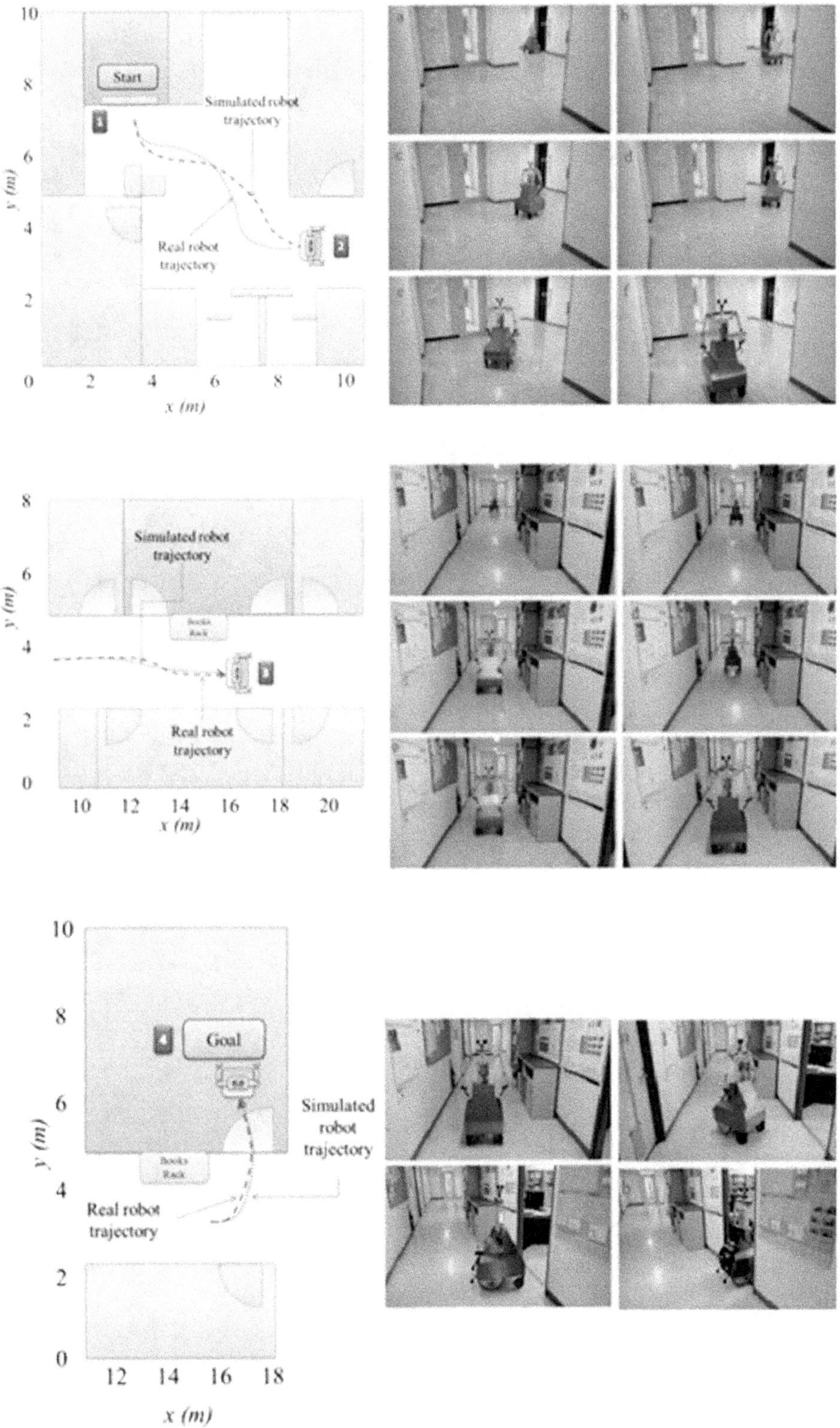

Figure 19. (a) Mobile robot approaching the table (b) Spray can picking and holding motion

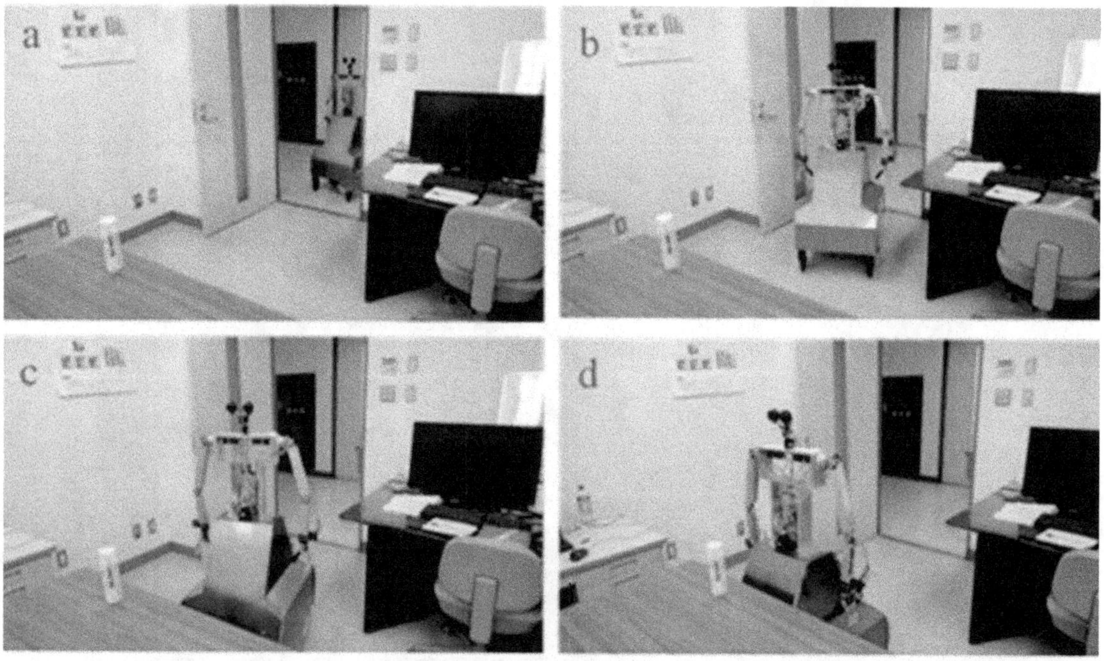

(a)

(b)

Figure 20. Comparison of the robot hand motion for x, y and z-axis (a) Goal Position 1 (b) Goal Position 2

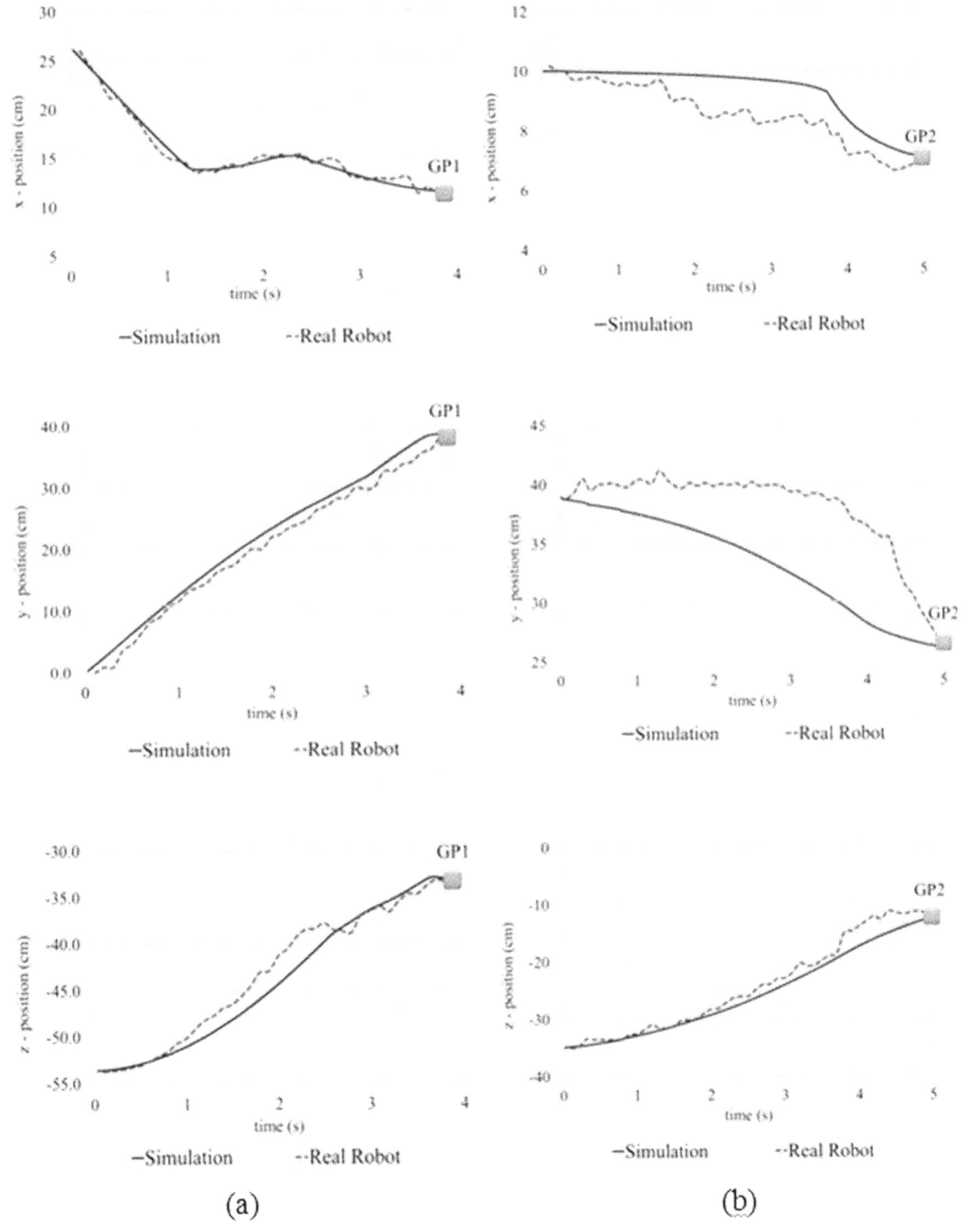

line trajectory toward the goal position when avoiding the table and the performance show similar result with the previous experiment. The robot successfully adapted the second neural controllers to complete the second task of picking the spray can. In this task, GP1 becomes the initial position for the robot hand to move to GP2. This shows that the generated neural controllers can adapt changes in initial and goal position without compromising the performance. Out of 20 trials, 90% the robot hand manage to reach the spray can successfully.

CONLCUSION AND DISCUSSION

In this chapter, the authors have presented an intelligent mobile humanoid robot for assistive tasks. The chapter is focused on robot navigation and humanoid robot arm motion generation. A new method for humanoid robot arm generation satisfying several objective functions is proposed. Three different objective functions are chosen for the robot arm motion generations which are minimum execution time, minimum distance and minimum acceleration. The advantage of the proposed algorithm is that the robot can generate the hand motion by the best neural controller based on the task it has to complete. In addition, the same neural controller can be employed to generate robot hand motion for different initial and goal positions. Our mobile humanoid robot shows good performance both in simulation and experiments. The robot was able to reach the table, select the neural controller for the robot arm motion generation to pick up the object. The robot motion is generated based on three different objective functions which are simultaneously optimized. Therefore, the humanoid robot can perform a wide range of tasks in real life environments, by selecting the appropriate motion.

For robot navigation, the authors have evolved neural networks that control the robot to reach the goal location. The robot was able to reach the goal by switching to the appropriate neural network. In addition, the navigation and robot arm motion are integrated showing a good performance. The robot reached the table and performed the spray can picking and holding task. Based on the task the specific neural controller was selected to generate the optimal motion. Results show that although there are some small differences between simulations and experiments the robot completed the task successfully.

REFERENCES

Asada, H., & Slotline, J.-J. (1986). *Robot Analysis and Control*. USA: John Wiley & Son.

Belding, T. (1995). The Distributed Genetic Algorithm Revisited. In *Proceedings of the 6th International Conference on Genetic Algorithms* (pp. 114–121).

Cantu-paz, E. (2000). Topologies, Migration Rates, and Multi-Population Parallel Genetic Algorithms. In *Proceedings of the Genetic and Evolutionary Computation Conference* (pp. 91–98).

Capi, G. (2007). Multiobjective Evolution of Neural Controllers and Task Complexity. *IEEE Transactions on Robotics*, *23*(6), 1225–1234. doi:10.1109/TRO.2007.910773

Chen, T. L., & Kemp, C. C. (2010). Lead me by the hand: Evaluation of a direct physical interface for nursing assistant robots. In *2010 5th ACM/IEEE International Conference on Human-Robot Interaction (HRI)* (pp. 367–374). IEEE. doi:10.1109/HRI.2010.5453162

Chen, T. L., & Kemp, C. C. (2011). A Direct Physical Interface for Navigation and Positioning of a Robotic Nursing Assistant. *Advanced Robotics*, *25*(5), 605–627. doi:10.1163/016918611X558243

Dias, A. H. F., & de Vasconcelos, J. (2002). Multiobjective genetic algorithms applied to solve optimization problems. *IEEE Transactions on Magnetics*, *38*(2), 1133–1136. doi:10.1109/20.996290

Flash, T., & Hogans, N. (1985). The Coordination of Arm Movements: Mathematical Model'. *The Journal of Neuroscience*, *5*(7), 1688–1703. PMID:4020415

Holz, D., Nieuwenhuisen, M., & Droeschel, D. (2013). Active Recognition and Manipulation for Mobile Robot Bin Picking.

Jazar, R. N. (2007). *Theory of Applied Robotics : Kinematics, Dynamics and Control*. New York: Springer. doi:10.1007/978-0-387-68964-7

Kawato, M., Maeda, Y., Uno, Y., & Suzuki, R. (1990). Trajectory Formation of Arm Movement by Cascade Neural Network Model Based on Minimum Torque Change. *Biological Cybernetics*, *62*(4), 275–288. doi:10.1007/BF00201442 PMID:2310782

Liu, Y., Jiang, Y., & Li, L. (2011). Multi-objective performance optimization of redundant robots using differential evolution. *Proceedings of 2011 6th International Forum on Strategic Technology*, (2), 410–414. doi:10.1109/IFOST.2011.6021052

Mohamed, Z., & Capi, G. (2012). Development of a New Mobile Humanoid Robot for Assisting Elderly People. *Procedia Engineering*, *41*(Iris), 345–351. doi:10.1016/j.proeng.2012.07.183

Mohamed, Z., Kitani, M., & Capi, G. (2013a). Adaptive Arm Motion Generation of Humanoid Robot Operating in Dynamic Environments. *Industrial Robot. International Journal (Toronto, Ont.)*.

Mohamed, Z., Kitani, M., & Capi, G. (2013b). Optimization of Robot Arm Motion in Human Environment. *International Journal of Enhanced Research Publication*, *2*(10), 1–7.

Mohamed, Z., Mano, M., Kitani, M., & Capi, G. (2013). Adaptive Humanoid Robot Arm Motion Generation by Evolved Neural Controllers. *International Journal of Control Automation and System*, 1–6.

Nakano, E., Imamizu, H., Osu, R., Uno, Y., Gomi, H., Yoshioka, T., & Kawato, M. (1999). Quantitative Examinations of Internal Representations for Arm Trajectory Planning : Minimum Commanded Torque Change Model Quantitative Examinations of Internal Representations for Arm Trajectory Planning : Minimum Commanded Torque Change Model. *Journal of Neurophysiology*, *81*, 2140–2155. PMID:10322055

Pires, E. J. S., de Moura Oliveira, P. B., & Machado, J. T. (2007). Manipulator trajectory planning using a MOEA. *Applied Soft Computing*, *7*(3), 659–667. doi:10.1016/j.asoc.2005.06.009

Pollack, M. E., Engberg, S., Matthews, J. T., Dunbar-jacob, J., Mccarthy, C. E., & Thrun, S. (2002). *Pearl : A Mobile Robotic Assistant for the Elderly*.

Ramabalan, S., Saravanan, R., & Balamurugan, C. (2008). Multi-objective dynamic optimal trajectory planning of robot manipulators in the presence of obstacles. *International Journal of Advanced Manufacturing Technology*, *41*(5-6), 580–594. doi:10.1007/s00170-008-1506-5

Rosenbaum, D. A., Loukopoulos, L. D., Meulenbroek, R. G. J., Vaughan, J., & Engelbrecht, S. E. (1995). Planning reached by evaluation stored posture.pdf. *Psychological Review*, *102*(1), 28–67. doi:10.1037/0033-295X.102.1.28 PMID:7878161

Sahar, G., & Hollerbach, J. M. (1986). Planning of Minimum - Time Trajectories for Robot Arms. *The International Journal of Robotics Research*, *5*(3), 90–100. doi:10.1177/027836498600500305

Sciavicco, L., & Siciliano, B. (2001). *Modelling and Control of Robot Manipulators*. London: Springer.

Spong, M., Lewis, F., & Abdallah, C. (1993). *Robot Control : Dynamics, Motion Planning, and Analysys*. New York: IEEE Press.

Stuckler, J., Grave, K., Klas, J., Muszynski, S., Schreiber, M., & Tischler, O., Behnke, S. (2009). Dynamaid : Towards a Personal Robot that Helps with Household Chores. In *Proceeding of RSS Workshop on Mobile Manipulation in Human Environments*.

Uno, Y., Kawato, M., & Suzuki, R. (1989). Formation and Control of Optimal Trajectory in Human Multijoint Arm Movement. *Biological Cybernetics*, *101*, 89–101. PMID:2742921

Ur-Rehman, R., Caro, S., Chablat, D., & Wenger, P. (2010). Multi-objective path placement optimization of parallel kinematics machines based on energy consumption, shaking forces and maximum actuator torques: Application to the Orthoglide. *Mechanism and Machine Theory*, *45*(8), 1125–1141. doi:10.1016/j.mechmachtheory.2010.03.008

Vahrenkamp, N., Kaiser, P., Asfour, T., & Dillmann, R. (2011). RDT+: A parameter-free algorithm for exact motion planning. *2011 IEEE International Conference on Robotics and Automation*, 715–722. doi:10.1109/ICRA.2011.5979777

Vahrenkamp, N., Scheurer, C., Asfour, T., Kuffner, J., & Str, H. (2008). Adaptive Motion Planning for Humanoid Robots. Proceedings of *2008 IEEE/RSJ International Conference on Intelligent Robots and Systems* (pp. 22–26).

Wada, Y., Kaneko, Y., Nakano, E., Osu, R., & Kawato, M. (2001). Quantitative examinations for multi joint arm trajectory planning--using a robust calculation algorithm of the minimum commanded torque change trajectory. *Neural Networks : The Official Journal of the International Neural Network Society*, *14*(4-5), 381–93. Retrieved from http://www.ncbi.nlm.nih.gov/pubmed/11411627

KEY TERMS AND DEFINITIONS

Arm Motion Generation: Generating robot arm motion from one point to another.

Artificial Neural Network (ANN): An information processing method inspired by human neuron system. It is consist of a large number of interconnected neurons to solve specific problems. ANNs, like human, it learns by example.

Genetic Algorithm (GA): Adaptive heuristic search method based on the evolutionary ideas of natural selection and genetics. As such they represent an intelligent exploitation of a random search used to solve optimization problems.

Human Environment: Our daily life environment where people and everyday life object is place such as laboratory or at home.

Mobile Humanoid Robot: A robot that consist of upper body (human like) and a mobile platform (wheel based).

Multi-Objective Evolutionary Algorithm: Optimization method utilizing NN and GA to search the best solution, optimizing more than one objective.

Robot Navigation: Robot movement from a starting position to a goal position while avoiding obstacles and mobbing object.

Chapter 2
Cooperative Robots

Pablo Sánchez-Sánchez
Universidad Autónoma de Puebla, Mexico & Universidad Nacional Autónoma de México, Mexico

Marco A. Arteaga-Pérez
Universidad Nacional Autónoma de México, Mexico

ABSTRACT

The interest in developing cooperative systems has increased due to the advantages they offer. Such systems can perform tasks that a single robot would be impossible to achieve. In this chapter, a summary of the cooperative robots's study, a classification of the type of grips, and path planning is presented. In addition, the properties and characteristics of the dynamic model, and the effects of torque and friction in contact tasks are shown. General considerations that should be made to analyze a cooperative system are introduced, and finally, the principle of orthogonalization, which separates the position and the force using a projection matrix which allows us to develop a control-observer scheme, is presented.

INTRODUCTION

The interest on developing cooperative systems has increased due to the advantages they offer against the single robot manipulators, since the cooperative systems can perform tasks which with a single robot would be impossible to achieve. In fact, in the industry many tasks are difficult or impossible to be executed by a single robot manipulator making it was necessary to use two or more manipulators in a cooperative way. Such tasks include handling heavy or large payloads, the assembly or disassembly of big or small pieces, and manipulating rigid or flexible objects.

The study of cooperative systems includes (Takase *et al.*, 1974) who implement the force/compliance control by using the back-drivability of actuators, without using force/torque's sensors. In (Nakano *et al.*, 1974) is proposed an approach of *master/slave forces control*, in order to coordinate two robot arms carrying an object cooperatively, and pointed out the necessity of force control for cooperative robots. In the work of (Fujii & Kurono, 1975) is used (Takase *et al.*, 1974) with the intention to adopt a concept of compliance control to coordinate multiple manipulators; defining a vector task respecting the object frame and controlled the fulfillment expressed in the same coordinate frame.

DOI: 10.4018/978-1-4666-8693-9.ch002

Based on the robot manipulator's theoretical results, there was a breakthrough in the 80s. Examples include the studies of (Dauchez & Zapata, 1984) concerning the definition of the task's vectors in relation with the object to be manipulated. Force control's issues, such as hybrid position/force control, were explored by (Uchiyama *et al.*, 1987). (Tarn *et al.*, 1988) proposed new nonlinear control algorithms for multiple robot arms using the dynamics and control of the closed kinematic chain formed by the multi-robot and the object.

A strong theoretical background for the control of multi-robots has been formed, providing the basis for research on more advanced topics from the 1990s to present day. Critical issues of the cooperative systems have been studied, like how to parameterize constraint forces/moments of the object based on the dynamic model of the whole system. Actually, this parameterization leads to define the task's variables, it allowing to control simultaneously the trajectory of the object, the mechanical stresses (internal forces/moments) acting on the object, load sharing among the arms, and even the external forces/moments of the object. The key for solving these problems may be the decomposition force according to (Walker *et al.*, 1991). Another important problem is developing a geometrically clear parameterization of the internal forces and momentum acting-on the object; (Williams & Khatib, 1993) have given a solution to this modeling of internal forces in multi-grasp manipulation. About the intersection of work areas, which occurs when the robots operate cooperatively, in (Chiacchio *et al.*, 1996) was proposed a scheme to regulate the workspace of two robots transporting coordinately a rigid object holding it firmly. (Caccavele *et al.*, 1999) described a cooperative system that consists of two robot manipulators proposing complementary roles performed during the execution of tasks, proposing for one of the robots, a position control and the other one a force/torque control with feedback, in order to reduce uncertainty in planning tasks.

In (Tinós *et al.*, 2002) is studied how increases dynamic load when working with cooperative manipulators, which is one of the most important advantages offered by such systems. In (Fonseca *et al.*, 2003) is developed a method of force/torque control taking into account the dynamics within the system by applying this control to a cooperative system consisting of two robot arms; this research is based-on (Pfeffer *et al.*, 1993) and (Woong & Beom, 1999).

Several cooperative control schemes based on the requested parameterization have been designed, including *motion control* and force-impedance/compliance control. Other approaches include adaptive control, kinematic control, task-space regulation, joint-space control, and coordinated control.

Generally, tasks performed by cooperative systems can be classified as: human-human, human-robot and robot-robot cooperative tasks. For a *robot-robot cooperative system*, performing a task is necessary to study the characteristics of the assignment performed by humans, which allows us to model and reproduce the interactions between *human-robot system* and *robot-robot system*. An example of cooperative tasks performed by the *human-robot system* is the medical applications, where a robotic system can assist the physician in a microsurgery which requires a fine manipulation and the surgeon's experience, Figure 1. In medical application, for safe and firm and handling a force/position control is used.

It should be noted when performing cooperative tasks between robot manipulators and human must be internal and external sensors, in order to avoid the user may suffer physical damage. Importantly, the tasks where it needed human intervention are applications where human qualities are needed, such as: skill, intelligence and judgment. The *robot-robot cooperative tasks* have a wide range of application in industry and used to transport materials that exceed the capacity of a single robot, the assembly and disassembly of products, the overall system performance, etc. being the main disadvantage of cooperative systems a considerable increase in the complexity of planning, control and execution of tasks, Figure 2.

Figure 1. Human-robot cooperative system (Instrumentalist robot)

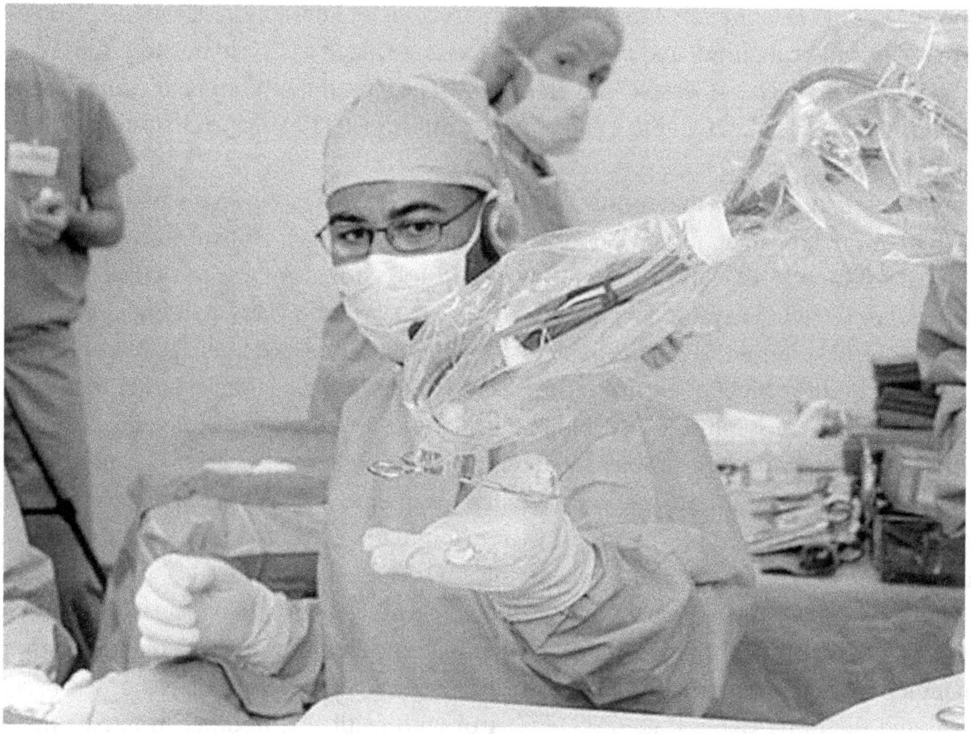

Figure 2. Robot-robot cooperative system

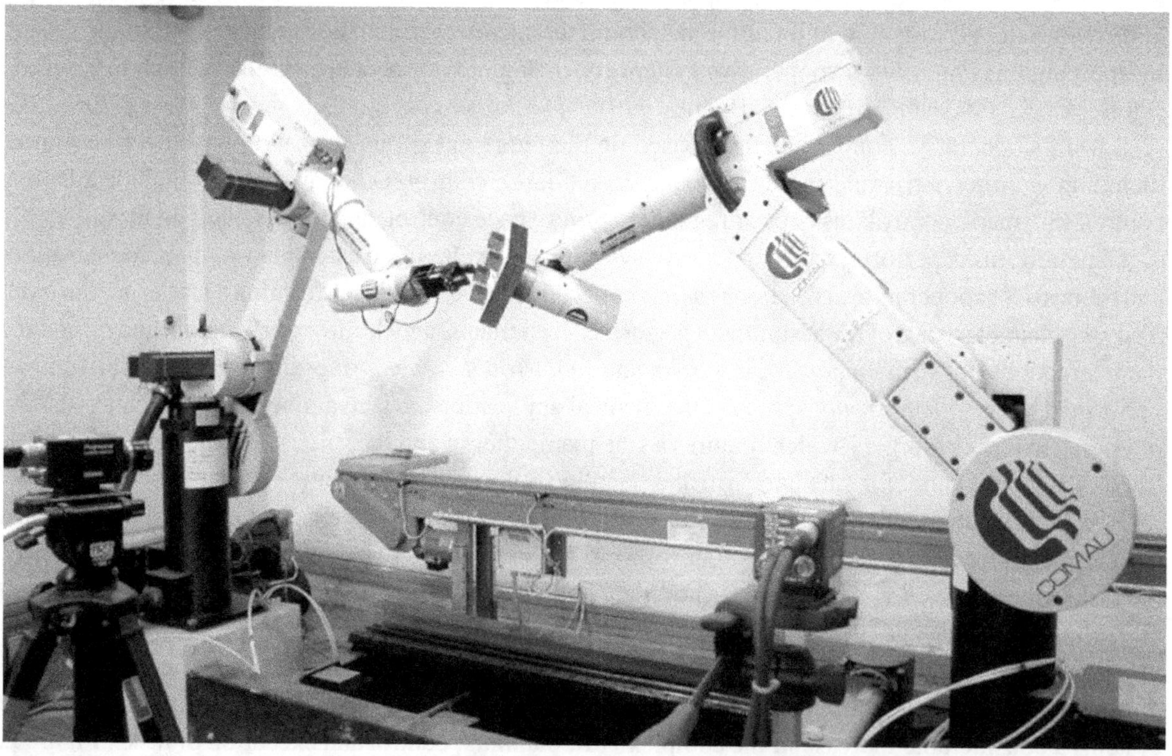

Considering that the systems where two or more manipulators work from a closed kinematic chain that creates the need to control the internal forces acting on the object. These internal forces give rise to the following classification:

- **Master-Slave**: This approach was used in the early work on the subject; this scheme is to apply position control in one of the manipulators (master) and to force control in the rest of the robots (slaves). The slave manipulators simply follow the movement and evolution of the master manipulator. This approach has a simple implementation by using simple control algorithms, but it leverages many of the benefits introduced through the use of multiple robots, such as: load sharing among robot manipulators.
- **Joint control**: In this approach, is considered the kinematic-chain as a whole, because of this for it is modeled and controlled the complex system. The main disadvantages of this method are that within the kinematic-chain manipulate the elements which have some distinct features and specifications, which makes losing generality and consistency to the method. The main advantage of this scheme is because the necessary coordination is naturally optimizing criteria.
- **Impedance control**: This scheme provides each of the robot's behavior a physical system consisting of a mass, a spring and a damper, Figure 3. In the case, no external forces are applied at the end-effector rated it continues its path, otherwise the robot modifies that path just as you are able to grab a mass subjected to these forces and linked to the nominal trajectory by a spring and

Figure 3. Nominal and real trajectories of a robot, Impedance control

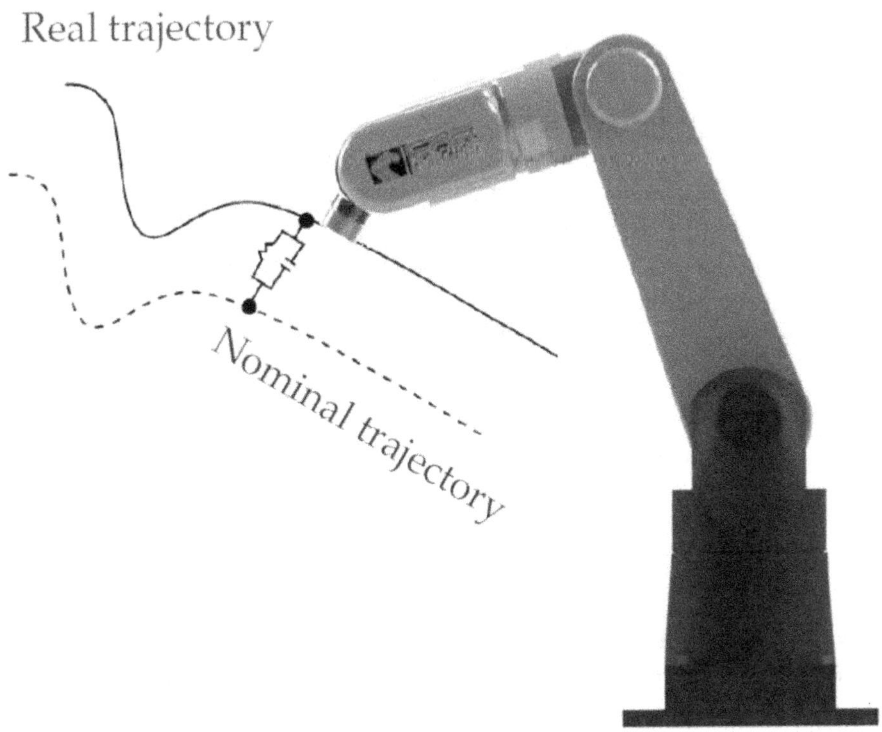

a damper. This scheme has the advantage the required system information is almost zero, so it is truly useful for unstructured systems or systems where there is very little information on its geometrical structure.

- **Force control**: This approach works directly on the manipulated object model, describing the exerted forces by the different manipulators on the body, which allows us to obtain the total force acting on the body and therefore, the trajectory it will follow.
- **Hybrid position/force control**: It consists on decomposing the force space and the position space which is intended to control. When the handling involves two or more manipulators, we seek to perform this decomposition, controlling in one side the path of the manipulated object and in another hand the internal forces acting on it.

According to the advantages of impedance control and the position/force control, these are the most used. The impedance control regulates the interaction between operators and environment, and the position/force control decomposes the forces applied to the manipulated object.

In (Szewczyk *et al.*, 1997) is presented a control strategy which consists on applying a distributed impedance control. Consequently, the robots are managed independently of each other. Each of them using its own impedance control developed separately, all that cooperatively performed are the specifications for the task, namely the position control and the forces that are applied in the manipulated object.

The strategy's advantage is applying to each robot a control system independent and autonomous. This facilitates the implementation of coordinated control, and in case the system topology changes (such as adding some device or other manipulator), not seems so affected the design of the controller. This provides flexibility and modularity by treating each system's component as a separate drive. Unlike other techniques or strategies that consider the robots and the object as a whole, leads to a complex controller design, since it must share information between manipulators and sensing the position in real-time. This leads to design a specific hardware for each work cell where the structure of the controller depends-on the topology of the system instantaneously.

Force control has the fundamental problem of decomposing the applied forces to the manipulated object; this fact produced internal forces and motion unwished. Another issue is the developing of the control algorithm. It should be noted that different approaches for the force control existed, as determining the forces' applied-on the body minimizing the norm of the torque vector, or dividing loads between the manipulators, or decreasing the forces applied taking into account the constraints of friction, among others.

In (Bonitz & Hsia, 1994) is proposed a control strategy based-on the decomposition of forces used to design the algorithm which regulates both the object's movement and the internal forces. This proposal considers a .. manipulator systems handling a common object where each manipulator holds for the object exerting their forces and momentums.

COOPERATIVE SYSTEM MODELING

A cooperative system consists of multiple robot manipulators who aim to hold an object, so that position on the end effector of each is limited geometrically. These constraints model object and caused a reduction in degrees of freedom of the cooperative system because the end effector of each of the manipulators must maintain contact with the object, so it cannot be moved in all directions. The motion degrees of freedom lost become in contact forces, so they should be included in the dynamics of each of the robots to form the cooperative system. The dynamic model for each manipulator with restricted movement is obtained by using the Lagrangian formulation:

$$\mathbf{H}_i(\mathbf{q}_i)\ddot{\mathbf{q}}_i + \mathbf{C}_i(\mathbf{q}_i, \dot{\mathbf{q}}_i)\dot{\mathbf{q}}_i + \mathbf{g}_i(\mathbf{q}_i) = \boldsymbol{\tau}_i \tag{1}$$

where $\mathbf{q}_i, \dot{\mathbf{q}}_i, \ddot{\mathbf{q}}_i \in \mathbb{R}^{n_i}$ are the position, velocity and acceleration in joint space, respectively; $\mathbf{H}_i(\mathbf{q}_i) \in \mathbb{R}^{n_i \times n_i}$ is a symmetric positive-definite inertia matrix; $\mathbf{C}_i(\mathbf{q}_i, \dot{\mathbf{q}}_i) \in \mathbb{R}^{n_i \times n_i}$ is a matrix containing the Coriolis and centripetal torque's effects; $\mathbf{g}_i(\mathbf{q}_i) \in \mathbb{R}^{n_i}$ is a vector of gravity torque obtained as a gradient result on the potential energy; and $\tau_i \in \mathbb{R}^{n_i}$ is the generalized torque.

PROPERTIES

It is essential to analyze the properties on the model to be able to apply them in the cooperative system.

Inertia Matrix Properties

The inertia matrix $\mathbf{H}_i(\mathbf{q}_i)$ has an important characteristic like its intimate relationship with kinetic energy $\mathcal{K}_i(\mathbf{q}_i, \dot{\mathbf{q}}_i)$,

$$\mathcal{K}_i(\mathbf{q}_i, \dot{\mathbf{q}}_i) = \frac{1}{2}\dot{\mathbf{q}}_i^T \mathbf{H}_i(\mathbf{q}_i)\dot{\mathbf{q}}_i. \tag{2}$$

The matrix is a symmetric positive definite matrix $\mathbf{H}_i(\mathbf{q}_i) = \mathbf{H}_i^T(\mathbf{q}_i) > \mathbf{0}$; the inertia matrix is considered full range $\rho[\mathbf{H}_i(\mathbf{q}_i)] = n$, so its inverse matrix exist, . Another important feature of the inertia matrix is the existence of a positive real constant α_i such as:

$$\mathbf{H}_i(\mathbf{q}_i) \geq \mathbf{I}_i\alpha_i \qquad \forall \mathbf{q}_i \in \mathbb{R}^{n_i} \tag{3}$$

where $\mathbf{I}_i$ is the identity matrix.

In the case of robots provided only rotational joints, there is a constant $\beta_i > 0$ such as:

$$\lambda_{\max}\{\mathbf{H}_i(\mathbf{q}_i)\} \leq \beta_i \qquad \forall \mathbf{q}_i \in \mathbb{R}^{n_i} \tag{4}$$

where β_i is described by:

$$\beta_i \geq n_i \left[\max_{i,j,q_i} \left| \mathbf{H}_{ij}\left(\mathbf{q}_i\right) \right| \right] \tag{5}$$

where $\mathbf{H}_{ij}\left(\mathbf{q}_i\right)$ being the j-th element of the matrix $\mathbf{H}_i\left(\mathbf{q}_i\right)$. In addition to the β_i constant, for robots with only revolute joints there must be a constant $k_m > 0$ such that:

$$\left\| \mathbf{H}_i\left(\mathbf{x}_i\right)\mathbf{z}_i - \mathbf{H}_i\left(\mathbf{y}_i\right)\mathbf{z}_i \right\| \leq k_{m_i} \left\| \mathbf{x}_i - \mathbf{y}_i \right\| \left\| \mathbf{z}_i \right\|, \tag{6}$$

for all vector $\mathbf{x}_i, \mathbf{y}_i, \mathbf{z}_i \in \mathbb{R}^{n_i}$. A simple way to determine k_{m_i} is:

$$k_{m_i} \geq n_i^2 \left[\max_{i,j,k,q_i} \left| \frac{\partial \mathbf{H}_{ij}\left(\mathbf{q}_i\right)}{\partial \mathbf{q}_k} \right| \right]. \tag{7}$$

Another constant used by robots equipped with rotational joints is $k'_{m_i} > 0$ defined as:

$$\left\| \mathbf{H}_i\left(\mathbf{x}_i\right)\mathbf{y}_i \right\| \leq k'_{m_i} \left\| \mathbf{y}_i \right\| \tag{8}$$

for all $\mathbf{x}_i, \mathbf{y}_i \in \mathbb{R}^{n_i}$. It should be noted that each inertia matrix $\mathbf{H}_i\left(\mathbf{q}_i\right)$ satisfies $\lambda_{\mathrm{h}_i} \left\|\mathbf{x}_i\right\|^2 \leq \mathbf{x}_i^T \mathbf{H}_i \mathbf{x}_i \leq \lambda_{\mathrm{H}_i} \left\|\mathbf{x}_i\right\|^2 \; \forall \mathbf{q}_i, \mathbf{x}_i \in \mathbb{R}^{n_i}$, where:

$$\lambda_{\mathrm{h}_i} \overset{\Delta}{=} \min_{\forall q_i \in \mathbf{R}^{n_i}} \lambda_{\min}\left(\mathbf{H}_i\right) \tag{9}$$

$$\lambda_{\mathrm{H}_i} \overset{\Delta}{=} \max_{\forall q_i \in \mathbf{R}^{n_i}} \lambda_{\max}\left(\mathbf{H}_i\right) \tag{10}$$

$$0 < \lambda_{\mathrm{h}_i} \leq \lambda_{\mathrm{H}_i} < \infty. \tag{11}$$

In addition, the inertia matrix satisfies the inequality $0 < \lambda_{\mathrm{h}_i} \leq \left\| \mathbf{H}_i\left(\mathbf{q}_i\right) \right\| \leq \lambda_{\mathrm{H}_i} < \infty$. The inverse matrix .. exist and satisfies $\sigma_{\mathrm{h}_i} \left\|\mathbf{x}_i\right\|^2 \leq \mathbf{x}_i^T \mathbf{H}_i^{-1}\left(\mathbf{q}_i\right) \mathbf{x}_i \leq \sigma_{H_i} \left\|\mathbf{x}_i\right\|^2$ for all $\mathbf{q}_i, \mathbf{x}_i \in \mathbb{R}^{n_i}$, where:

$$\sigma_{h_i} \overset{\Delta}{=} \min_{\forall q_i \in \mathbf{R}^{n_i}} \lambda_{\min}\left(\mathbf{H}_i^{-1}\right) \tag{12}$$

$$\sigma_{H_i} \stackrel{\Delta}{=} \max_{\forall q_i \in \mathbf{R}^{n_i}} \lambda_{\max}\left(\mathbf{H}_i^{-1}\right) \tag{13}$$

$$0 < \sigma_{h_i} \leq \sigma_{H_i} < \infty \tag{14}$$

where $\mathbf{H}_i^{-1}\left(\mathbf{q}_i\right)$ satisfies $0 < \sigma_{h_i} \leq \left\|\mathbf{H}_i^{-1}\left(\mathbf{q}_i\right)\right\| \leq \sigma_{H_i} < \infty$.

Coriolis and Centripetal Terms Properties

$\mathbf{C}_i\left(\mathbf{q}_i, \dot{\mathbf{q}}_i\right)$ matrix is a quadratic term related to generalized velocity $\dot{\mathbf{q}}_i$. Coriolis matrix and centripetal force is bounded as follows:

$$\left\|\mathbf{C}_i\left(\mathbf{q}_i, \dot{\mathbf{q}}_i\right)\right\| \leq v_{b_i}\left(\mathbf{q}_i\right)\left\|\dot{\mathbf{q}}_i\right\|^2, \tag{15}$$

where $v_{b_i}\left(\mathbf{q}_i\right)$ is a scalar constant for a joint revolution, generally is a scalar function of $\mathbf{q}_i$ for a prismatic joint. $\mathbf{C}_i\left(\mathbf{q}_i, \dot{\mathbf{q}}_i\right)$ matrix may not be unique, but the vector $\mathbf{C}_i\left(\mathbf{q}_i, \dot{\mathbf{q}}_i\right)\dot{\mathbf{q}}_i$ is unique. The matrix of Coriolis evaluated to a zero velocity, $\dot{\mathbf{q}}_i = \mathbf{0}$, is zero for every vector $\mathbf{q}_i \in \mathbb{R}^{n_i}$; whereas for all vector $\mathbf{q}_i, \mathbf{x}_i, \mathbf{y}_i, \mathbf{z}_i \in \mathbb{R}^{n_i}$ and the scale α_i we have:

$$\mathbf{C}_i\left(\mathbf{q}_i, \mathbf{x}_i\right)\mathbf{y}_i = \mathbf{C}_i\left(\mathbf{q}_i, \mathbf{y}_i\right)\mathbf{x}_i, \tag{16}$$

$$\mathbf{C}_i\left(\mathbf{q}_i, \mathbf{z}_i + \alpha_i \mathbf{x}_i\right)\mathbf{y}_i = \mathbf{C}_i\left(\mathbf{q}_i, \mathbf{z}_i\right)\mathbf{y}_i + \alpha_i \mathbf{C}_i\left(\mathbf{q}_i, \mathbf{x}_i\right)\mathbf{y}_i. \tag{17}$$

Vector $\mathbf{C}_i\left(\mathbf{q}_i, \mathbf{x}_i\right)\mathbf{y}_i$ can be expressed as follows:

$$\mathbf{C}_i\left(\mathbf{q}_i, \mathbf{x}_i\right)\mathbf{y}_i = \begin{bmatrix} \mathbf{x}_i^T \mathbf{C}_1\left(\mathbf{q}_i\right)\mathbf{y}_i \\ \mathbf{x}_i^T \mathbf{C}_2\left(\mathbf{q}_i\right)\mathbf{y}_i \\ \vdots \\ \mathbf{x}_i^T \mathbf{C}_n\left(\mathbf{q}_i\right)\mathbf{y}_i \end{bmatrix} \tag{18}$$

where $\mathbf{C}_k\left(\mathbf{q}_i\right)$ are symmetrical matrices of dimensions $n_i \times n_i$ for all $k = 1, 2, \ldots, n$. In fact the ij-th element $\mathbf{C}_{k_{ij}}\left(\mathbf{q}_i\right)$ of matrix $\mathbf{C}_k\left(\mathbf{q}_i\right)$ corresponds to the symbol of Chistoffel:

$$\mathbf{C}_{ijk}(\mathbf{q}) = \frac{1}{2}\left[\frac{\partial \mathbf{H}_{kj}(\mathbf{q})}{\partial \mathbf{q}_i} + \frac{\partial \mathbf{H}_{ki}(\mathbf{q})}{\partial \mathbf{q}_j} - \frac{\partial \mathbf{H}_{ij}(\mathbf{q})}{\partial \mathbf{q}_k}\right] \tag{19}$$

where $\mathbf{H}_{ij}(\mathbf{q})$ denotes the j-th element of the inertia matrix $\mathbf{H}_i(\mathbf{q}_i)$. The kj-th element of $\mathbf{C}_{kj}(\mathbf{q}_i, \dot{\mathbf{q}}_i)$ of matrix $\mathbf{C}_i(\mathbf{q}_i, \dot{\mathbf{q}}_i)$ can be obtained as follows:

$$\mathbf{C}_{kj}(\mathbf{q}_i, \dot{\mathbf{q}}_i) = \begin{bmatrix} c_{1kj}(\mathbf{q}_i) \\ c_{2kj}(\mathbf{q}_i) \\ \vdots \\ c_{nkj}(\mathbf{q}_i) \end{bmatrix}^T \dot{\mathbf{q}}_i . \tag{20}$$

In the case of robots provided only of rotational joints, constant $k_{C_1} > 0$ exists like:

$$\left\|\mathbf{C}(\mathbf{q}_i, \mathbf{x}_i)\mathbf{y}_i\right\| \le k_{c_1} \left\|\mathbf{x}_i\right\| \left\|\mathbf{y}_i\right\| \quad \forall \quad \mathbf{q}_i, \mathbf{x}_i, \mathbf{y}_i \in \mathbb{R}^{n_i} . \tag{21}$$

In the case of robots provided only of rotational joints, constants $k_{C_1} > 0$ and $k_{C_2} > 0$ exist such as:

$$\left\|\mathbf{C}(\mathbf{x}_i, \mathbf{z}_i)\mathbf{w}_i - \mathbf{C}(\mathbf{y}_i, \mathbf{v}_i)\mathbf{w}_i\right\| \le k_{c_1} \left\|\mathbf{z}_i - \mathbf{v}_i\right\| \left\|\mathbf{w}_i\right\| + k_{c_2} \left\|\mathbf{x}_i - \mathbf{y}_i\right\| \left\|\mathbf{w}_i\right\| \left\|\mathbf{z}_i\right\| \tag{22}$$

for all $\mathbf{x}_i, \mathbf{y}_i, \mathbf{z}_i, \mathbf{u}_i, \mathbf{w}_i \in \mathbb{R}^{n_i}$. $\mathbf{C}_i(\mathbf{q}_i, \dot{\mathbf{q}}_i)$ matrix is related with the inertia matrix $\mathbf{H}_i(\mathbf{q}_i)$ by the expression:

$$\mathbf{C}_i(\mathbf{q}_i, \dot{\mathbf{q}}_i)\dot{\mathbf{q}}_i = \dot{\mathbf{H}}_i(\mathbf{q}_i)\dot{\mathbf{q}}_i - \frac{1}{2}\frac{\partial}{\partial \mathbf{q}_i}\left[\dot{\mathbf{q}}_i^T \mathbf{H}_i(\mathbf{q}_i)\dot{\mathbf{q}}_i\right]. \tag{23}$$

$\dot{\mathbf{H}}_i(\mathbf{q}_i)$ is related to the Coriolis matrix as follow:

$$\dot{\mathbf{H}}_i(\mathbf{q}_i) = \mathbf{C}_i(\mathbf{q}_i, \dot{\mathbf{q}})_i + \mathbf{C}_i^T(\mathbf{q}_i, \dot{\mathbf{q}}_i) \tag{24}$$

where $\mathbf{C}_i^T(\mathbf{q}_i, \dot{\mathbf{q}}_i)$ is defined as follow:

$$\mathbf{C}_i^T(\mathbf{q}_i, \dot{\mathbf{q}}_i) = \frac{1}{2}\frac{\partial}{\partial \mathbf{q}_i}\left[\dot{\mathbf{q}}_i^T \mathbf{H}_i(\mathbf{q}_i)\dot{\mathbf{q}}_i\right]. \tag{25}$$

Since $\mathbf{C}_i(\mathbf{q}_i, \dot{\mathbf{q}}_i)$ is not a unique matrix, must be carefully selected to satisfy the following property:

$$\frac{1}{2}\dot{\mathbf{q}}_i^{\,T}\left[\dot{\mathbf{H}}_i\left(\mathbf{q}_i\right)-\mathbf{C}_i\left(\mathbf{q}_i,\dot{\mathbf{q}}_i\right)\right]\dot{\mathbf{q}}_i\equiv 0 \text{ Skew-symmetric matrix.} \tag{26}$$

Gravity Terms Properties

The gravity vector is situated in robots which have not been designed mechanically with compensation of gravity, so, without counterbalances; or for robots assigned to move outside the horizontal plane. The gravity terms only depends on joint positions $\mathbf{q}_i$; and the gravity terms can be related with the joint velocity $\dot{\mathbf{q}}_i$ this means:

$$\int_0^T \mathbf{g}_i\left(\mathbf{q}_i\right)^T \dot{\mathbf{q}}_i dt=\mathcal{U}_i\left(\mathbf{q}_i\left(T\right)\right)-\mathcal{U}_i\left(\mathbf{q}_i\left(0\right)\right) \text{ for all } T\in\mathbb{R}_+. \tag{27}$$

In the case of robots provided solely of rotational joints, a constant $k_{\mathcal{U}}$ exists like:

$$\int_0^T \mathbf{g}_i\left(\mathbf{q}_i\right)^T \dot{\mathbf{q}}_i dt+\mathcal{U}_i\left(\mathbf{q}_i\left(0\right)\right)\geq k_{\mathcal{U}} \text{ for all } T\in\mathbb{R}_+ \text{ and where } k_{\mathcal{U}}=\min_q\left\{\mathcal{U}\left(\mathbf{q}_i\right)\right\}. \tag{28}$$

In the case of robots provided solely of rotational joints the gravity vector $\mathbf{g}_i\left(\mathbf{q}_i\right)$ is Lipschitz, this means that a constant $k_{g_i}>0$ exists like:

$$\left\|\mathbf{g}_i\left(\mathbf{x}_i\right)-\mathbf{g}_i\left(\mathbf{y}_i\right)\right\|\leq k_{g_i}\left\|\mathbf{x}_i-\mathbf{y}_i\right\| \text{ for all } \mathbf{x}_i,\mathbf{y}_i\in\mathbb{R}^{n_i}. \tag{29}$$

A simple form to calculate k_{g_i} is:

$$k_{g_i}\geq n_i\left[\max_{i,j,q}\left|\frac{\partial \mathbf{g}_i\left(\mathbf{q}_i\right)}{\partial \mathbf{q}_j}\right|\right]. \tag{30}$$

In addition k_{g_i} satisfies:

$$k_{g_i}\geq\left\|\frac{\partial \mathbf{g}\left(\mathbf{q}_i\right)}{\partial \mathbf{q}_i}\right\|\geq\lambda_{\max}\left\{\frac{\partial \mathbf{g}\left(\mathbf{q}_i\right)}{\partial \mathbf{q}_i}\right\}. \tag{31}$$

The gravity term $\mathbf{g}\left(\mathbf{q}_i\right)$ is bounded only if $\mathbf{q}_i$ is bounded:

$$\left\|\mathbf{g}\left(\mathbf{q}_i\right)\right\|\leq g_{b_i} \tag{32}$$

where g_{b_i} is a scalar constant for revolute arms and a scalar function of $\mathbf{q}_i$ for arms containing revolute joints.

EFFECTS OF TORQUE AND FRICTION IN CONTACT TASKS

Torque τ_i, applied to the joints of the robot, is provided by a direct-drive motor or by gear drive. The following torque contributions are in each joint:

$$\tau_i = \tau_{a_i} - \tau_{f_i} - \tau_{e_i} \tag{33}$$

where $\tau_i \in \mathbb{R}^{n_i}$ is the generalized torque; $\tau_{a_i} \in \mathbb{R}^{n_i}$ represents the torque applied to the system, which is responsible for moving the joints; $\tau_{f_i} \in \mathbb{R}^{n_i}$ is the torque due to friction; and $\tau_{e_i} \in \mathbb{R}^{n_i}$ is the torque caused by the forces and moments exerted by the end-effector, when exist contact with the environment. In view of the contributions made by the torque, the dynamic model described in **(1)** can be rewritten as:

$$\mathbf{H}_i\left(\mathbf{q}_i\right)\ddot{\mathbf{q}}_i + \mathbf{C}_i\left(\mathbf{q}_i,\dot{\mathbf{q}}_i\right)\dot{\mathbf{q}}_i + \mathbf{g}_i\left(\mathbf{q}_i\right) + \tau_{f_i} = \tau_{a_i} - \tau_{e_i}\,. \tag{34}$$

The complete dynamic model is described in (34), which provides the possibility of representing the system behavior to the friction phenomena and the interaction between the end-effector and environment. On contact with the environment, the torque caused by the forces and momentums applied by the end-effector, is defined as:

$$\tau_{e_i} = \mathbf{J}^T\left(\mathbf{q}_i\right)\lambda_i \tag{35}$$

where $\mathbf{J}\left(\mathbf{q}_i\right) \in \mathbb{R}^{n_i \times n_i}$ is the Jacobian matrix, which is considered full-rank; and $\lambda_i \in \mathbb{R}^{6\times1}$ is a vector of external forces, and momentums present in the end-effector defined as:

$$\lambda_i = \begin{bmatrix} \mathbf{f}_i \\ \mu_i \end{bmatrix} \tag{36}$$

where $\mathbf{f}_i \in \mathbb{R}^{3\times1}$ represents the external force at the end-effector; and $\mu_i \in \mathbb{R}^{3\times1}$ is the external momentum in the end-effector. In order to define if the end-effector is or is not interacting with their environment, this can be determined through vector λ_i. If $\lambda_i = 0$ there is not an interaction between the end-effector and the environment, and the dynamic model is defined as follows:

$$\mathbf{H}_i\left(\mathbf{q}_i\right)\ddot{\mathbf{q}}_i+\mathbf{C}_i\left(\mathbf{q}_i,\dot{\mathbf{q}}_i\right)\dot{\mathbf{q}}_i+\mathbf{g}_i\left(\mathbf{q}_i\right)+\boldsymbol{\tau}_{f_i}=\boldsymbol{\tau}_{a_i}, \tag{37}$$

whereas that if $\lambda_i \neq 0$ is considered that exists the interaction with the environment and is described by:

$$\mathbf{H}_i\left(\mathbf{q}_i\right)\ddot{\mathbf{q}}_i+\mathbf{C}_i\left(\mathbf{q}_i,\dot{\mathbf{q}}_i\right)\dot{\mathbf{q}}_i+\mathbf{g}_i\left(\mathbf{q}_i\right)+\boldsymbol{\tau}_{f_i}=\boldsymbol{\tau}_{a_i}-\mathbf{J}^T\left(\mathbf{q}_i\right)\begin{bmatrix}\mathbf{f}_i\\ \boldsymbol{\mu}_i\end{bmatrix}. \tag{38}$$

The effects of friction in the joints are difficult to be modeled, as a result; a simplified representation only used the viscous friction defined as:

$$\boldsymbol{\tau}_{f_i}=\mathbf{D}_i\dot{\mathbf{q}}_i \tag{39}$$

where $\mathbf{D}_i \in \mathbb{R}^{n_i x n_i}$ is a positive definite diagonal matrix containing the viscous friction coefficients set in the joints; and $\dot{\mathbf{q}}_i \in \mathbb{R}^{n_i}$ is the joint velocity vector. Finally, the dynamic model is defined as:

$$\mathbf{H}_i\left(\mathbf{q}_i\right)\ddot{\mathbf{q}}_i+\mathbf{C}_i\left(\mathbf{q}_i,\dot{\mathbf{q}}_i\right)\dot{\mathbf{q}}_i+\mathbf{g}_i\left(\mathbf{q}_i\right)+\mathbf{D}_i\dot{\mathbf{q}}_i=\imath_{a_i}-\mathbf{J}^T\left(\mathbf{q}_i\right)\begin{bmatrix}\mathbf{f}_i\\ \boldsymbol{\mu}_i\end{bmatrix}. \tag{40}$$

HOLONOMIC HOMOGENEOUS COOPERATIVE SYSTEM

The interaction with environment makes no sense when the end-effector of the manipulator is in contact with any surface, otherwise there is free movement. Mechanically constrained robots, or simply robots restricted, are those interact with its environment. The main advantage of a restricted system is the ability to apply force on any surface using the end-effector. This change in mobility by force is a quality of the hybrid control for robot. When a cooperative system holds an object in m_i autonomous points and the body is not deformed, a set of m_i independent constraint is imposed on every touch point, Figure 4.

The manipulation of a rigid object using a cooperative system imposes some kinematic and dynamic constraints. The constraint is a characteristic which limits the geometry and system movement; it also must be applied to achieve a strong grip, quick control, proper load distribution, compliance with the planned trajectory, stable transition to perform the task and efficient closed-loop energy balance. Such requirements provide greater dexterity, flexibility in handling and assembly tasks. Consequently, in order to manipulate a body using a cooperative system, the goals will simultaneously control the position and velocity of each robot's end-effectors under the internal forces applied over the body; ensure the grip; and control the external forces which move the object. In other words, the target in this task is to hold a

Figure 4. Manipulation of an object using cooperative robots (constraint representations)

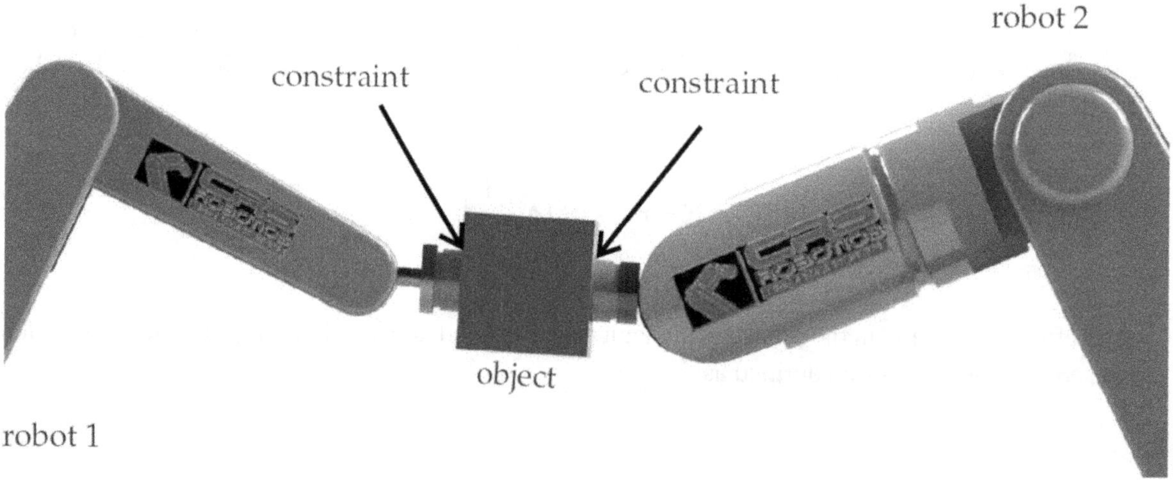

body on a support, which let you do the grip with enough force to prevent the body from turning, slipping, falling or uncontrolled movements, allowing movement within your workspace. In order to perform these tasks there exist some possibilities, but the most-used control structure is the position/force control.

CONSTRAINTS

The force control used in a manipulator in order to interact with the environment and perform a task, can be discussed in terms of constraints. A constraint limits the geometrical configuration and the system's movement. The restriction is imposed by the environment. When a manipulator moves freely in the workspace has no restrictions, and therefore, no reaction force produced by their environment, according to **(35)** the value of the vector λ_i would be zero. A force sensor at the wrist, this sensor only registered inertial forces due to acceleration of the end-effector.

In contrast, when the manipulator is in contact with its environment, being $\lambda_i \neq 0$, one or more movement degrees of freedom may be lost, consequently, the manipulator cannot go through the surface, However, the manipulator will be exerting a normal force $\mathbf{N}$ at the surface, Figure 5.

In Figure 5, we show that the movement of the end-effector can be performed in the direction tangent to the constrained surface, while the force can be applied perpendicular to the surface. Thus, position control and force control are mutually exclusive and cannot control a force and a position along the same direction simultaneously.

It should be noted that the differential kinematic establishes the velocity equation of the robot manipulator, relating the linear $\mathbf{v}$ and the angular ω_i velocities of the end-effectors, that is to say, the joint velocities. This mapping is described by the Jacobian matrix, which is one of the most important tools in motion control analysis. Jacobian matrix provides information about the robot's characteristics in different aspects such as: the planning and execution of smooth trajectories, in the determination of singular configurations, in the analysis of redundancy, in the execution of anthropomorphic movements,

Figure 5. Robot end-effector in contact with a surface (appearance of the normal force)

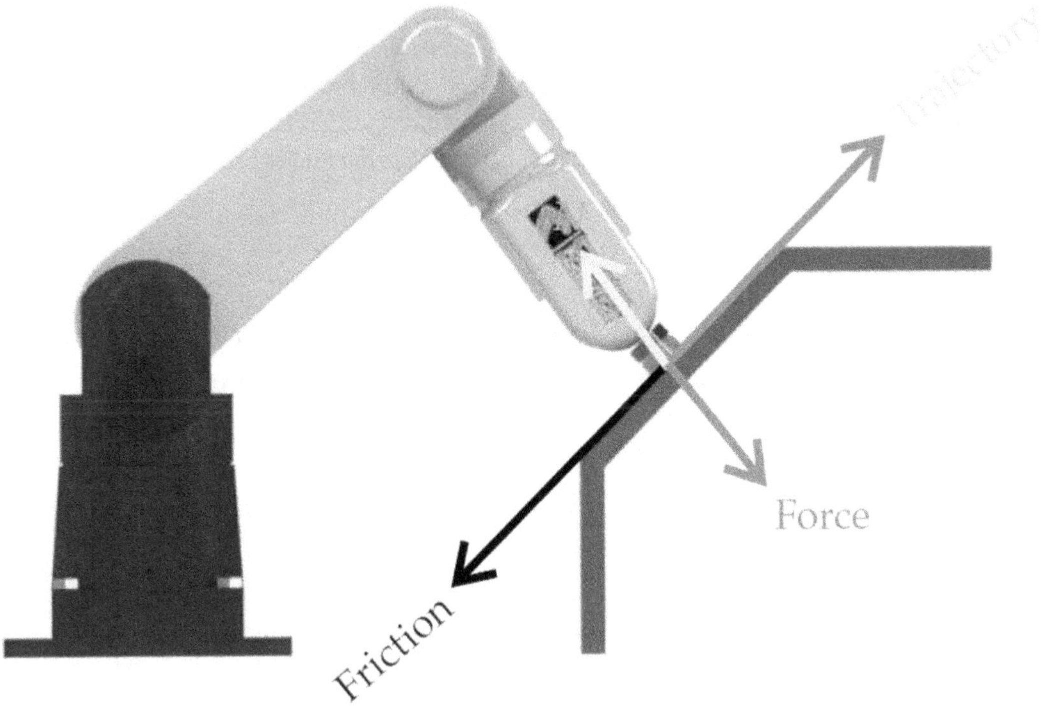

in the derivation of the dynamic equations of motion and transformation of forces and torques at each joint end-effector. The relationship between the angular and linear velocities of the end effector is set by the differential equation:

$$\nu_i = \begin{bmatrix} \mathbf{v}_i \\ \omega_i \end{bmatrix} = \mathbf{J}_i\left(\mathbf{q}_i\right)\dot{\mathbf{q}}_i \tag{41}$$

where $\mathbf{v}_i \in \mathbb{R}^{3\times1}$ is the linear velocity, $\omega_i \in \mathbb{R}^{3\times1}$ represent the angular velocity; and $\mathbf{J}_i\left(\mathbf{q}_i\right) \in \mathbb{R}^{6\times n_i}$ is the Jacobian matrix. Obtaining the Jacobian matrix usually follows a geometric method based on joint contributions of velocity (linear and angular), by this fact, the matrix $\mathbf{J}_i\left(\mathbf{q}_i\right)$ is known as geometric Jacobian of the manipulator.

Type of Constraints

The constraints can be classified as:

- **Holonomic constraints:** are those constraints which can be written as a function of the coordinates of the particles, $f\left(r_1, r_2, \ldots, r_n, t\right) = 0$. This means that the different degrees of freedom are independent, i.e. are decoupled. When the number of controllable degrees of freedom is equal to the total number of actuators is said to be a *holonomic robot.*

- **Non-holonomic constraints:** as the name implies, they are those which cannot be written as a function between the coordinates of the particles. This means that the different degrees of freedom are dependent, i.e. are coupled. If the number of controllable degrees of freedom is smaller than the total number of actuators is said to be a *non-holonomic robot*.
- **Rheonomic constraints:** a constraint it is said rheonomic if it is time-dependent, i.e. the time variable t appears explicitly. This restriction is time-varying and is represented as:

$$\varphi_i\left(\mathbf{q}_i, t\right) = 0 \tag{42}$$

where $\varphi_i \in \mathbb{R}^{m_i}$ and $\mathbf{q}_i \in \mathbb{R}^{n_i}$ are the vector of generalized coordinates of the i-th manipulator.

- **Scleronomic constraint**: a constraint is scleronomic if it is time-independent, i.e. this constraint is invariant in time and its general representation is given by a function in the joint space:

$$\varphi_i\left(\mathbf{q}_i\right) = 0 \tag{43}$$

where $\varphi_i \in \mathbb{R}^{m_i}$ and $\mathbf{q}_i \in \mathbb{R}^{n_i}$ are the vectors of generalized coordinates of the i-th manipulator.

For holonomic constraints is appropriate to introduce a set of independent coordinates called *generalized coordinates* $q_1, q_2, \ldots, q_f$ where $f = 3n - k$, n is the number of particles ($3n$ coordinates) and k the number of constraints on the system. Then, there exist k functions on the form:

$$f_1\left(r_1, r_2, \ldots, r_n, t\right) = 0$$

$$\vdots$$

$$f_k\left(r_1, r_2, \ldots, r_n, t\right) = 0 \tag{44}$$

which can be written as:

$$r_1 = g_1\left(q_1, q_2, \ldots, q_f, t\right)$$

$$\vdots$$

$$r_n = g_n\left(q_1, q_2, \ldots, q_f, t\right) \tag{45}$$

where f is the number of system's degrees of freedom. In order to develop equations by describing the dynamics of each robot manipulator which includes cooperative system, it is necessary to incorporate a set of level holonomic constraints imposed on kinematic manipulators. Then there is the whole dynamic of the cooperative system described by nonlinear ordinary differential equations over another set of algebraic equations, which represent the constraints, thus resulting in a system of differential algebraic equations. In order to define the cooperative system's characteristics, robot manipulators are considered, each robot has n_i degrees of freedom and m_i constraints caused by contact with the manipulated object. Therefore, the total number of joints is given by and the total number of constraints is given by $m = \sum_{i=1}^{l} m_i$, with $n_i > m_i$.

Definition 1. If the restrictions of movement, in a cooperative system form by two or more robotic manipulators, are holonomic, we said that cooperation was holonomic. ■

Definition 2. It is said that the movement constraints imposed on the cooperative system are homogeneous if the scleronomic constraint of movement of a cooperative system given by **(43)**, we can be written as follows:

$$\varphi\left(q_1, q_2, \ldots, q_l\right) = \sum_{i=1}^{l} \varphi_i\left(\mathbf{q}_i\right) = 0 \tag{46}$$

where $\varphi\left(\mathbf{q}_i\right)$ depends only on the joint variables of the i-th robot. ■

The object to be manipulated by the cooperative system should be modeled by using a set of homogeneous holonomic constraints, which restrict the movement of the robot manipulators at each contact point. It should be noted that with the appropriate definition of homogeneous holonomic constraints, the dynamics of the object does not have to be explicitly calculated for controlling a cooperative system. However, we implicitly must consider its size, weight, etc. Using kinematics relationship, which is based on the joint variables $\mathbf{q}_i \in \mathbb{R}^{n_i}$, defined as:

$$\mathbf{x}_i = f_i\left(\mathbf{q}_i\right) \tag{47}$$

where $\mathbf{x}_i \in \mathbb{R}^{k_i}$ denotes the position; k is the dimension of the task space, and $f_i\left(\mathbf{q}_i\right)$ is a scalar function. Holonomic constraint described by (42) can be expressed as:

$$\varphi_i\left(\mathbf{x}_i\right) = 0. \tag{48}$$

Consequently, we have a kinematic level, the joint and Cartesian velocities related by the Jacobian of the manipulator $\mathbf{J}_i\left(\mathbf{q}_i\right) \in \mathbb{R}^{3 \times n_i}$:

$$\mathbf{J}_i\left(\mathbf{q}_i\right) \in \mathbb{R}^{3 \times n_i}. \tag{49}$$

Furthermore, by a cooperative system formed by k robot manipulators with holonomic constraints, movement restrictions can be expressed generally as:

$$\varphi_i\left(q_1, q_2, \ldots, q_l\right) = 0 \tag{50}$$

where $\boldsymbol{\varphi}_i \in \mathbb{R}^{m_i}$ is a vector of functions, which depends on the joint variables of the cooperative system, m is the number of movement constraints on the system cooperative, $\mathbf{q}_i \in \mathbb{R}^{n_i}$; $i = 1, \ldots, l$ are the variables of the i-th joint robot. If it is one considered that each robot holds the object in a single point of contact, then the number of cooperative system constraints equals the number of manipulators who form the system, i.e. $m = l$.Generally, the Jacobian of restrictions is defined as follows:

$$\mathbf{J}_{\varphi_i}\left(\mathbf{q}_i\right) = \frac{\partial \boldsymbol{\varphi}_i\left(\mathbf{q}_i\right)}{\partial \mathbf{q}_i}. \tag{51}$$

Deriving (46), we obtain the movement restriction at the velocity level defined as:

$$\sum_{i=1}^{l} \mathbf{J}_{\varphi_i}\left(\mathbf{q}_i\right)\dot{\mathbf{q}}_i = \sum_{i=1}^{l} L_{\dot{q}_i}\boldsymbol{\varphi}_i\left(\mathbf{q}_i\right) = 0 \tag{52}$$

where $L_{\dot{q}_i}\boldsymbol{\varphi}_i\left(\mathbf{q}_i\right)$ is the directional derivative of the scalar function representing the i holonomic constraints of joint movement in the direction of the vector field $\dot{\mathbf{q}}_i$.

POSITION AND VELOCITY RESTRICTED

The positions and velocities restricted, provide us information about the cooperative system's behavior at the kinematic level. In this kind of systems, the position of the end-effector of each robot is in contact with the object, which is supposed to rigid, and the movement restrictions appear. It should be noted that the restrictions imposed on each robot are homogeneous, that is, each robot depends on its own joint variables and not the others. Since there is no relative movement between each manipulator's end-effectors and the object, we were establishing geometrically constants, connecting endpoints of the manipulators. Therefore, end-effector of the i-th manipulator can be defined in terms of the positions of the remaining manipulators. The same applies for velocity. This is a logical consequence, since otherwise, we could lose the contact point between any of the manipulators and the object. The position and velocity restricted to have already been introduced in some control schemes for cooperative systems, but these were used only to raise the kinematic problem. The definitions of position and velocity restricted used within the control law to provide more information on restricted kinematics were introduced by (Liu *et al.*, 1997).

Definition 3. Given the holonomic homogeneous constraint (43), the restricted velocity of the i-th manipulator is defined as:

$$\dot{\mathbf{p}}_i = J_{\varphi_i}\left(\mathbf{q}_i\right)\dot{\mathbf{q}}_i = L_{\dot{q}_i}\varphi_i\left(\mathbf{q}_i\right). \tag{53}$$

In view of (46), we have:

$$\sum_{i=1}^{l}\dot{\mathbf{p}}_i = \sum_{i=1}^{l}\mathbf{J}_{\varphi_i}\left(\mathbf{q}_i\right)\dot{\mathbf{q}}_i = \sum_{i=1}^{l}L_{\dot{q}_i}\varphi_i\left(\mathbf{q}_i\right) = 0. \blacksquare \tag{54}$$

Definition 4. Restricted position variable is given by the integral of the restricted velocity, that is:

$$\mathbf{p}_i = \int_0^t \mathbf{J}_{\varphi_i}\left(\mathbf{q}_i\right)\dot{\mathbf{q}}_i dt = \int_0^t L_{\dot{q}_i}\varphi_i\left(\mathbf{q}_i\right)dt = 0 \tag{55}$$

with $\mathbf{p}_i\left(0\right) = 0$. ■

The restricted position $\mathbf{p}_i$ of a cooperative system is calculated directly as a function of desired position $\mathbf{q}_{d_i}$ as follow:

$$\sum_{i-1}^{l}\mathbf{p}_{d_i} = \sum_{i-1}^{l}\int_0^t \mathbf{J}_{\varphi_i}\left(\mathbf{q}_{d_i}\right)\dot{\mathbf{q}}_{d_i} dt = \int_0^t L_{\dot{q}_i}\varphi_i\left(\mathbf{q}_i\right)dt = 0 \text{ with } \mathbf{p}_{d_i}(0) = 0. \tag{56}$$

The same applies to calculate the restricted velocity $\dot{\mathbf{p}}_i$:

$$\sum_{i=1}^{l}\dot{\mathbf{p}}_{d_i} = \sum_{i=1}^{l}\mathbf{J}_{\varphi_i}\left(\mathbf{q}_{d_i}\right)\dot{\mathbf{q}}_{d_i} = \sum_{i=1}^{l}L_{\dot{q}_i}\varphi_i\left(\mathbf{q}_i\right) = 0. \tag{57}$$

In order to control the position and velocity restricted, the following errors are defined:

$$\tilde{\mathbf{p}}_i = \Delta\mathbf{p}_i \stackrel{\Delta}{=} \mathbf{p}_i - \mathbf{p}_{d_i} \tag{58}$$

$$\dot{\tilde{\mathbf{p}}}_i = \Delta\dot{\mathbf{p}}_i \stackrel{\Delta}{=} \dot{\mathbf{p}}_i - \dot{\mathbf{p}}_{d_i}, \tag{59}$$

satisfying the equalities:

$$\sum_{i=1}^{l} \tilde{\mathbf{p}}_i = 0 \tag{60}$$

$$\sum_{i=1}^{l} \dot{\tilde{\mathbf{p}}}_i = 0\,. \tag{61}$$

The desired positions must comply the restriction at any instant of time in order to ensure the contact is maintained between the end-effector and the object. For this reason, the position and velocity errors are included in the control law and in the nominal reference signal, which ensures the rigid object is in contact with the end-effectors of the robots involved in the cooperative system.

ASSUMPTIONS ON COOPERATIVE SYSTEM

Dynamic model of each robot of the cooperative system is obtained from an Euler-Lagrange formulation. In order to facilitate the analysis, the following assumptions are made.

Assumptions on Manipulators

Assumption 1. The robots are formed by rotational joints. ■
Assumption 2. The links of the cooperative manipulators are rigid. ■
Assumption 3. The robot arms does not enter at singular configurations throughout the task. ■
Assumption 4. Cooperative robots are non-redundant. ■

Assumptions on Manipulated Object

Assumption 5. The robot manipulators rigidly holding the object, therefore, no relative movement between the end-effectors and the object (stable grip condition). ■
Assumption 6. As it doesn't exist a relative movement between the end-effector and the object, the effects due to the tangential friction between the effectors of robotic manipulators, and object are null. ■
Assumption 7. The manipulated object is rigid and does not undergo deformation when it is gripped. ■
Assumption 8. We know the forward kinematic of the rigid object. ■

Assumptions on Constraints

Assumption 9. Considering a single contact point between each robot manipulator and the object, the movement constraints' number on the cooperative system is equal to the number of robot manipulators. ■

Assumption 10. The movement constraints imposed on the cooperative system are holonomic and homogeneous. ■

Assumption 11. The robots of the cooperative system satisfy the scleronomic constraints. ■

Assumption 12. $\mathbf{J}_{\varphi_i}$ matrix is Lipschitz continuous, i.e.

$$\left\| \mathbf{J}_{\varphi_i}\left(\mathbf{q}_i\right) - \mathbf{J}_{\varphi_i}\left(\mathbf{q}_{d_i}\right) \right\| \leq L_{p_i} \left\| \mathbf{q}_i - \mathbf{q}_{d_i} \right\| \tag{62}$$

for a positive constant L_{p_i} and for all $\mathbf{q}_i, \mathbf{q}_{d_i} \in \mathbb{R}^{n_i}$. Exist finite positive constants c_{0_i} and c_{1_i}, satisfying:

$$c_{0_i} = \max_{\forall q_i \in \mathbf{R}^{n_i}} \left\| \mathbf{J}^{+}_{\varphi_i}\left(\mathbf{q}_i\right) \right\| \tag{63}$$

$$c_{1_i} = \max_{\forall q_i \in \mathbf{R}^{n_i}} \left\| \mathbf{J}_{\varphi_i}\left(\mathbf{q}_i\right) \right\|. \blacksquare \tag{64}$$

DYNAMIC MODEL OF A RIGID BODY

The movement of the cooperative system formed by the two robots arms are dynamically coupled by contact forces caused by the interaction with the rigid body. In order to describe this interaction, it is necessary to understand the dynamics of the object, so raises the free body diagram below.

Considering Assumption 2, 5-7 and the free body diagram, Figure 6, the Newton equation of motion is:

$$\mathbf{m}_i \ddot{\mathbf{x}}_i - \mathbf{m}_i g_i = \mathbf{f}_{1_i} - \mathbf{f}_{2_i} \tag{65}$$

where $\mathbf{m}_i \in \mathbb{R}^{3\times3}$ is the diagonal matrix of mass of the object, $\ddot{\mathbf{x}}_i \in \mathbb{R}^{3\times1}$ is the vector which describes the translational acceleration on the center of mass of the object, $\mathbf{f}_{1_i}$ and $\mathbf{f}_{2_i} \in \mathbb{R}^{3\times3}$ are the forces exerted by the robots. All vectors are expressed concerning to the inertial coordinate frame. The contact force is given by:

$$\mathbf{f}_i = n_i \lambda_i \tag{66}$$

where n_i represents the direction of the normal forces respecting to the restriction, and λ_i is a vector of external forces, and momentums set in the end-effector defined in **(36)**.

Figure 6. Free body diagram

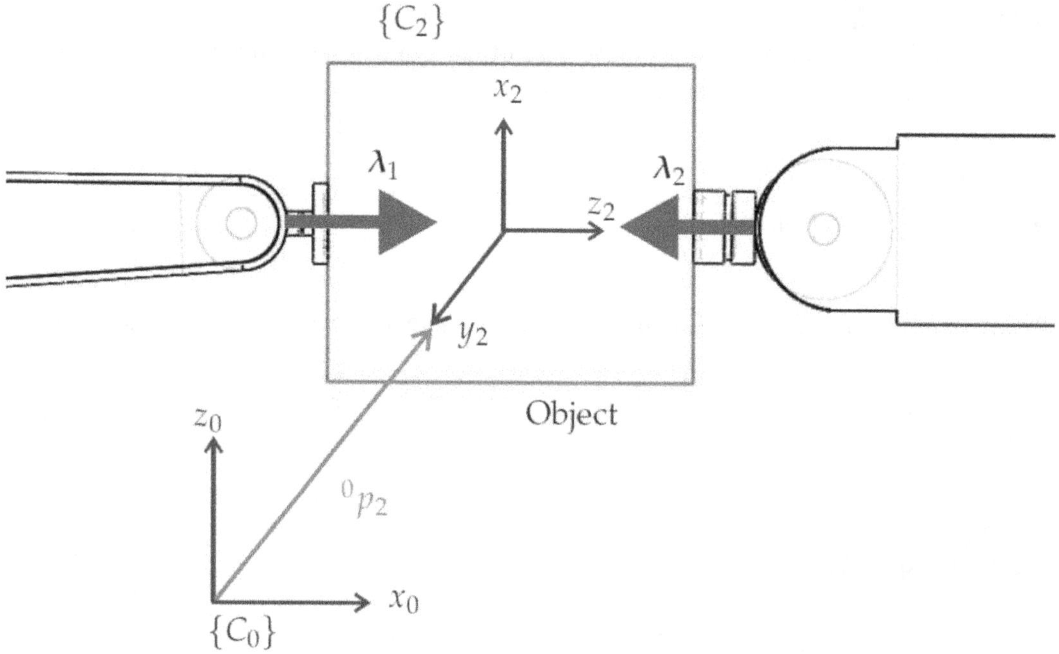

DYNAMIC COUPLING

The position, velocity and acceleration of the center of mass of the object respecting to the inertia coordinate system $\{C_0\}$ are given in Cartesian coordinates by:

$$\mathbf{x} = \mathbf{f}_{cm_1}\left(\mathbf{q}_i\right)\ i = 1,2 \tag{67}$$

$$\dot{\mathbf{x}} = \frac{\partial \mathbf{f}_{cm_1}\left(\mathbf{q}_i\right)}{\partial \mathbf{q}_i}\dot{\mathbf{q}}_i = L_{\dot{q}_i}\mathbf{f}_{cm_1}\left(\mathbf{q}_i\right) = \mathbf{J}_{cm_i}\dot{\mathbf{q}}_i\ i = 1,2 \tag{68}$$

$$\ddot{\mathbf{x}} = \frac{\partial^2 \mathbf{f}_{cm_1}\left(\mathbf{q}_i\right)}{\partial^2 \mathbf{q}_i}\dot{\mathbf{q}}_i = L^2_{\dot{q}_i}\mathbf{f}_{cm_1}\left(\mathbf{q}_i\right) = \mathbf{J}_{cm_i}\ddot{\mathbf{q}}_i + \dot{\mathbf{J}}_{cm_i}\dot{\mathbf{q}}_i\ i = 1,2 \tag{69}$$

where $\mathbf{f}_{cm_i}\left(\mathbf{q}_i\right) \in \mathbb{R}^{3\times 1}$ is the forward kinematics of the center of mass of the rigid object expressed in the coordinate system $\{C_0\}$ and $\mathbf{J}_{cm_i} \in \mathbb{R}^{3\times 3}$ corresponds to the Jacobian matrix of $\mathbf{f}_{cm_i}\left(\mathbf{q}_i\right)$. Substituting (69) on (65), we get the value of $\mathbf{f}_i$:

$$\mathbf{m}_i\ddot{\mathbf{x}}_i - \mathbf{m}_i g_i = \mathbf{f}_i - \mathbf{f}_{i+1}$$

$$\mathbf{m}_i\left[\mathbf{J}_{cm_i}\left(\mathbf{q}_i\right)\ddot{\mathbf{q}}_i + \dot{\mathbf{J}}_{cm_i}\left(\mathbf{q}_i\right)\dot{\mathbf{q}}_i\right] - \mathbf{m}_i g_i = \mathbf{f}_i - \mathbf{f}_{i+1}$$

$$\mathbf{m}_i\mathbf{J}_{cm_i}\left(\mathbf{q}_i\right)\ddot{\mathbf{q}}_i + \mathbf{m}_i\ddot{\mathbf{J}}_{cm_i}\left(\mathbf{q}_i\right)\dot{\mathbf{q}}_i - \mathbf{m}_i g_i + \mathbf{f}_{i+1} = \mathbf{f}_i\,. \tag{70}$$

By definition it follows that the Jacobian is related with constraints φ_i, which have the following equivalence with analytical Jacobian:

$$\mathbf{J}_{\varphi_i}^T\left(\mathbf{q}_i\right) = \mathbf{J}_{a_i}^T\left(\mathbf{q}_i\right)n_i = \mathbf{J}_i^T\left(\mathbf{q}_i\right)n_i\,. \tag{71}$$

The analytic Jacobian is obtained by deriving the forward kinematics function (end-effector location expressed by a minimum operational representation space) respecting to the joint variables. Considering the contact forces $\mathbf{f}_i$, we have:

$$\mathbf{f}_i = n_i\lambda_i$$

$$\mathbf{m}_i\mathbf{J}_{cm_i}\left(\mathbf{q}_i\right)\ddot{\mathbf{q}}_i + \mathbf{m}_i\dot{\mathbf{J}}_{cm_i}\left(\mathbf{q}_i\right)\dot{\mathbf{q}}_i - \mathbf{m}_i g_i + \mathbf{f}_{i+1} = n_i\lambda_i\,. \tag{72}$$

Substituting **(71)** in **(40)**, we have:

$$\mathbf{H}_i\left(\mathbf{q}_i\right)\ddot{\mathbf{q}}_i + \mathbf{C}_i\left(\mathbf{q}_i,\dot{\mathbf{q}}_i\right)\dot{\mathbf{q}}_i + g_i\left(\mathbf{q}_i\right) + \mathbf{F}_{v_i}\dot{\mathbf{q}}_i = \tau_{a_i} - \mathbf{J}_{\varphi_i}^T\left(\mathbf{q}\right)\lambda_i$$

$$= \tau_{a_i} - \mathbf{J}_{a_i}^T\left(\mathbf{q}_i\right)n_i\lambda_i \tag{73}$$

allowing replacement **(72)** in **(73)**, we get:

$$\mathbf{H}_i\left(\mathbf{q}_i\right)\ddot{\mathbf{q}}_i + \mathbf{C}_i\left(\mathbf{q}_i,\dot{\mathbf{q}}_i\right)\dot{\mathbf{q}}_i + g_i\left(\mathbf{q}_i\right) + \mathbf{D}_i\dot{\mathbf{q}}_i = \tau_{a_i} - \mathbf{J}_{\varphi_i}^T\left(\mathbf{q}\right)\mathbf{f}_i$$

$$= \tau_{a_i} - \mathbf{J}_{a_i}^T\left(\mathbf{q}_i\right)\left[\mathbf{m}_i\mathbf{J}_{cm_i}\left(\mathbf{q}_i\right)\ddot{\mathbf{q}}_i + \mathbf{m}_i\ddot{\mathbf{J}}_{cm_i}\left(\mathbf{q}_i\right)\dot{\mathbf{q}}_i - \mathbf{m}_i g_i + \mathbf{f}_{i+1}\right] \tag{74}$$

Consequently, we have:

$$\begin{aligned}\tau_{a_i} - \mathbf{J}_{a_i}^T\left(\mathbf{q}_i\right)\mathbf{f}_{i+1} = &\left(\mathbf{H}_i\left(\mathbf{q}_i\right) + \tau_{a_i}\mathbf{J}_{a_i}^T\left(\mathbf{q}_i\right)\mathbf{m}_i\mathbf{J}_{cm_i}\right)\ddot{\mathbf{q}}_i + \mathbf{g}_i\left(\mathbf{q}_i\right) - \tau_{a_i}\mathbf{J}_{a_i}^T\left(\mathbf{q}_i\right)\mathbf{m}_i g_i \\ &+ \left(\mathbf{C}_i\left(\mathbf{q}_i,\dot{\mathbf{q}}_i\right) + \tau_{a_i}\mathbf{J}_{a_i}^T\left(\mathbf{q}_i\right)\mathbf{m}_i\dot{\mathbf{J}}_{cm_i} + \mathbf{D}_i\right)\dot{\mathbf{q}}_i\end{aligned} \tag{75}$$

In order to facilitate handling of terms is performed the following assignment:

$$\mathbf{H}_{T_i}\left(\mathbf{q}_i\right) = \mathbf{H}_i\left(\mathbf{q}_i\right) + \tau_{a_i}\mathbf{J}_{a_i}^T\left(\mathbf{q}_i\right)\mathbf{m}_i\mathbf{J}_{cm_i} \tag{76}$$

$$\mathbf{C}_{T_i}\left(\mathbf{q}_i, \dot{\mathbf{q}}_i\right) = \mathbf{C}_i\left(\mathbf{q}_i, \dot{\mathbf{q}}_i\right) + \tau_{a_i}\mathbf{J}_{a_i}^T\left(\mathbf{q}_i\right)\mathbf{m}_i\dot{\mathbf{J}}_{cm_i} + \mathbf{D}_i \tag{77}$$

$$\mathbf{g}_{T_i}\left(\mathbf{q}_i\right) = g_i\left(\mathbf{q}_i\right) - \tau_{a_i}\mathbf{J}_{a_i}^T\left(\mathbf{q}_i\right)\mathbf{m}_i g_i \tag{78}$$

Therefore, we get:

$$\mathbf{H}_{T_i}\left(\mathbf{q}_i\right)\ddot{\mathbf{q}}_i + \mathbf{C}_{T_i}\left(\mathbf{q}_i, \dot{\mathbf{q}}_i\right)\dot{\mathbf{q}}_i + \mathbf{g}_{T_i}\left(\mathbf{q}_i\right) = \tau_{a_i} - \mathbf{J}_{a_i}^T\left(\mathbf{q}_i\right)\mathbf{f}_{i+1}. \tag{79}$$

CONTACT MODELS

Contact between an end-effector and an object can be described as mapping between the forces exerted by the end-effector at a point of contact and reaction forces in some reference point on the object. Contact models are used to classify common contact configurations, which can be extended to consider effects of friction.

CONTACT POINT FRICTIONLESS

The contact point frictionless is obtained when it is not considered the friction between the end-effector and object. In this case, the forces can be applied only in the normal direction to the contact surface, the implementation of pushing force can be represented as:

$$\mathbf{F}_{c_i} = \begin{bmatrix} 0 \\ 0 \\ 1 \\ 0 \\ 0 \\ 0 \end{bmatrix}\mathbf{f}_{c_i} \tag{80}$$

where $\mathbf{f}_{c_i} \geq 0 \in \mathbb{R}$ is the magnitude of the force applied by the end-effector in the direction normal to the contact. The requirement of a positive value of $\mathbf{f}_{c_i}$ in the pattern starts from the fact that contacts of this type push the object being pulled. The frictionless contact never occurs in practical situations, but this approach can be used in applications in which the influence of friction is small enough to be ignored. In applications which require holding an object make their appearance frictional forces; therefore, they are set to be considered in the model. As a first approximation, we can simply use a model that contains only the Coulomb friction which provides an empirical model which states that the tangential forces are proportional to the normal force application, while the constant of proportionality is a function of the materials in contact.

It should be noted that $\mathbf{f}^t \in \mathbb{R}$ is defined as the magnitude of the tangential force, and $\mathbf{f}^n \in \mathbb{R}$ as the magnitude of the normal force, Coulomb's law states that the sliding contact begins when $\left|\mathbf{f}^t\right| > \mu \mathbf{f}^n$ where $\mu > 0$ is the coefficient of static friction. This implies that the range of forces that can be applied tangent to the contact is given by:

$$\left|\mathbf{f}^t\right| > \mu \mathbf{f}^n \quad \mu < 1 \text{ (Typical values).} \tag{81}$$

It shows that $\mathbf{f}^n$ must be positive because this relationship is valid for nonzero values of $\mathbf{f}^t$. We can represent (81) geometrically as Figure 7(b). The set of forces which can be applied in a contact must be in a cone centered on the normal surface. This cone is called *friction cone*, Figure 7(a). The taper angle with the normal one is given by:

$$\alpha = \tan^{-1}(\mu) \tag{82}$$

by the *friction cone* the angles are typically less than 45°.

Figure 7. Geometric interpretation of the Coulomb friction model

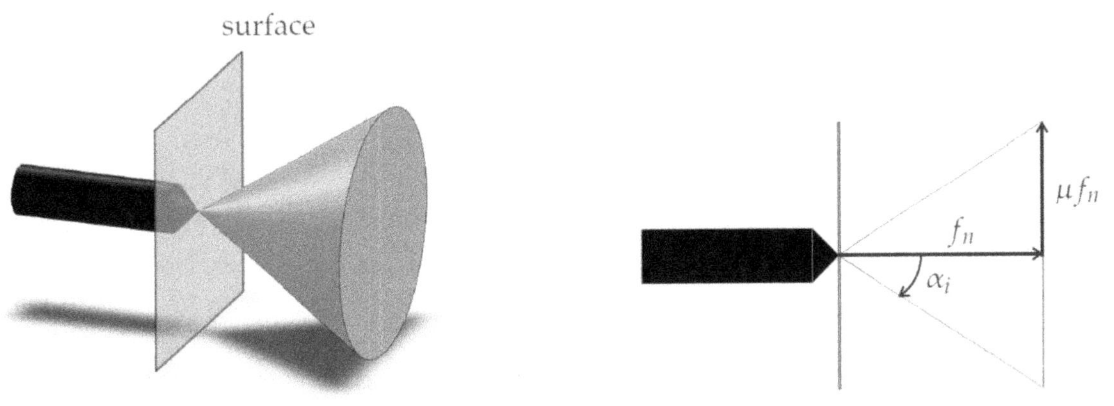

CONTACT POINT WITH FRICTION

The contact point with friction is used when there is friction between the tip of the end-effector and object, in this case the forces can be exerted in any direction within the friction cone. It may represent the application of the thrust exerted on the object respecting to a base of directions, which are consistent with the friction model:

$$\mathbf{F}_{c_i} = \begin{bmatrix} 1 & 0 & 0 \\ 0 & 1 & 0 \\ 0 & 0 & 1 \\ 0 & 0 & 0 \\ 0 & 0 & 0 \\ 0 & 0 & 0 \end{bmatrix} \mathbf{f}_{c_i} \tag{83}$$

where $\mathbf{f}_{c_i} \in C_f$ and the friction cone are defined as $C_f = \left(\mathbf{f} \in \mathbb{R}^{3\times 1}, \sqrt{f_1^2 + f_2^2} \leq \mu f_3, f_3 \geq 0\right)$.

SOFT-FINGER MODEL

Soft-finger model is a more realistic model which not only consider the applied forces in the friction cone on the normal surface but also on the torques which are considered normal. The model presented considered a limit to the initial torque by friction torque γ. Soft-finger model is described by:

$$\mathbf{F}_{c_i} = \begin{bmatrix} 1 & 0 & 0 & 0 \\ 0 & 1 & 0 & 0 \\ 0 & 0 & 1 & 0 \\ 0 & 0 & 0 & 0 \\ 0 & 0 & 0 & 0 \\ 0 & 0 & 0 & 1 \end{bmatrix} \mathbf{f}_{c_i} \tag{84}$$

where $\mathbf{f}_{c_i} \in C_f = \left(\mathbf{f} \in R^{4\times 1}, \sqrt{f_1^2 + f_2^2} \leq \mu f_3, f_3 \geq 0, \left|f_4\right| \leq \gamma f_3\right)$ and $\gamma > 0$ is the *friction torque*.

In general, contact is modeled by using thrust bases, $\mathbf{B}_{c_i} \in \mathbb{R}^{p\times m_i}$, and the friction cone C_f. The dimension at the thrust base m_i indicates that the number of independent forces can be applied at the moment of contact and which requires the friction cone C_f satisfies the following properties:

Property 1. The friction cone C_f is a closed subset of $\mathbb{R}^{m_i}$ with non-empty interior. ■

Property 2. $f_1, f_2 \in C_f \rightarrow \alpha f_1 + \beta f_2 \in C_f$ for $\alpha, \beta > 0$. ■

The set of contact forces applied to meet the properties above for a given contact is:

$$\mathbf{F}_{c_i} = \mathbf{B}_{c_i}\mathbf{f}_{c_i} \quad (85)$$

where $\mathbf{B}_{c_i} \in \mathbb{R}^{p \times m_i}$ models the contact by using thrust bases; y $\mathbf{f}_{c_i} \in C_f$.

PRINCIPLE OF ORTHOGONALIZATION

The principle of orthogonalization was proposed by (Arimoto *et al.*, 1994) as an extended notion of position-force control for robot manipulators under geometric constraints. This principle separates the position-feedback signals from the force-feedback signals by using a projection matrix defined to project the error signals of position and velocity onto the tangent plane to the surface at each instantaneous contact point.

Consequently, the residual position and the velocity error become perpendicular to the force vector, which is normal to the surface. In order to preserve this property, the constraint is manipulated to obtain two orthogonal subspaces. The principle of orthogonalization is related to two types of control: hybrid trajectory tracking and model-based adaptive control. The first type assumes that exact dynamic models with friction characteristics are available. In addition, it is reasonably assumed that the acceleration signal is not accessible. However, position, velocity and momentum signals are considered measurable in real-time. On the other hand, the model-based adaptive control is based on parameterization of the manipulator dynamics via the vector θ, whose components are unknown or uncertain functions of masses and inertia moments of the links. The control structure obtained by using the principle of orthogonalization can also be applied to the cooperative robot system; that is, a set of manipulators performing a coordinated task. This kind of systems is employed in activities where the use of a single robot is insufficient.

A typical example is a flexible assembly, where robots connect two or more parts of a product. Cooperative manipulators can also be used in material handling, for example, transporting objects beyond the load carrying capacity of a single robot. Furthermore, their employment allows to improve the quality of tasks in the manufacturer industry that require of great precision. On the other hand, cooperative robots are indispensable for grasping and skillful manipulation of objects. When force is applied on a rigid surface (restriction) and at the same time there exist a displacement on it, the velocity vector is in the plane tangent to the point of contact, whereas the application of force perpendicular to the plane, Figure 8.

In order to preserve this property, the restriction is manipulated to obtain two orthogonal subspaces. These subspaces give a raise two transformations which are used to get a suitable representation which will eliminate cross terms in the analysis of stability, so they can be made simple controller with simple stability tests, and convergence properties of position and force. These transformations are used to make the orthogonal projection of the vector $\mathbf{x}$ on a space dimension $(n-1)$ (tangent space) orthogonal to the Jacobian vector $\mathbf{J}_{\varphi x}^{T}(\mathbf{q})$, the tangent plane can be represented in matrix form by:

$$\mathbf{Q}_x = \mathbf{I} - \mathbf{J}_{\varphi x}^{T}(\mathbf{q})\left(\mathbf{J}_{\varphi x}(\mathbf{q})\mathbf{J}_{\varphi x}^{T}(\mathbf{q})\right)\mathbf{J}_{\varphi x}^{T}(\mathbf{q}) \quad (86)$$

Figure 8. Geometric decomposition point of contact

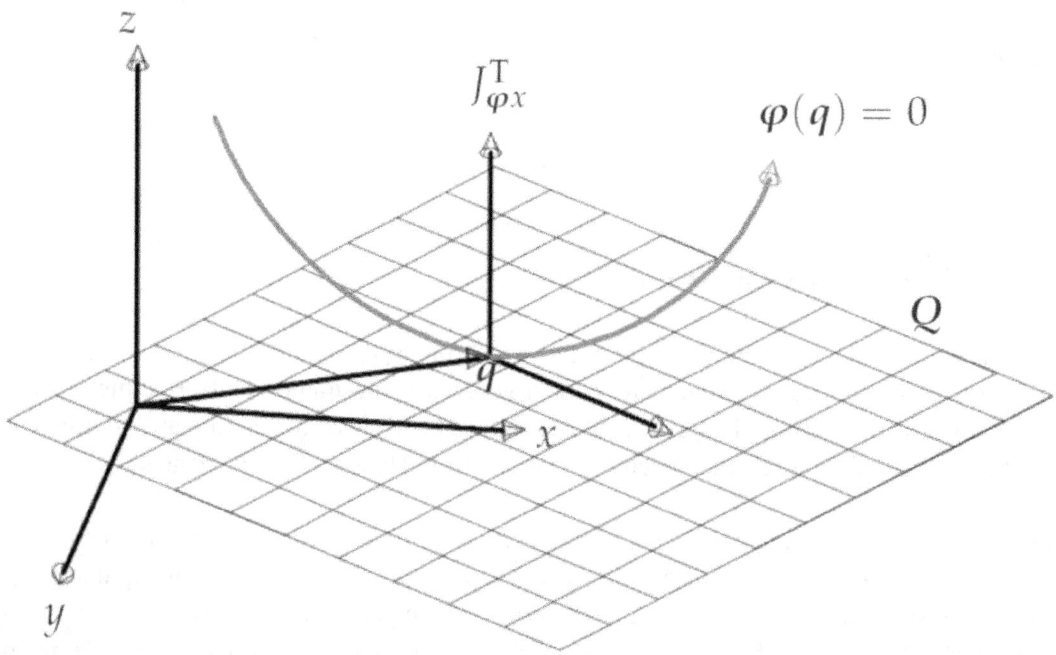

where $\mathbf{I}$ is an identity matrix. While the manipulator to move over the surface is satisfied:

$$\mathbf{Q}_x\dot{\mathbf{x}}=\dot{\mathbf{x}} \tag{87}$$

Property 3. The vector $\dot{\mathbf{q}}$ can be written as follows:

$$\dot{\mathbf{q}}_i=\mathbf{Q}_i\left(\mathbf{q}_i\right)\dot{\mathbf{q}}_i+\mathbf{P}_i\left(\mathbf{q}_i\right)\dot{\mathbf{q}}_i \tag{88}$$

where $\mathbf{P}_i\left(\mathbf{q}_i\right)=\mathbf{J}^{+}_{\varphi_i}\left(\mathbf{q}_i\right)\mathbf{J}_{\varphi_i}\left(\mathbf{q}_i\right),\mathbf{Q}_i\left(\mathbf{q}_i\right)=\mathbf{I}-\mathbf{P}_i\left(\mathbf{q}_i\right)\in\mathbb{R}^{n_i\times n_i}$ with $\operatorname{rank}\left[\mathbf{Q}_i\left(\mathbf{q}_i\right)\right]=n_i-m_i$, I is the identity matrix; $\mathbf{J}^{+}_{\varphi_i}\left(\mathbf{q}_i\right)=\mathbf{J}_{\varphi_i}{}^{T}\left(\mathbf{q}_i\right)\left[\mathbf{J}_{\varphi_i}\left(\mathbf{q}_i\right)\mathbf{J}_{\varphi_i}{}^{T}\left(\mathbf{q}_i\right)\right]^{-1}$ is Moore-Penrose pseudo invers. $\mathbf{J}^{+}_{\varphi_i}\left(\mathbf{q}_i\right)$ and $\mathbf{Q}_i\left(\mathbf{q}_i\right)$ are orthogonal, i.e., $\mathbf{Q}_i\left(\mathbf{q}_i\right)\mathbf{J}^{+}_{\varphi_i}\left(\mathbf{q}_i\right)=0$ and $\mathbf{Q}_i\left(\mathbf{q}_i\right)\mathbf{J}^{T}_{\varphi_i}\left(\mathbf{q}_i\right)=0$. ■

Property 4. While the end-effector of the manipulators are in contact with the surface, that is $\varphi_i\left(\mathbf{q}_i\right)=0$, are fulfilled:

$$\mathbf{J}_{\varphi_i}\left(\mathbf{q}_i\right)\dot{\mathbf{q}}_i=0 \tag{89}$$

therefore,

$$\mathbf{Q}_i\left(\mathbf{q}_i\right)\dot{\mathbf{q}}_i = \dot{\mathbf{q}}_i \tag{90}$$

where $\mathbf{Q}_i\left(\mathbf{q}_i\right)\mathbf{J}_{\varphi_i}^T\left(\mathbf{q}_i\right) = 0$. ■

Remark 1. Suppose a bounded trajectory fulfills the constraint in $\varphi_i\left(\mathbf{q}_i\right) = 0$ for $\mathbf{q}_i$ and $\mathbf{q}_{di}$. In this case, the tracking error can be defined as:

$$\begin{aligned}\tilde{\mathbf{q}}_i &= \mathbf{q}_i - \mathbf{q}_{\mathrm{d}i} \\ &= \mathbf{Q}_i\left(\mathbf{q}_i\right)\dot{\mathbf{q}}_i + \mathbf{J}_{\varphi_i}^{+}\left(\mathbf{q}_i\right)\mathbf{J}_{\varphi_i}\left(\mathbf{q}_i\right)\mathbf{q}_i - \mathbf{Q}_i\left(\mathbf{q}_{\mathrm{d}i}\right)\dot{\mathbf{q}}_{\mathrm{d}i} + \mathbf{J}_{\varphi_i}^{+}\left(\mathbf{q}_{\mathrm{d}i}\right)\mathbf{J}_{\varphi_i}\left(\mathbf{q}_{\mathrm{d}i}\right)\mathbf{q}_{\mathrm{d}i}\end{aligned} \tag{91}$$

In Figure 9(a), the case in which the tracking error is large is shown, and in Figure 9(b), we can see that $\tilde{\mathbf{q}}_i$ tends to be tangent to the surface as the error decreases.

By considering the surface smoothness and the small position error, we can do the following relationship:

$$\tilde{\mathbf{q}}_i = \mathbf{Q}_i\left(\mathbf{q}_i\right)\left[\mathbf{q}_i - \mathbf{q}_{\mathrm{d}i}\right] = \mathbf{Q}_i\left(\mathbf{q}_i\right)\tilde{\mathbf{q}}_i\,. \tag{92}$$

Using Property 3 and Property 4, we have:

$$\dot{\tilde{\mathbf{q}}}_i = \mathbf{Q}_i\left(\mathbf{q}_i\right)\left[\dot{\mathbf{q}}_i - \dot{\mathbf{q}}_{\mathrm{d}i}\right] = \mathbf{Q}_i\left(\mathbf{q}_i\right)\dot{\tilde{\mathbf{q}}}_i\,. \tag{93}$$

Figure 9. Surface tangent plane (tracking error). (left) Large tracking error (right) Small tracking error

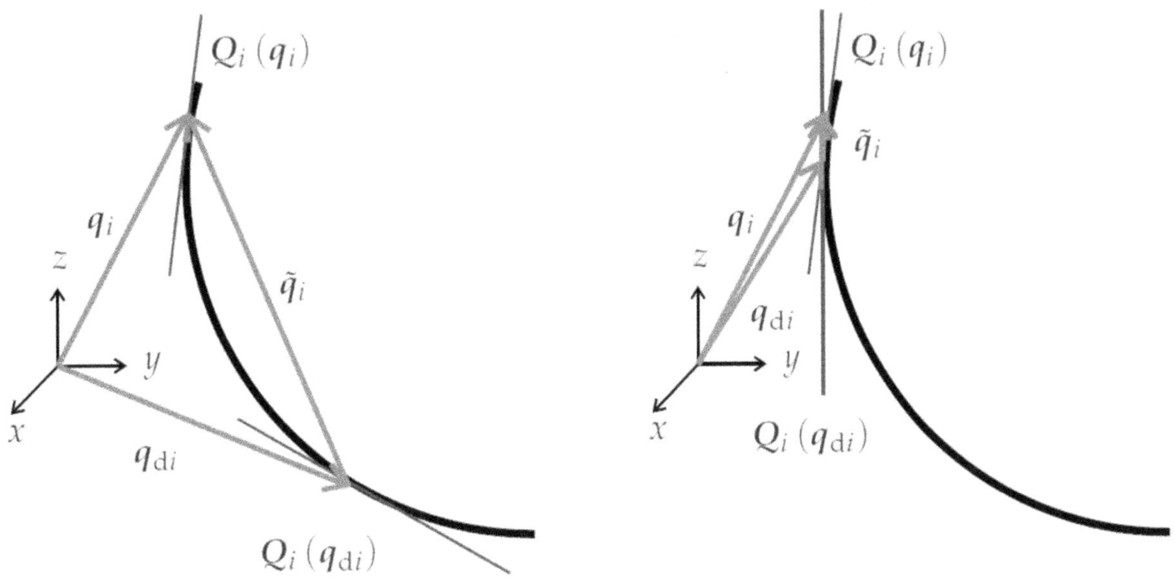

We can perform a local stability analysis for a small enough region around $\tilde{\mathbf{q}}_i = 0$. ■

Assumption 13. $\mathbf{J}_{\varphi_i}(\mathbf{q}_i) \in \mathbb{R}^{m_i \times n_i}$ is Lipschitz continuous, that is to say:

$$\left\| \mathbf{J}_{\varphi_i}(\mathbf{q}_i) - \mathbf{J}_{\varphi_i}(\mathbf{q}_{di}) \right\| \leq L_i \left\| \mathbf{q}_i - \mathbf{q}_{di} \right\| \tag{94}$$

where L_i is a positive constant and $\mathbf{q}_i, \mathbf{q}_{di} \in \mathbb{R}^{n_i}$. Besides, there exists a positive-finite constant C_{0i}, such as:

$$C_{0i} = \max_{\forall \mathbf{q}_i \in \mathbb{R}^{n_i}} \left\| \mathbf{J}^{+}_{\varphi_i}(\mathbf{q}_i) \right\| \blacksquare \tag{95}$$

By considering the use of robots with rotational joints, Assumption 13 is quite reasonable, since the elements of qi appear as an argument to the functions $\sin(\cdot)$ and $\cos(\cdot)$. For this reason, (95) is valid. It is should be noted that $\mathbf{J}^{+}_{\varphi_i}(\mathbf{q}_i)$, and $\mathbf{J}_{\varphi_i}(\mathbf{q}_i)$ are bounded for any $\mathbf{q}_i$. In addition, we consider that by applying the appropriate contact force on the surface, the effects of friction disappear, simplifying the dynamic model.

COOPERATIVE SYSTEM DESCRIPTION

The cooperative system consists in two robots: CRS A255 (5 DOF) and CRS A465 (6 DOF), each equipped with a force sensor JR3 and an impact guard mounted on the wrist and end-effector, Figure 10. The generalized coordinates of the cooperative system are the position, velocity and acceleration, but also the forces occurring from contact end-effector on the rigid object, and input generalized forces and torques exerted by the joints. In order to describe the kinematic relationship between the robots and the object, we select a stationary coordinate system $\{C_0\}$ as a frame of reference located at the base of the robot arm CRS A255, which has been regarded as the inertia coordinates frame of the entire system.

The object coordinate system $\{C_2\}$ is located in the center of mass. The origin of the coordinated frame $\{C_1\}$ is located at the Center Point of the A225 robot arm end-effector. Similarly, the origin of the coordinate system $\{C_3\}$ is located in the Center Point of the A465 robot arm end-effector. ${}^0p_2$ is the position vector of the center of mass on the object expressed in the coordinate system $\{C_0\}$. ${}^0p_1$ and ${}^0p_3$ are vectors that describe the position of the contact points between the end-effectors robot's arms A465, A255 and the object respectively, all expressed in the coordinate system $\{C_0\}$. Moreover, the cooperative system of Figure 10 is subject to $m = 2$ holonomic constraints given by:

$$\varphi(q_1, q_2) = 0 \tag{96}$$

Figure 10. Cooperative system diagram

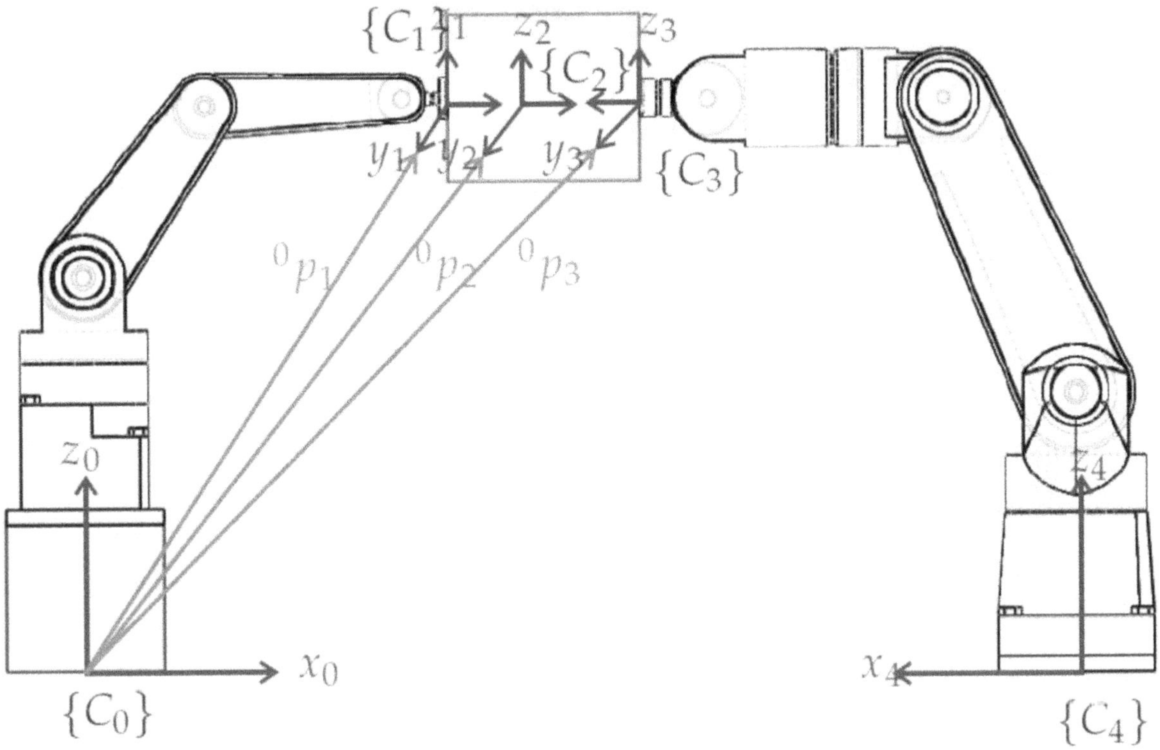

This means that the object is being manipulated, and the environment is described by restrictions (96). If the holonomic constraints are calculated correctly there is no guarantee that the robot arms will not drop the object. If one considers that the homogeneous holonomic constraints can be written position, velocity and acceleration as follows:

$$\varphi_i(\mathbf{q}_i) = \mathbf{f}(\mathbf{q}_i) = 0 \tag{97}$$

$$\ddot{\varphi}_i(\mathbf{q}_i) = \mathbf{J}_{\varphi_i}(\mathbf{q}_i)\ddot{\mathbf{q}} + \dot{\mathbf{J}}_{\varphi_i}(\mathbf{q}_i)\dot{\mathbf{q}} = L^2_{\dot{q}_i}\varphi_i(\mathbf{q}_i) = 0 \tag{98}$$

$$\dot{\varphi}_i(\mathbf{q}_i) = \mathbf{J}_{\varphi_i}(\mathbf{q}_i)\dot{\mathbf{q}} = L_{\dot{q}_i}\varphi_i(\mathbf{q}_i) = 0 \tag{99}$$

The relationship between the velocity restrictions and joint acceleration is described in (97)-(99). Since a limiting constraint is that geometry and motion of the system and being imposed by the environment, we can talk about defining a restricted space variables work, i.e. work space environment dependent. From (98), the joint velocity can be represented in terms of constraints as follows:

$$\dot{\mathbf{q}}_i = \mathbf{J}_{\varphi_i}^{-1}\left(\mathbf{q}_i\right)\dot{\varphi}_i \tag{100}$$

and joint acceleration $\ddot{\mathbf{q}}$ of the form:

$$\ddot{\varphi}_i = \mathbf{J}_{\varphi_i}\left(\mathbf{q}_i\right)\ddot{\mathbf{q}} + \dot{\mathbf{J}}_{\varphi_i}\left(\mathbf{q}_i\right)\dot{\mathbf{q}}$$

$$\mathbf{J}_{\varphi_i}\left(\mathbf{q}_i\right)\ddot{\mathbf{q}} = \ddot{\varphi}_i\left(\mathbf{q}_i\right) - \dot{\mathbf{J}}_{\varphi_i}\left(\mathbf{q}_i\right)\dot{\mathbf{q}}$$

$$\ddot{\mathbf{q}} = \mathbf{J}_{\varphi_i}^{-1}\left(\mathbf{q}_i\right)\ddot{\varphi}_i - \mathbf{J}_{\varphi_i}^{-1}\left(\mathbf{q}_i\right)\dot{\mathbf{J}}_{\varphi_i}\left(\mathbf{q}_i\right)\dot{\mathbf{q}} \tag{101}$$

substituting (100) in (101) is obtained the acceleration joint define as:

$$\ddot{\mathbf{q}}_i = \mathbf{J}_{\varphi_i}^{-1}\left(\mathbf{q}_i\right)\ddot{\varphi}_i - \mathbf{J}_{\varphi_i}^{-1}\left(\mathbf{q}_i\right)\dot{\mathbf{J}}_{\varphi_i}\left(\mathbf{q}_i\right)\mathbf{J}_{\varphi_i}^{-1}\left(\mathbf{q}_i\right)\dot{\varphi}_i \,. \tag{102}$$

SYMMETRIC FORMULATION

Considering a system of two cooperative robots manipulating a common object, Figure 10, where $\left\{C_2\right\}$ is a fixed point of the object, which is the *center of mass*, its position in the base frame $\left\{C_0\right\}$ is given by the vector ${}^0p_2$. Let $p_i \in \mathbb{R}^{n\times 1}$ denote the vector of the position of the i-th end-effector coordinate frame. Let $\Theta_i \in \mathbb{R}^{3\times 3}$ is a matrix which expresses the orientation of $\left\{C_2\right\}$ respecting to the base frame $\left\{C_0\right\}$. End-effector orientation may be expressed in Euler angles ϕ_i terms. Forward kinematics can be expressed by using an operational space vector x_i defined in (47). The external forces described in (36) represents the generalized forces causing the object's motion. Before making contact with the body, the manipulators are considered open kinematic chains, Figure 11.

If we describe the movement of robots in open kinematic chain, is observed that robots at in adequate position created an intersection between workspaces, Figure 12.

Considering the structural geometry and degrees of freedom of the robots CRS A255 and CRS A465, when the manipulators come into contact with the object, the cooperative system becomes into a closed kinematic chain, which has movement restrictions. The restrictions are due to the need to maintain contact with the body, allowing movement and handling. Figure 6 shows that in the robots are applied opposite forces in direction but equals in magnitude in order to maintain contact with the object. In this example, the robot manipulators who form the closed kinematic chain perform symmetrical joint displacements in magnitude for purpose of maintain contact with the body, Figure 13.

Figure 11. Cooperative system (open kinematic chains)

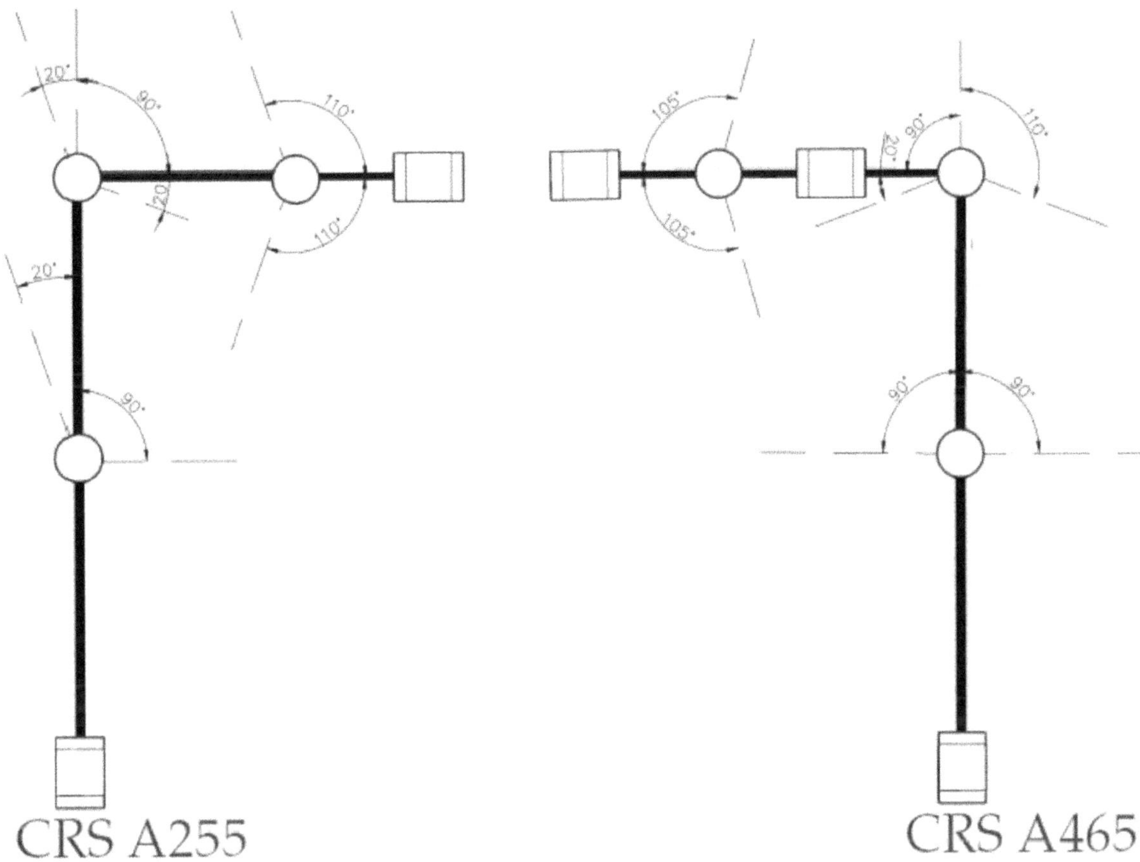

The workspace on the robot CRS A255 is defined as Ω_1 and the workspace on the robot CRS A465 is defined as Ω_2. We can observe in Figure 13 that workspaces of each robot manipulator are intersected and create the workspace of the cooperative system ($\Omega_3 = \Omega_1 \cap \Omega_2$). If the body is in the intersection Ω_3, we can say that the body can be manipulated by the cooperative system. However, we need to keep in mind that when the end-effectors of the robot manipulators are making contact with the body to be handled are created *movement restrictions* that limit the movement of the cooperative system to a small workspace. This fact generates inside the workspace of the cooperative system another smaller space where the closed kinematic chain can moves freely without losing contact with the object. In Figure 14(a) we can observe the workspace intersection of two robot manipulators, in this workspace they can move freely before the contact with the object, but when the contact is made between the object and the robots, the manipulators have a *restriction of movement,* and the workspace is reduce, as we can see in Figure 14(b).

As noted, we must first ensure that the manipulators reach the vicinity of the object, then we must make a smooth contact, and finally we must apply the necessary force in order to manipulate the body.

Figure 12. Workspace intersection

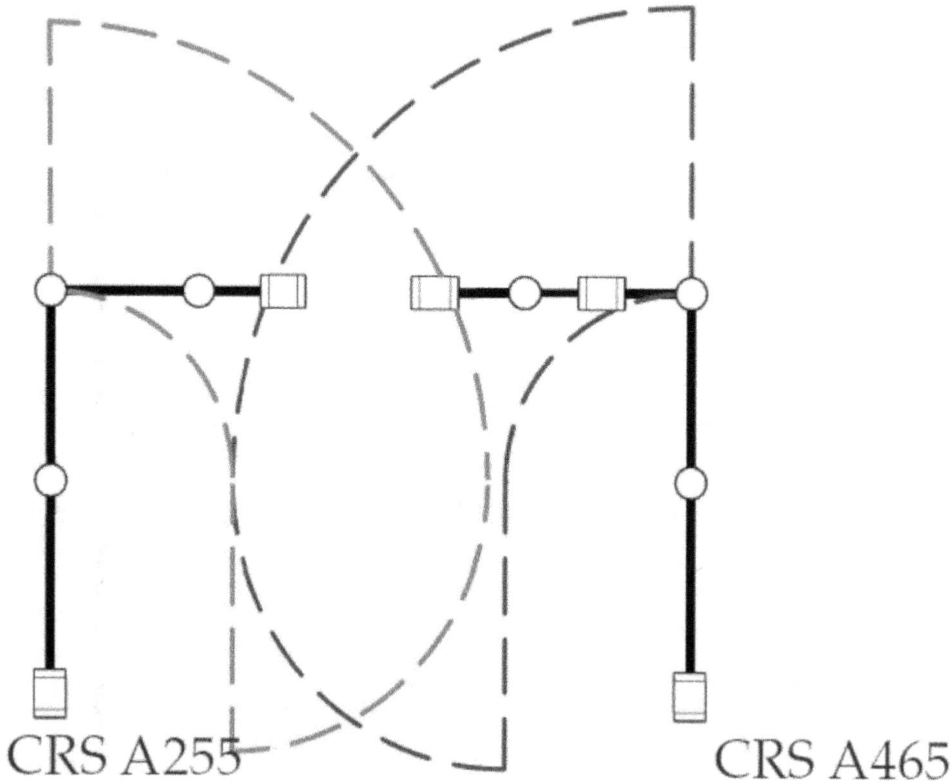

CLASSIFICATION AND GRIP TYPES

Generally, holding an object can be defined as *the application of functional forces to an object to perform a certain task*. In robotics, the grip is understood within *a system where an object is gripped by a set of robot manipulators or by the fingers of a robot*. In order to make a grip, we need to define a set of contact points or regions on the surface of an object and define the forces to be applied over them to satisfy certain criteria of immobility and handling. There are a variety of ways to hold an object, which are obtained by changing the kinematics and kinetics of the end-effector. We can handle the force applied at each end-effectors and change grip kinematics, by using different positions to hold an object. The grip type classification is performed based-on the morphology of the hand. Skinner has classified the grips in: *cylindrical, punctual, palmar, side, spherical* and *hook*, as we can see in Figure 15.

Napier's grab classification is more useful when considering functional aspects, while Skinner's classification is more extensive. Napier classified the grips into two categories: *force grip* and *precision grip*, as we can see in Figure 16.

Furthermore, Napier defines opposed grip, which is a movement in which the contact surface is placed at right angles or diametrically opposed to the object. Grip in opposition let us describe three basic directions through which forces can be applied.

- **Punctual opposed grip**: it occurs between the fingertips along a direction which is generally parallel to the palm, Figure 17(a).

Figure 13. Cooperative system (closed kinematic chains)

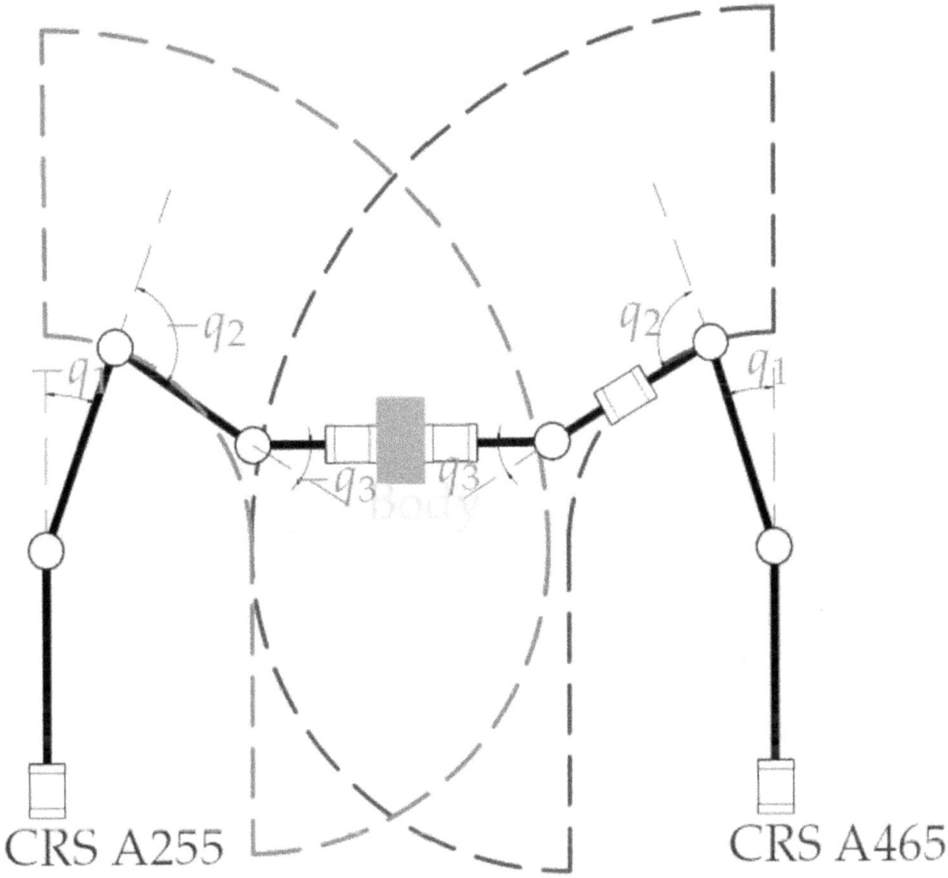

Figure 14. Workspaces intersection

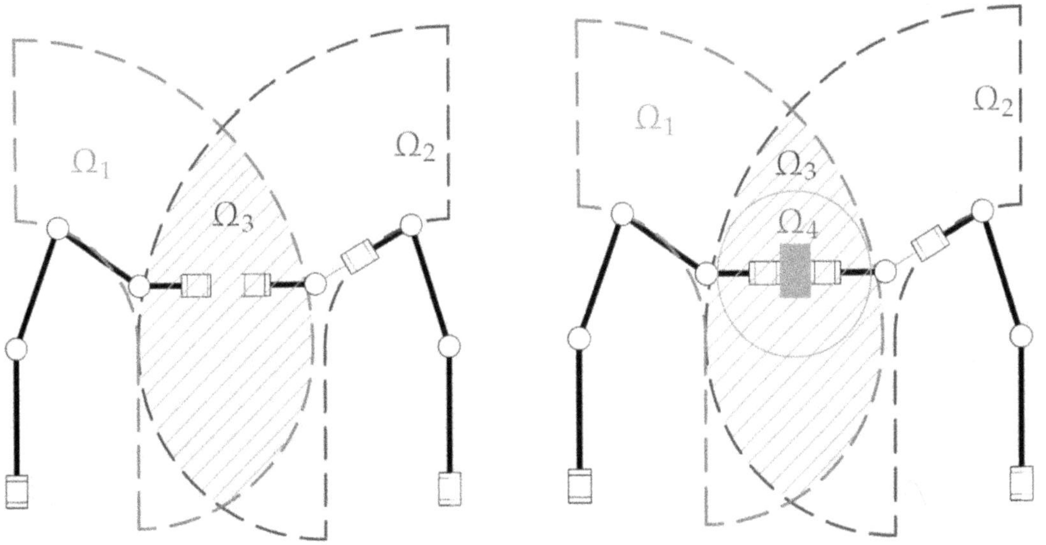

Figure 15. Skinner grip classification. (a) cylindrical grip, (b) punctual grip, (c) palmar grip, (d) side grip, (e) spherical grip, (f) hook grip

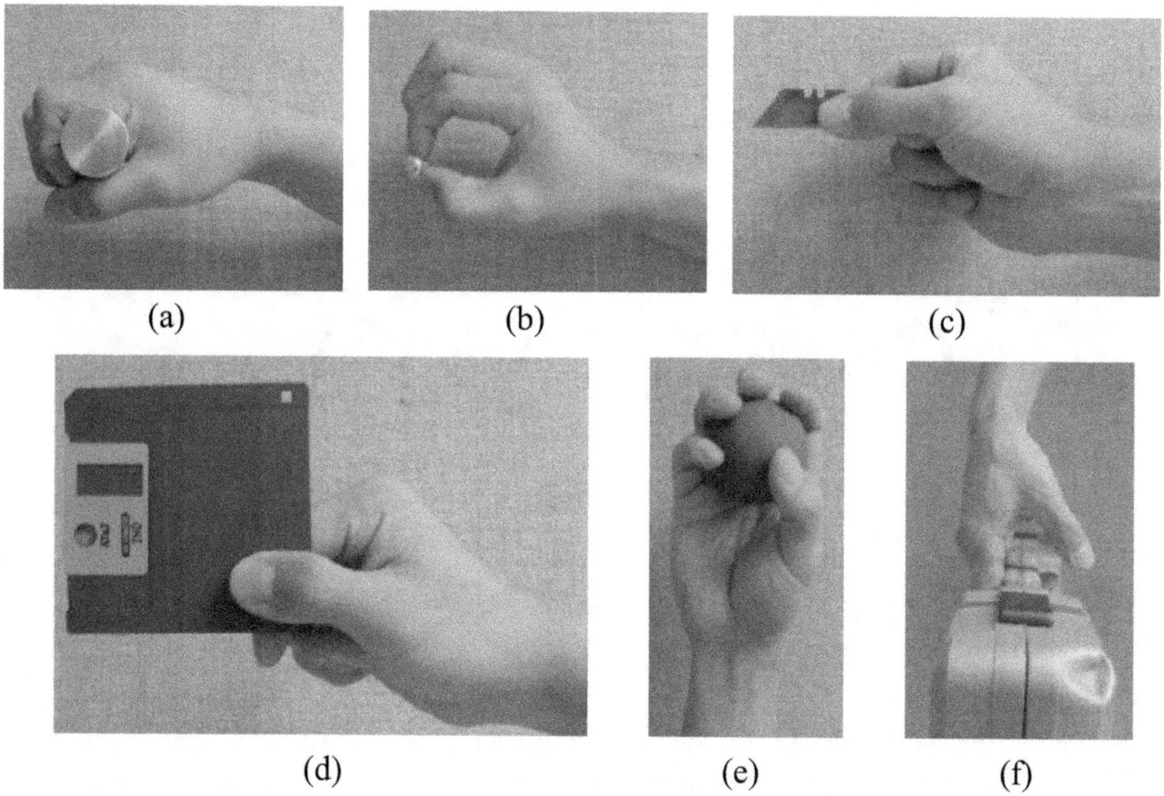

Figure 16. Napier's grab classification (left) force grip (right) precision grip

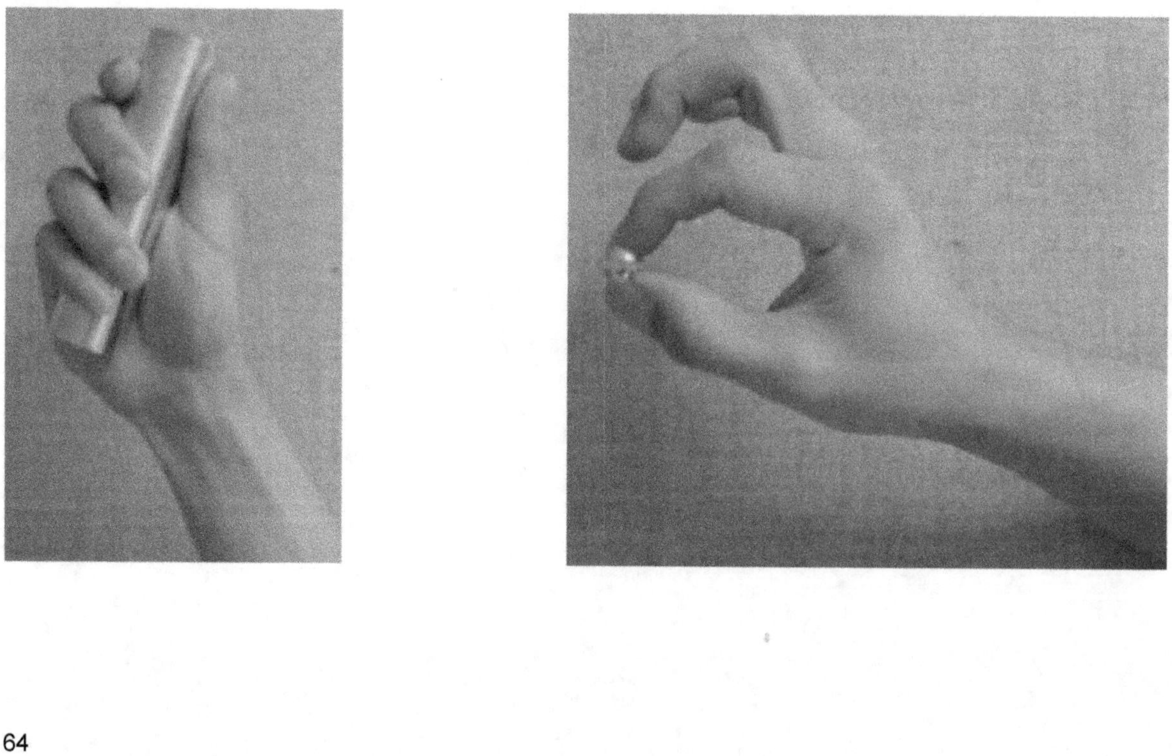

Figure 17. Napier's opposed grab classification (left) Punctual opposed grip (center) Palmar opposed grip (right) Side opposed grip

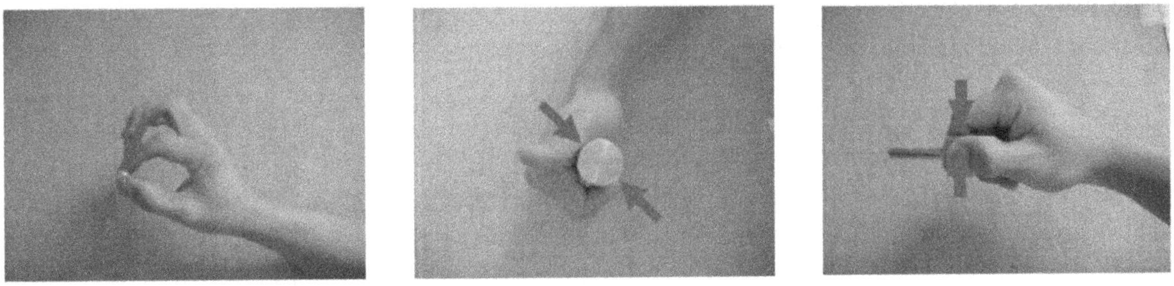

- **Palmar opposed grip**: it occurs between the surfaces of the fingers and the palm, along a direction which is generally perpendicular to the palm, Figure 17(b).
- **Side opposed grip**: it occurs between the surface of the index finger and thumb through a direction generally transverse to the palm, Figure 17(c).

Now we can define the grip due to the nature of the end-effectors of the robot manipulators CRS A255 and CRS A465, which is a *precision grip with a punctual opposite point*. In order to make a gripping task it must be resolved first *gripping planning problem* and then the *grip control problem*. Within the *gripping planning problem* are two stages, the first is the *synthesis of the grip*, and the second is the *analysis of grip*. The *synthesis of the grip* is the stage where we determine the contact points and the necessary forces to ensure a grip that completely restricts the movement of the object considering the geometry of the object, the model of type of contact and external forces. Moreover, given the contact points, the gripping forces and model type of contact, the objective of the grip analysis is determining its kinematic and dynamic properties. The most important properties of the grip are the equilibrium, dexterity, stability, strength lock and dynamic behavior. *Grip is in equilibrium* when the sums of forces and moments acting on the objects are zero. The *dexterity of a grip* is defined as the ability to achieve one or more secondary tasks while maintaining contact with the object. We say that a *grip is stable* when any errors in the position and force trajectories are caused by disturbances, vanishes with time. That is, if a force perturbs the object out of its balance, once the disturbing force disappears, the object returns to its equilibrium position.

We say that a grip has the property of *force lock* if it can withstand any external force or torque applied to the object. A grip with this property can generate forces of suitable magnitude in any direction in order to withstand the effect of external forces and maintain the stability of the grip. The dynamic behavior is defined as the response of the grip on time before changes or disturbances in the paths of position and force applied to the contact point on the object surface. It is noteworthy that this property is desirable but not essential.

PLANNING TRAJECTORIES

Planning is to select the appropriate trajectory in order to bring the robot end-effector at the vicinity of the body and choosing properly the contact points between the object and the end-effectors. In order to ensure a right grip between the body and the end-effectors of the robot manipulators. We first must be planning a soft trajectory that carries the end-effectors of all robot manipulators at the vicinity of the body, then we must be planning the contact by determining the properly contact points, and finally we must hold the object ensuring the sufficient force so that no relative motion exists.

Trajectories for Point to Point Motion

The main problem is to find a trajectory which connects an initial to a final configuration while satisfying other specified constraints at the endpoints. We will concern ourselves with the problem of determining $\mathbf{q}(t)$, where $\mathbf{q}(t)$ is a scalar joint variable. In order to obtain a continuous position, velocities and accelerations at the start and end point's times we require a fifth order polynomial and its first two derivatives. Expressing the fifth order polynomial considering $t = t_0$ we have:

$$\mathbf{q}(t_0) = \mathbf{q}_0 = a_0 + a_1 t_0 + a_2 t_0^2 + a_3 t_0^3 + a_4 t_0^4 + a_5 t_0^5 \tag{103}$$

$$\dot{\mathbf{q}}(t_0) = \dot{\mathbf{q}}_0 = a_1 + 2a_2 t_0 + 3a_3 t_0^2 + 4a_4 t_0^3 + 5a_5 t_0^4 \tag{104}$$

$$\ddot{\mathbf{q}}(t_0) = \ddot{\mathbf{q}}_0 = 2a_2 + 6a_3 t_0 + 12a_4 t_0^2 + 20a_5 t_0^3 \tag{105}$$

Now defining the fifth order polynomial according to $t = t_f$, we have:

$$\mathbf{q}(t_f) = \mathbf{q}_f = a_0 + a_1 t_f + a_2 t_f^2 + a_3 t_f^3 + a_4 t_f^4 + a_5 t_f^5 \tag{106}$$

$$\dot{\mathbf{q}}(t_f) = \dot{\mathbf{q}}_f = a_1 + 2a_2 t_f + 3a_3 t_f^2 + 4a_4 t_f^3 + 5a_5 t_f^4 \tag{107}$$

$$\ddot{\mathbf{q}}(t_f) = \ddot{\mathbf{q}}_f = 2a_2 + 6a_3 t_f + 12a_4 t_f^2 + 20a_5 t_f^3 \tag{108}$$

Evaluating **(103)-(105)** for $t_0 = 0$, we have:

$$\mathbf{q}_0 = a_0 + a_1(0) + a_2(0)^2 + a_3(0)^3 + a_4(0)^4 + a_5(0)^5 = a_0 \tag{109}$$

$$\dot{\mathbf{q}}_0 = a_1 + 2a_2(0) + 3a_3(0)^2 + 4a_4(0)^3 + 5a_5(0)^4 = a_1 \tag{110}$$

$$\ddot{\mathbf{q}}_0 = 2a_2 + 6a_3(0) + 12a_4(0)^2 + 20a_5(0)^3 = 2a_2 \tag{111}$$

Defining the following initial conditions:

$$\mathbf{q}(t_0) = \mathbf{q}_0 = \mathbf{q} \tag{112}$$

$$\dot{\mathbf{q}}(t_0) = \dot{\mathbf{q}}_0 = 0 \tag{113}$$

$$\ddot{\mathbf{q}}(t_0) = \ddot{\mathbf{q}}_0 = 0 \tag{114}$$

from **(109)-(111)**, we have:

$$a_0 = \mathbf{q} \tag{115}$$

$$a_1 = 0 \tag{116}$$

$$a_2 = 0 \tag{117}$$

Substituting **(115)-(117)** in **(109)-(111)**, we get:

$$\mathbf{q}_f = \mathbf{q} + (0)t_f + (0)t_f^2 + a_3t_f^3 + a_4t_f^4 + a_5t_f^5 = \mathbf{q} + a_3t_f^3 + a_4t_f^4 + a_5t_f^5 \tag{118}$$

$$\dot{\mathbf{q}}_f = (0) + 2(0)t_f + 3a_3t_f^2 + 4a_4t_f^3 + 5a_5t_f^4 = 3a_3t_f^2 + 4a_4t_f^3 + 5a_5t_f^4 \tag{119}$$

$$\ddot{\mathbf{q}}_f = 2(0) + 6a_3t_f + 12a_4t_f^2 + 20a_5t_f^3 = 6a_3t_f + 12a_4t_f^2 + 20a_5t_f^3 \tag{120}$$

Defining the following final conditions:

$$\mathbf{q}(t_f) = \mathbf{q}_f = \mathbf{q}_d \tag{121}$$

$$\dot{\mathbf{q}}\left(t_f\right) = \dot{\mathbf{q}}_f = 0 \tag{122}$$

$$\ddot{\mathbf{q}}\left(t_f\right) = \ddot{\mathbf{q}}_f = 0 \tag{123}$$

Forming the following system:

$$\mathbf{q}_d = \mathbf{q} + a_3 t_f^3 + a_4 t_f^4 + a_5 t_f^5 \tag{124}$$

$$0 = 3a_3 t_f^2 + 4a_4 t_f^3 + 5a_5 t_f^4 \tag{125}$$

$$0 = 6a_3 t_f + 12a_4 t_f^2 + 20a_5 t_f^3 \tag{126}$$

The system can be written as follows:

$$\begin{bmatrix} t_f^3 & t_f^4 & t_f^5 \\ 3t_f^2 & 4t_f^3 & 5t_f^4 \\ 6t_f & 12t_f^2 & 20t_f^3 \end{bmatrix} \begin{bmatrix} a_3 \\ a_4 \\ a_5 \end{bmatrix} = \begin{bmatrix} \mathbf{q}_d - \mathbf{q} \\ 0 \\ 0 \end{bmatrix} = \begin{bmatrix} \tilde{\mathbf{q}} \\ 0 \\ 0 \end{bmatrix} \tag{127}$$

Applying Cramer's rule yields the following solutions:

$$a_3 = -\frac{10}{t_f^3}\tilde{\mathbf{q}} \tag{128}$$

$$a_4 = \frac{15}{t_f^4}\tilde{\mathbf{q}} \tag{129}$$

$$a_5 = -\frac{6}{t_f^5}\tilde{\mathbf{q}} \tag{130}$$

We need to keep in mind that $t_f > 0$. This polynomial trajectory ensures the end-effectors place in the vicinity of the object within a smooth way in a finite time, Figure 18.

Figure 18. End-effectors in the vicinity of the object

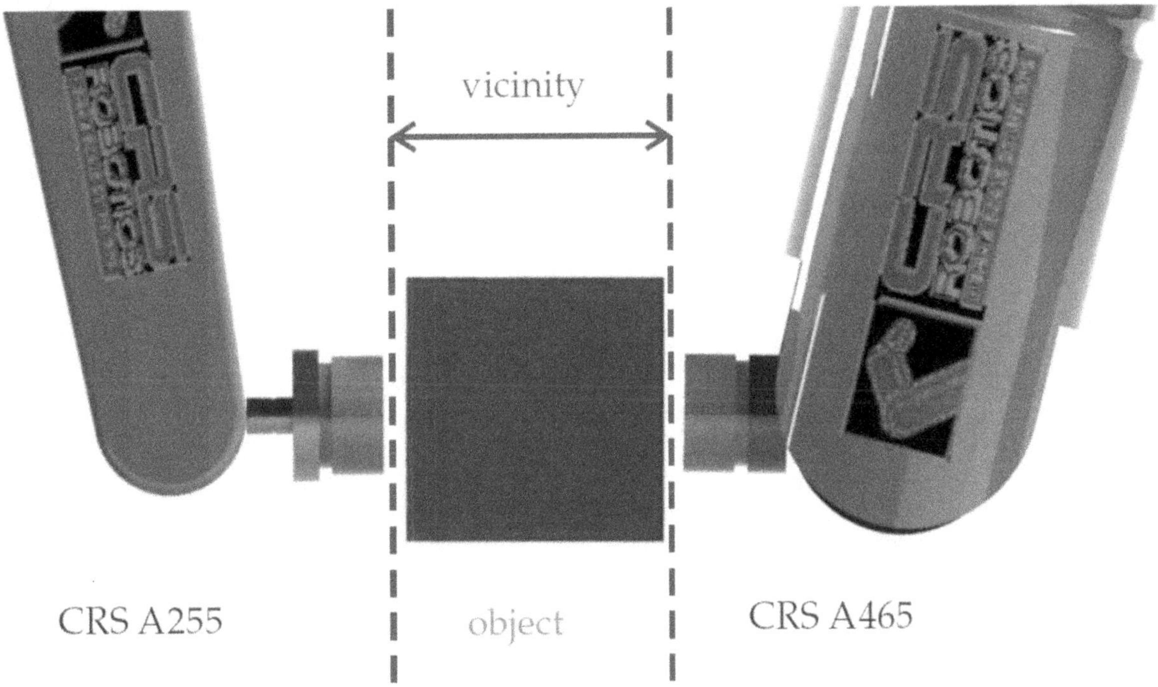

Control with Velocity Estimation

After placing the robots close to the object, the next step is to contact the object and manipulate it. In order to design the control law, consider the dynamic model described in (40) writing as:

$$\mathbf{H}_i\left(\mathbf{q}_i\right)\ddot{\mathbf{q}}_i+\mathbf{C}_i\left(\mathbf{q}_i,\dot{\mathbf{q}}_i\right)\dot{\mathbf{q}}_i+\mathbf{g}_i\left(\mathbf{q}_i\right)+\mathbf{D}_i\dot{\mathbf{q}}_i=\tau_{a_i}-\mathbf{J}_{\varphi_i}^T\left(\mathbf{q}\right)\boldsymbol{\lambda}_i \tag{131}$$

and define the tracking, observation, force, and constrained position errors as follow:

$$\tilde{\mathbf{q}}_i \stackrel{\Delta}{=} \mathbf{q}_i-\mathbf{q}_{d_i} \tag{132}$$

$$z_i \stackrel{\Delta}{=} \mathbf{q}_i-\hat{\mathbf{q}}_i \tag{133}$$

$$\Delta\boldsymbol{\lambda}_i \stackrel{\Delta}{=} \boldsymbol{\lambda}_i-\boldsymbol{\lambda}_{d_i} \tag{134}$$

$$\Delta \mathbf{p}_i \overset{\Delta}{=} \mathbf{p}_i - \mathbf{p}_{d_i} \tag{135}$$

where $\mathbf{q}_{d_i}$ is a desired smooth bounded trajectory, $\mathbf{p}_{d_i}$ is the desired constrained position which satisfies (60) and (61), $\hat{\mathbf{q}}_i$ represents the estimated value of $\mathbf{q}_i$ and $\boldsymbol{\lambda}_{d_i}$ is the desired force to be applied by each end-effector on the constrained surface. $\hat{\boldsymbol{\lambda}}_i$ represents the estimated value of $\boldsymbol{\lambda}_i$. The corresponding observation error is given by:

$$\Delta \bar{\boldsymbol{\lambda}}_i \overset{\Delta}{=} \boldsymbol{\lambda}_i - \hat{\boldsymbol{\lambda}}_i \tag{136}$$

Now, using the following definitions, we have:

$$\mathbf{q}_{r_i} \overset{\Delta}{=} \mathbf{Q}_i(\mathbf{q}_i)\left[\dot{\mathbf{q}}_{d_i} - \Lambda_i\left(\hat{\mathbf{q}}_i - \mathbf{q}_{d_i}\right)\right] + \mathbf{J}^+_{\varphi_i}(\mathbf{q}_i)\left[\dot{\mathbf{p}}_{d_i} - \beta_i \Delta \mathbf{p}_i + \xi_i \Delta \mathbf{F}_i\right] \tag{137}$$

$$s_i \overset{\Delta}{=} \dot{\mathbf{q}}_i - \dot{\mathbf{q}}_{r_i} = \mathbf{Q}_i(\mathbf{q}_i)\left[\dot{\tilde{\mathbf{q}}}_i - \Lambda_i \tilde{\mathbf{q}}_i - \Lambda_i z_i\right] + \mathbf{J}^+_{\varphi_i}(\mathbf{q}_i)\left[\Delta \dot{\mathbf{p}}_i + \beta_i \Delta \mathbf{p}_i - \xi_i \Delta \mathbf{F}_i\right] \tag{138}$$

$$\Delta \mathbf{F}_i \overset{\Delta}{=} \int_0^t \Delta \lambda_i d\vartheta . \tag{139}$$

where $\boldsymbol{\lambda}_i$ is an auxiliary force error. $\Lambda_i = k_i \mathbf{I} \in \mathbb{R}^{n_i x n_i}$ with $k_i > 0$, and $\xi_i \in \mathbb{R}^{m_i x m_i}$ are diagonal positive definite matrixes, and β_i is a positive constant. Note that s_{p_i} and s_{f_i} are orthogonal vectors. Analyzing $\ddot{\mathbf{q}}_{r_i}$, quantity given by:

$$\begin{aligned} \ddot{\mathbf{q}}_{r_i} \overset{\Delta}{=} & \dot{\mathbf{Q}}_i(\mathbf{q}_i)\left[\dot{\mathbf{q}}_{d_i} - \Lambda_i\left(\hat{\mathbf{q}}_i - \mathbf{q}_{d_i}\right)\right] + \dot{\mathbf{J}}^+_{\varphi_i}(\mathbf{q}_i)\left[\dot{\mathbf{p}}_{d_i} - \beta_i \Delta \mathbf{p}_i + \xi_i \Delta \mathbf{F}_i\right] \\ & + \mathbf{Q}_i(\mathbf{q}_i)\left[\ddot{\mathbf{q}}_{d_i} - \Lambda_i\left(\hat{\mathbf{q}}_i - \mathbf{q}_{d_i}\right)\right] + \mathbf{J}^+_{\varphi_i}(\mathbf{q}_i)\left[\ddot{\mathbf{p}}_{d_i} - \beta_i\left(\dot{\mathbf{p}}_i - \dot{\mathbf{p}}_{d_i}\right) + \xi_i \Delta \boldsymbol{\lambda}_i\right] \end{aligned} \tag{140}$$

As it will be shown later, $\ddot{\mathbf{q}}_{r_i}$ is necessary to implement the controller and the observer. However, this quantity is not available since $\mathbf{q}_i$ is not measurable. In order to overcome this drawback, let us consider $\mathbf{Q}_i(\mathbf{q}_i) \in \mathbb{R}^{n_i x n_i}$. We have:

$$\dot{\mathbf{Q}}_i\left(\mathbf{q}_i\right)=\begin{bmatrix} \frac{\partial a_{11}\left(\mathbf{q}_i\right)}{\partial \mathbf{q}_i}\dot{\mathbf{q}}_i & \cdots & \frac{\partial a_{1n_i}\left(\mathbf{q}_i\right)}{\partial \mathbf{q}_i}\dot{\mathbf{q}}_i \\ \vdots & \ddots & \vdots \\ \frac{\partial a_{n_i1}\left(\mathbf{q}_i\right)}{\partial \mathbf{q}_i}\dot{\mathbf{q}}_i & \cdots & \frac{\partial a_{n_in_i}\left(\mathbf{q}_i\right)}{\partial \mathbf{q}_i}\dot{\mathbf{q}}_i \end{bmatrix} \tag{141}$$

where $a_{\alpha_i\beta_i}$ is the $\alpha_i\beta_i$ element of $\mathbf{Q}_i\left(\mathbf{q}_i\right)$.

Based on (141) we consider the following definition:

$$\dot{\hat{\mathbf{Q}}}_i\left(\mathbf{q}_{oi}\right)=\begin{bmatrix} \frac{\partial a_{11}\left(\mathbf{q}_i\right)}{\partial \mathbf{q}_i}\dot{\mathbf{q}}_{oi} & \cdots & \frac{\partial a_{1n_i}\left(\mathbf{q}_i\right)}{\partial \mathbf{q}_i}\dot{\mathbf{q}}_{oi} \\ \vdots & \ddots & \vdots \\ \frac{\partial a_{n_i1}\left(\mathbf{q}_i\right)}{\partial \mathbf{q}_i}\dot{\mathbf{q}}_{oi} & \cdots & \frac{\partial a_{n_in_i}\left(\mathbf{q}_i\right)}{\partial \mathbf{q}_i}\dot{\mathbf{q}}_{oi} \end{bmatrix} \tag{142}$$

with

$$\dot{\mathbf{q}}_{oi}\overset{\Delta}{=}\dot{\hat{\mathbf{q}}}_i-\Lambda_i z_i \tag{143}$$

Then, we can compute:

$$\dot{\tilde{\mathbf{Q}}}_i\left(r_i\right)=\begin{bmatrix} \frac{\partial a_{11}\left(\mathbf{q}_i\right)}{\partial \mathbf{q}_i}r_i & \cdots & \frac{\partial a_{1n_i}\left(\mathbf{q}_i\right)}{\partial \mathbf{q}_i}r_i \\ \vdots & \ddots & \vdots \\ \frac{\partial a_{n_i1}\left(\mathbf{q}_i\right)}{\partial \mathbf{q}_i}r_i & \cdots & \frac{\partial a_{n_in_i}\left(\mathbf{q}_i\right)}{\partial \mathbf{q}_i}r_i \end{bmatrix} \tag{144}$$

with $\dot{\tilde{\mathbf{Q}}}_i\left(r_i\right)\overset{\Delta}{=}\dot{\mathbf{Q}}_i\left(\mathbf{q}_i\right)-\dot{\hat{\mathbf{Q}}}_i\left(\mathbf{q}_{oi}\right)$, and

$$r_i\overset{\Delta}{=}\dot{\mathbf{q}}_i-\dot{\mathbf{q}}_{oi}=\dot{z}_i+\Lambda_i z_i \tag{145}$$

From (142) the following substitution for $\mathbf{q}_{r_i}$ is used:

$$\ddot{\mathbf{q}}_{r_i} \stackrel{\Delta}{=} \dot{\hat{\mathbf{Q}}}_i(\mathbf{q}_i)\left[\dot{\mathbf{q}}_{d_i} - \Lambda_i\left(\hat{\mathbf{q}}_i - \mathbf{q}_{d_i}\right)\right] + \dot{\hat{\mathbf{J}}}^{+}_{\varphi_i}(\mathbf{q}_i)\left[\dot{\mathbf{p}}_{d_i} - \beta_i \Delta \mathbf{p}_i + \xi_i \Delta \mathbf{F}_i\right] \\ + \mathbf{Q}_i(\mathbf{q}_i)\left[\ddot{\mathbf{q}}_{d_i} - \Lambda_i\left(\dot{\hat{\mathbf{q}}}_i - \dot{\mathbf{q}}_{d_i}\right)\right] + \mathbf{J}^{+}_{\varphi_i}(\mathbf{q}_i)\left[\ddot{\mathbf{p}}_{d_i} - \beta_i \Delta\left(\dot{\mathbf{p}}_i - \dot{\mathbf{p}}_{d_i}\right) + \xi_i \Delta \boldsymbol{\lambda}_i\right] \tag{146}$$

where $\dot{\hat{\mathbf{J}}}^{+}_{\varphi_i}(\mathbf{q}_i)$ is defined in the very same fashion as in (142). Note that $\dot{\mathbf{p}}_i$ is still used since this value is known from (60) and (61). After some manipulation, it is possible to get:

$$\ddot{\ddot{\mathbf{q}}}_{r_i} = \ddot{\mathbf{q}}_{r_i} + e_i(r_i) \tag{147}$$

where

$$e_i(r_i) \stackrel{\Delta}{=} -\dot{\tilde{\mathbf{Q}}}_i(r_i)\left[\dot{\mathbf{q}}_{d_i} - \Lambda_i\left(\hat{\mathbf{q}}_i - \mathbf{q}_{d_i}\right)\right] - \dot{\tilde{\mathbf{J}}}^{+}_{\varphi_i}(r_i)\left[\dot{\mathbf{p}}_{d_i} - \beta_i \Delta \mathbf{p}_i + \xi_i \Delta \mathbf{F}_i\right] \tag{148}$$

The controller is given for each single input by:

$$\tau_i \stackrel{\Delta}{=} \mathbf{H}_i(\mathbf{q}_i)\ddot{\mathbf{q}}_{r_i} + \mathbf{H}_i(\mathbf{q}_i)\mathbf{e}_i(\mathbf{r}_i) + \mathbf{C}_i\left(\mathbf{q}_i, \dot{\mathbf{q}}_{r_i}\right)\dot{\mathbf{q}}_{r_i} + \mathbf{D}_i\dot{\mathbf{q}}_{r_i} + \\ g_i(\mathbf{q}_i) - K_R\left(\dot{\mathbf{q}}_{o_i} - \dot{\mathbf{q}}_{r_i}\right) - \mathbf{J}^{T}_{\varphi_i}(\mathbf{q}_i)\left[\boldsymbol{\lambda}_{d_i} - k_{F_i}\Delta \mathbf{F}_i\right] \tag{149}$$

where $K_{R_i} \in \mathbb{R}^{n_i \times n_i}$ is a diagonal positive definite matrix and K_{F_i} is a positive constant. From (147), we get:

$$\tau_i \stackrel{\Delta}{=} \mathbf{H}_i(\mathbf{q}_i)\left[\ddot{\mathbf{q}}_{r_i} - e_i\right] + \mathbf{C}_i\left(\mathbf{q}_i, \dot{\hat{\mathbf{q}}}_{r_i}\right)\dot{\hat{\mathbf{q}}}_{r_i} + \\ \mathbf{F}_{v_i}\dot{\mathbf{q}}_{r_i} + g_i - K_{R_i}\left(s_i - r_i\right) - \mathbf{J}^{T}_{\varphi_i}(\mathbf{q}_i)\left[\boldsymbol{\lambda}_{d_i} - K_{F_i}\Delta \mathbf{F}_i\right] \tag{150}$$

Now, we compute the closed loop dynamics:

$$\mathbf{H}_i(\mathbf{q}_i)\ddot{\mathbf{q}}_i + \mathbf{C}_i(\mathbf{q}_i, \dot{\mathbf{q}}_i)\dot{\mathbf{q}}_i + \mathbf{D}_i\dot{\mathbf{q}}_i + \mathbf{g}_i(\mathbf{q}_i) = \\ \mathbf{H}_i(\mathbf{q}_i)\ddot{\mathbf{q}}_{r_i} + \mathbf{H}_i(\mathbf{q}_i)\mathbf{e}_i(\mathbf{r}_i) + \mathbf{C}_i\left(\mathbf{q}_i, \dot{\mathbf{q}}_{r_i}\right)\dot{\mathbf{q}}_{r_i} + \mathbf{D}_i\dot{\mathbf{q}}_{r_i}$$

$$+\mathbf{g}_i(\mathbf{q}_i) - \mathbf{K}_{R_i}\left(\dot{\mathbf{q}}_{0i} - \dot{\mathbf{q}}_{r_i}\right) - \mathbf{J}^{T}_{\varphi_i}(\mathbf{q}_i)\left[\boldsymbol{\lambda}_{d_i} - k_{F_i}\Delta \mathbf{F}_i\right]$$

$$+\mathbf{J}^{T}_{\varphi_i}(\mathbf{q}_i)\boldsymbol{\lambda}_i \tag{151}$$

Note that:

$$\dot{\mathbf{q}}_{oi} - \dot{\mathbf{q}}_{r_i} = \dot{\mathbf{q}}_i - \dot{\mathbf{q}}_{r_i} + \dot{\mathbf{q}}_{oi} - \dot{\mathbf{q}}_i = \mathbf{s}_i - \mathbf{r}_i. \tag{152}$$

Thus we have:

$$\begin{aligned}&\mathbf{H}_i\left(\mathbf{q}_i\right)\dot{\mathbf{s}}_i + \mathbf{C}_i\left(\mathbf{q}_i,\dot{\mathbf{q}}_i\right)\dot{\mathbf{q}}_i - \mathbf{C}_i\left(\mathbf{q}_i,\dot{\mathbf{q}}_{r_i}\right)\dot{\mathbf{q}}_{r_i} + \mathbf{D}_i\mathbf{s}_i + \mathbf{K}_{R_i}\mathbf{s}_i = \\ &\mathbf{J}_{\varphi_i}^T\left(\mathbf{q}_i\right)\left[\Delta\boldsymbol{\lambda}_i - k_{F_i}\Delta\mathbf{F}_i\right] + \mathbf{K}_{R_i}\mathbf{r}_i + \mathbf{H}_i\left(\mathbf{q}_i\right)\mathbf{e}_i\left(\mathbf{r}_i\right)\end{aligned} \tag{153}$$

Now, we compute

$$\begin{aligned}&\mathbf{C}_i\left(\mathbf{q}_i,\dot{\mathbf{q}}_i\right)\dot{\mathbf{q}}_i - \mathbf{C}_i\left(\mathbf{q}_i,\dot{\mathbf{q}}_{r_i}\right)\dot{\mathbf{q}}_{r_i} = \\ &\mathbf{C}_i\left(\mathbf{q}_i,\dot{\mathbf{q}}_i\right)\dot{\mathbf{q}}_i - \mathbf{C}_i\left(\mathbf{q}_i,\dot{\mathbf{q}}_{r_i}\right)\dot{\mathbf{q}}_{r_i} + \mathbf{C}_i\left(\mathbf{q}_i,\dot{\mathbf{q}}_i\right)\dot{\mathbf{q}}_{r_i} - \mathbf{C}_i\left(\mathbf{q}_i,\dot{\mathbf{q}}_i\right)\dot{\mathbf{q}}_{r_i} \\ &= \mathbf{C}_i\left(\mathbf{q}_i,\dot{\mathbf{q}}_i\right)\left[\dot{\mathbf{q}}_i - \dot{\mathbf{q}}_{r_i}\right] + \mathbf{C}_i\left(\mathbf{q}_i,\dot{\mathbf{q}}_i\right)\dot{\mathbf{q}}_{r_i} - \mathbf{C}_i\left(\mathbf{q}_i,\dot{\mathbf{q}}_{r_i}\right)\dot{\mathbf{q}}_{r_i} \\ &= \mathbf{C}_i\left(\mathbf{q}_i,\dot{\mathbf{q}}_i\right)\mathbf{s}_i + \mathbf{C}_i\left(\mathbf{q}_i,\mathbf{q}_i - \dot{\mathbf{q}}_{r_i}\right)\dot{\mathbf{q}}_{r_i}.\end{aligned} \tag{154}$$

Applying (16), we have:

$$\begin{aligned}&\mathbf{C}_i\left(\mathbf{q}_i,\dot{\mathbf{q}}_i\right)\dot{\mathbf{q}}_i - \mathbf{C}_i\left(\mathbf{q}_i,\dot{\mathbf{q}}_{r_i}\right)\dot{\mathbf{q}}_{r_i} = \mathbf{C}_i\left(\mathbf{q}_i,\dot{\mathbf{q}}_i\right)\mathbf{s}_i + \mathbf{C}_i\left(\mathbf{q}_i,\mathbf{q}_i - \dot{\mathbf{q}}_{r_i}\right)\dot{\mathbf{q}}_{r_i} \\ &= \mathbf{C}_i\left(\mathbf{q}_i,\dot{\mathbf{q}}_i\right)\mathbf{s}_i + \mathbf{C}_i\left(\mathbf{q}_i,\mathbf{s}_i\right)\dot{\mathbf{q}}_{r_i} \\ &= \mathbf{C}_i\left(\mathbf{q}_i,\dot{\mathbf{q}}_i\right)\mathbf{s}_i + \mathbf{C}_i\left(\mathbf{q}_i,\dot{\mathbf{q}}_{r_i}\right)\mathbf{s}_i\end{aligned} \tag{155}$$

This means that the close loop dynamics is given by:

$$\begin{aligned}&\mathbf{H}_i\left(\mathbf{q}_i\right)\dot{\mathbf{s}}_i = \mathbf{K}_{R_i}\mathbf{r}_i + \mathbf{J}_{\varphi_i}^T\left(\mathbf{q}_i\right)\left[\Delta\boldsymbol{\lambda}_i - k_{F_i}\Delta\mathbf{F}_i\right] - \mathbf{C}_i\left(\mathbf{q}_i,\mathbf{s}_i\right)\dot{\mathbf{q}}_{r_i} + \mathbf{H}_i\left(\mathbf{q}_i\right)\mathbf{e}_i\left(\mathbf{r}_i\right) \\ &-\mathbf{C}_i\left(\mathbf{q}_i,\dot{\mathbf{q}}_i\right)\mathbf{s}_i - \mathbf{D}_i\mathbf{s}_i - \mathbf{K}_{R_i}\mathbf{s}_i \\ &= \mathbf{K}_{R_i}\mathbf{r}_i + \mathbf{J}_{\varphi_i}^T\left(\mathbf{q}_i\right)\left[\Delta\boldsymbol{\lambda}_i - k_{F_i}\Delta\mathbf{F}_i\right] - \mathbf{C}_i\left(\mathbf{q}_i,\mathbf{s}_i\right)\dot{\mathbf{q}}_{r_i} \\ &+\mathbf{H}_i\left(\mathbf{q}_i\right)\mathbf{e}_i\left(\mathbf{r}_i\right) - \mathbf{C}_i\left(\mathbf{q}_i,\dot{\mathbf{q}}_i\right)\mathbf{s}_i - \mathbf{K}_{DR}\mathbf{s}_i\end{aligned} \tag{156}$$

where $\mathbf{K}_{DR} = \mathbf{D}_i + \mathbf{K}_{R_i}$.

Observer Definition

The observer is given by:

$$\dot{\hat{\mathbf{q}}}_i = \dot{\hat{\mathbf{q}}}_{o_i} + \Lambda_i \mathbf{z}_i + k_{d_i} \mathbf{z}_i \quad (157)$$

$$\ddot{\hat{\mathbf{q}}}_{o_i} = \ddot{\hat{\mathbf{q}}}_{r_i} + k_{d_i} \Lambda_i \mathbf{z}_i + \mathbf{H}_i^{-1}(\mathbf{q}_i) \mathbf{J}_{\varphi_i}^T(\mathbf{q}_i) \left[\Delta\boldsymbol{\lambda}_i + k_{F_i} \Delta\mathbf{F}_i \right]. \quad (158)$$

To compute the close loop dynamics, we have from (157):

$$\dot{\hat{\mathbf{q}}}_{o_i} = \dot{\hat{\mathbf{q}}}_i - \Lambda_i \mathbf{z}_i - k_{d_i} \mathbf{z}_i \quad (159)$$

Deriving (159) and substituting in (158) we get:

$$\ddot{\hat{\mathbf{q}}}_i - \Lambda_i \dot{\mathbf{z}}_i - k_{\mathrm{d}i} \dot{\mathbf{z}}_i = \ddot{\hat{\mathbf{q}}}_{r_i} + k_{\mathrm{d}i} \Lambda_i \mathbf{z}_i + \mathbf{H}_i^{-1}(\mathbf{q}_i) \mathbf{J}_{\varphi_i}^{\mathrm{T}}(\mathbf{q}_i) \left[\Delta\boldsymbol{\lambda}_i + k_{Fi} \Delta\mathbf{F}_i \right] \quad (160)$$

Substituting (175) in (160), we have:

$$\ddot{\hat{\mathbf{q}}}_i - \Lambda_i \dot{\mathbf{z}}_i - k_{\mathrm{d}i} \dot{\mathbf{z}}_i = \ddot{\mathbf{q}}_{r_i} + \mathbf{e}(\mathbf{r}_i) + k_{\mathrm{d}i} \Lambda_i \mathbf{z}_i + \mathbf{H}_i^{-1}(\mathbf{q}_i) \mathbf{J}_{\varphi_i}^{\mathrm{T}}(\mathbf{q}_i) \left[\Delta\boldsymbol{\lambda}_i + k_{Fi} \Delta\mathbf{F}_i \right] \quad (161)$$

or:

$$\ddot{\mathbf{q}}_i - \ddot{\mathbf{q}}_{r_i} = \ddot{\mathbf{q}}_i - \ddot{\hat{\mathbf{q}}}_i + \Lambda_i \dot{\mathbf{z}}_i + k_{di} \left(\dot{\mathbf{z}}_i + \Lambda_i \mathbf{z}_i \right) + \mathbf{e}(\mathbf{r}_i) + \mathbf{H}_i^{-1}(\mathbf{q}_i) \mathbf{J}_{\varphi_i}^{\mathrm{T}}(\mathbf{q}_i) \left[\Delta\boldsymbol{\lambda}_i + k_{Fi} \Delta\mathbf{F}_i \right] \quad (162)$$

Then, we have:

$$\dot{\mathbf{s}}_i = \dot{\mathbf{r}}_i + k_{di} \mathbf{r}_i + \mathbf{e}(\mathbf{r}_i) + \mathbf{H}_i^{-1}(\mathbf{q}_i) \mathbf{J}_{\varphi_i}^{\mathrm{T}}(\mathbf{q}_i) \left[\Delta\boldsymbol{\lambda}_i + k_{Fi} \Delta\mathbf{F}_i \right] \quad (163)$$

Multiplying by $\mathbf{H}_i(\mathbf{q}_i)$, we get:

$$\mathbf{H}_i(\mathbf{q}_i) \dot{\mathbf{s}}_i = \mathbf{H}_i(\mathbf{q}_i) \dot{\mathbf{r}}_i + \mathbf{H}_i(\mathbf{q}_i) k_{di} \mathbf{r}_i + \mathbf{H}_i(\mathbf{q}_i) \mathbf{e}(\mathbf{r}_i)$$

$$+ \mathbf{J}_{\varphi_i}^{\mathrm{T}}(\mathbf{q}_i) \left[\Delta\boldsymbol{\lambda}_i + k_{Fi} \Delta\mathbf{F}_i \right]$$

$$\mathbf{H}_i(\mathbf{q}_i)\dot{\mathbf{s}}_i - \mathbf{H}_i(\mathbf{q}_i)\mathbf{e}(\mathbf{r}_i) - \mathbf{J}^{\mathrm{T}}_{\varphi_i}(\mathbf{q}_i)\left[\Delta\boldsymbol{\lambda}_i + k_{Fi}\Delta\mathbf{F}_i\right] = \mathbf{H}_i(\mathbf{q}_i)\dot{\mathbf{r}}_i + \mathbf{H}_i(\mathbf{q}_i)k_{di}\mathbf{r}_i \tag{164}$$

Substituting (156) in (164), we have:

$$\begin{aligned}
&\mathbf{H}_i(\mathbf{q}_i)\dot{\mathbf{r}}_i + \mathbf{H}_i(\mathbf{q}_i)k_{di}\mathbf{r}_i = \mathbf{K}_{R_i}\mathbf{r}_i + \\
&\cancel{\mathbf{J}^{\mathrm{T}}_{\varphi_i}(\mathbf{q}_i)\left[\Delta\boldsymbol{\lambda}_i + k_{Fi}\Delta\mathbf{F}_i\right]} - \mathbf{C}_i(\mathbf{q}_i,\mathbf{s}_i)\dot{\mathbf{q}}_{r_i} + \cancel{\mathbf{H}_i(\mathbf{q}_i)\mathbf{e}(\mathbf{r}_i)} \\
&-\mathbf{C}_i(\mathbf{q}_i,\dot{\mathbf{q}}_i)\mathbf{s}_i - \mathbf{K}_{DR_i}\mathbf{s}_i - \cancel{\mathbf{H}_i(\mathbf{q}_i)\mathbf{e}(\mathbf{r}_i)} - \cancel{\mathbf{J}^{\mathrm{T}}_{\varphi_i}(\mathbf{q}_i)\left[\Delta\boldsymbol{\lambda}_i + k_{Fi}\Delta\mathbf{F}_i\right]} \\
&= \mathbf{K}_{R_i}\mathbf{r}_i - \mathbf{C}_i(\mathbf{q}_i,\mathbf{s}_i)\dot{\mathbf{q}}_{r_i} - \mathbf{C}_i(\mathbf{q}_i,\dot{\mathbf{q}}_i)\mathbf{s}_i - \mathbf{K}_{DR_i}\mathbf{s}_i
\end{aligned} \tag{165}$$

Applying the property (16), we have:

$$\mathbf{H}_i(\mathbf{q}_i)\dot{\mathbf{r}}_i + \mathbf{H}_i(\mathbf{q}_i)k_{di}\mathbf{r}_i = \mathbf{K}_{R_i}\mathbf{r}_i - \mathbf{C}_i(\mathbf{q}_i,\dot{\mathbf{q}}_{r_i})\mathbf{s}_i - \mathbf{C}_i(\mathbf{q}_i,\dot{\mathbf{q}}_i)\mathbf{s}_i - \mathbf{K}_{DR_i}\mathbf{s}_i \tag{166}$$

We make the following definition:

$$\mathbf{H}_{Rdi} = \mathbf{H}_i(\mathbf{q}_i)k_{di} - \mathbf{K}_{R_i} \tag{167}$$

then, we have:

$$\mathbf{H}_i(\mathbf{q}_i)\dot{\mathbf{r}}_i = -\mathbf{H}_{Rdi}\mathbf{r}_i - \mathbf{C}_i(\mathbf{q}_i,\dot{\mathbf{q}}_{r_i})\mathbf{s}_i - \mathbf{C}_i(\mathbf{q}_i,\dot{\mathbf{q}}_i)\mathbf{s}_i - \mathbf{K}_{DR_i}\mathbf{s}_i \tag{168}$$

Simplify with $\dot{\mathbf{q}}_i = \mathbf{s}_i + \dot{\mathbf{q}}_{r_i}$, we have:

$$\begin{aligned}
&\mathbf{C}_i(\mathbf{q}_i,\dot{\mathbf{q}}_{r_i})\mathbf{s}_i + \mathbf{C}_i(\mathbf{q}_i,\dot{\mathbf{q}}_i)\mathbf{s}_i = \mathbf{C}_i(\mathbf{q}_i,\dot{\mathbf{q}}_i)\mathbf{r}_i - \\
&\mathbf{C}_i(\mathbf{q}_i,\dot{\mathbf{q}}_i)\mathbf{r}_i + \mathbf{C}_i(\mathbf{q}_i,\dot{\mathbf{q}}_{r_i})\mathbf{s}_i + \mathbf{C}_i(\mathbf{q}_i,\dot{\mathbf{q}}_i)\mathbf{s}_i \\
&= \mathbf{C}_i(\mathbf{q}_i,\dot{\mathbf{q}}_i)\mathbf{r}_i - \mathbf{C}_i(\mathbf{q}_i,\mathbf{s}_i + \dot{\mathbf{q}}_{r_i})\mathbf{r}_i + \mathbf{C}_i(\mathbf{q}_i,\mathbf{s}_i + \dot{\mathbf{q}}_{r_i})\mathbf{s}_i + \mathbf{C}_i(\mathbf{q}_i,\dot{\mathbf{q}}_{r_i})\mathbf{s}_i \\
&= \mathbf{C}_i(\mathbf{q}_i,\dot{\mathbf{q}}_i)\mathbf{r}_i - \mathbf{C}_i(\mathbf{q}_i,\mathbf{s}_i + \dot{\mathbf{q}}_{r_i})\mathbf{r}_i + \mathbf{C}_i(\mathbf{q}_i,\mathbf{s}_i)\left[\mathbf{s}_i + \dot{\mathbf{q}}_{r_i}\right] + \mathbf{C}_i(\mathbf{q}_i,\dot{\mathbf{q}}_{r_i})\mathbf{s}_i \\
&= \mathbf{C}_i(\mathbf{q}_i,\dot{\mathbf{q}}_i)\mathbf{r}_i - \mathbf{C}_i(\mathbf{q}_i,\mathbf{s}_i + \dot{\mathbf{q}}_{r_i})\mathbf{r}_i + \mathbf{C}_i(\mathbf{q}_i,\mathbf{s}_i)\left[\mathbf{s}_i + \dot{\mathbf{q}}_{r_i}\right] + \mathbf{C}_i(\mathbf{q}_i,\mathbf{s}_i)\dot{\mathbf{q}}_{r_i}
\end{aligned}$$

$$= \mathbf{C}_i(\mathbf{q}_i, \dot{\mathbf{q}}_i)\mathbf{r}_i - \mathbf{C}_i(\mathbf{q}_i, \mathbf{s}_i + \dot{\mathbf{q}}_{r_i})\mathbf{r}_i + \mathbf{C}_i(\mathbf{q}_i, \mathbf{s}_i)\left[\mathbf{s}_i + 2\dot{\mathbf{q}}_{r_i}\right] \tag{169}$$

Thus, we get:

$$\begin{aligned}\mathbf{H}_i(\mathbf{q}_i)\dot{\mathbf{r}}_i &= \mathbf{C}_i(\mathbf{q}_i, \mathbf{s}_i + \dot{\mathbf{q}}_{r_i})\mathbf{r}_i - \mathbf{C}_i(\mathbf{q}_i, \dot{\mathbf{q}}_i)\mathbf{r}_i - \\ &\mathbf{C}_i(\mathbf{q}_i, \mathbf{s}_i)\left[\mathbf{s}_i + 2\dot{\mathbf{q}}_{r_i}\right] - \mathbf{H}_i(\mathbf{q}_i)k_{di}\mathbf{r}_i - \mathbf{K}_{DR_i}\mathbf{s}_i\end{aligned} \tag{170}$$

Now, defining the variable $\mathbf{x}_i$, we have:

$$\mathbf{x}_i = \begin{bmatrix}\mathbf{s}_i \\ \mathbf{r}_i \\ \Delta\mathbf{F}_i\end{bmatrix} \tag{171}$$

as state for the cooperative system. The main idea of the control-observer design is to show that whenever $\|\mathbf{x}_i\|$ tends to zero, the tracking errors $\tilde{\mathbf{q}}_i, \dot{\tilde{\mathbf{q}}}_i, \Delta\mathbf{p}_i, \Delta\dot{\mathbf{p}}_i$ and $\Delta\boldsymbol{\lambda}_i$ the observation errors $\mathbf{z}_i$ and $\dot{\mathbf{z}}_i$ will tend all to zero as well. From (173), it is rather obvious that if $\mathbf{r}_i$ is bounded and tends to zero, so will do $\mathbf{z}_i$ and $\dot{\mathbf{z}}_i$. However, this is not clear for the other variables. Following **Lemma 1** showing this case is indeed under some conditions.

Lemma 1. If $\mathbf{x}_i$ is bounded by $\mathbf{x}_{\max i}$ and tends to zero, then the following facts hold:

- $\Delta\mathbf{p}_i$ and $\Delta\dot{\mathbf{p}}_i$ remain bounded and tend to zero.
- $\tilde{\mathbf{q}}_i$ and $\dot{\tilde{\mathbf{q}}}_i$ remain bounded. If the bound $\mathbf{x}_{\max i}$ for $\|\mathbf{x}_i\|$ is selected sufficiently small to guarantee that $\|\tilde{\mathbf{q}}_i\| < \eta_i$ for all t, with η_i a positive and small enough constant, subsequently both $\tilde{\mathbf{q}}_i$ and $\dot{\tilde{\mathbf{q}}}_i$ will tend to zero as well.
- In addition, if the velocity vector $\dot{\mathbf{q}}_i$ is bounded, then $\Delta\boldsymbol{\lambda}_i$ will remain bounded and tend to zero. ■

Recall that the state for (156) and (170) is given by (171). We define:

$$\mathbf{M}_i(\mathbf{q}_i) = \begin{bmatrix}\mathbf{H}_i(\mathbf{q}_i) & 0 & 0 \\ 0 & \mathbf{H}_i(\mathbf{q}_i) & 0 \\ 0 & 0 & \xi_i\end{bmatrix} \tag{172}$$

and

$$V_i\left(\mathbf{x}_i\right)=\frac{1}{2}\mathbf{x}_i^{\mathrm{T}}\mathbf{M}_i\left(\mathbf{q}_i\right)\mathbf{x}_i \tag{173}$$

Clearly, we have:

$$\lambda_{1i}\left\|\mathbf{x}_i\right\|^2 \leq V_i\left(\mathbf{x}_i\right) \leq \lambda_{2i}\left\|\mathbf{x}_i\right\|^2 \tag{174}$$

where:

$$\lambda_{1_i}=\frac{1}{2}\min_{\forall q_i\in\mathbb{R}^{n_i}}\lambda_{\min}\left(\mathbf{M}_i\right) \tag{175}$$

$$\lambda_{2_i}=\frac{1}{2}\max_{\forall q_i\in\mathbb{R}^{n_i}}\lambda_{\max}\left(\mathbf{M}_i\right) \tag{176}$$

Suppose we define the following region of attraction:

$$S_{a_i}=\left\{\mathbf{x}_i:\left\|\mathbf{x}_i\right\|\leq a_i\right\} \tag{177}$$

Furthermore, suppose that $\dot{V}_i(\mathbf{x}_i) - 0$. In that case we have:

$$\lambda_{1i}\left\|\mathbf{x}_i\right\|^2 \leq V_i\left(\mathbf{x}_i\right) < \lambda_{2i}\left\|\mathbf{x}_i\left(0\right)\right\|^2 \leq \lambda_{2i}a_i^2 \tag{178}$$

or,

$$\left\|\mathbf{x}_i\right\| \leq \mathbf{x}_{max_i} \qquad \forall t\geq 0 \tag{179}$$

with

$$\mathbf{x}_{maxi}=a_i\sqrt{\frac{\lambda_{2i}}{\lambda_{1i}}} \tag{180}$$

Now, we are going to prove that for the proposed controller-observer scheme $\dot{V}\left(\mathbf{x}_i\right)=0$ where $\mathbf{x}_i\equiv 0$ and otherwise $\dot{V}\left(\mathbf{x}_i\right)<0$, so that $\mathbf{x}_i \rightarrow 0$.

In order to define how large can be a_i, we require that

$$\left\|\tilde{\mathbf{q}}_i\right\| \le \eta_i \qquad \forall t \ge 0 \tag{181}$$

Then, we have:

$$\mathbf{x}_{maxi} = a_i \sqrt{\frac{\boldsymbol{\lambda}_{2i}}{\boldsymbol{\lambda}_{1i}}} \le \frac{\eta_i \alpha_i}{1 + \bar{\xi}_i c_{oi} + \sqrt{n_i}} \tag{182}$$

or

$$\mathbf{x}_{maxi} = a_i \sqrt{\frac{\boldsymbol{\lambda}_{2i}}{\boldsymbol{\lambda}_{1i}}} \le \frac{\eta_i \alpha_i}{1 + \bar{\xi}_i c_{oi} + \sqrt{n_i}} \tag{183}$$

Assuming that a_i satisfies (183), we can prove that $\left\|x_i\right\| \to 0$, with a proper selection of gains. However, if $\left\|x_i\right\| \le x_{max_i}$ we must have:

$$\left\|\mathbf{e}_i\left(\mathbf{r}_i\right)\right\| \le \mathbf{M}_{e_i}\left(\mathbf{x}_{\max i}\right)\left\|\mathbf{r}_i\right\| \qquad \mathbf{M}_{e_i} > 0 \tag{184}$$

where

$$\mathbf{M}_{e_i}\left(\mathbf{x}_{\max i}\right) = c_{1i}\left(\nu_{m_i} + k_i \eta_i + \sqrt{n_i}\mathbf{x}_{\max i}\right) + c_{2i}\bar{\xi}_i \mathbf{x}_{\max i} \tag{185}$$

To prove that $\dot{V}_i\left(\mathbf{x}_i\right) \le 0$, rewrite $V_i\left(\mathbf{x}_i\right)$ as:

$$V_i\left(\mathbf{x}_i\right) = \frac{1}{2}\mathbf{s}_i^T \mathbf{H}_i\left(\mathbf{q}_i\right)\mathbf{s}_i + \frac{1}{2}\mathbf{r}_i^T \mathbf{H}_i\left(\mathbf{q}_i\right)\mathbf{r}_i + \frac{1}{2}\Delta\mathbf{F}_i^T \xi_i\left(\mathbf{q}_i\right)\Delta\mathbf{F}_i \tag{186}$$

Then we have:

$$\dot{V}_i\left(\mathbf{x}_i\right) = \frac{1}{2}\mathbf{s}_i^T \dot{\mathbf{H}}_i\left(q_i\right)\mathbf{s}_i + \frac{1}{2}\mathbf{r}_i^T \dot{\mathbf{H}}_i\left(q_i\right)\mathbf{r}_i + \mathbf{s}_i^T \mathbf{H}_i\left(q_i\right)\dot{\mathbf{s}}_i + \mathbf{r}_i^T \mathbf{H}_i\left(q_i\right)\dot{\mathbf{r}}_i + \Delta\mathbf{F}_i^T \xi_i \Delta\boldsymbol{\lambda}_i \tag{187}$$

Substituting (156) and (170) in (187) we get:

$$\begin{aligned}\dot{V}_i\left(\mathbf{x}_i\right) &= \frac{1}{2}\mathbf{s}_i^T \dot{\mathbf{H}}_i\left(\mathbf{q}_i\right)\mathbf{s}_i + \frac{1}{2}\mathbf{r}_i^T \dot{\mathbf{H}}_i\left(\mathbf{q}_i\right)\mathbf{r}_i + \\ &\mathbf{s}_i^T \mathbf{K}_{R_i}\mathbf{r}_i + \mathbf{s}_i^T \mathbf{J}_{\varphi i}^T\left(\mathbf{q}_i\right)\left[\Delta\boldsymbol{\lambda}_i + k_{Fi}\Delta\mathbf{F}_i\right] - \mathbf{s}_i^T \mathbf{C}_i\left(\mathbf{q}_i, \mathbf{s}_i\right)\dot{\mathbf{q}}_{r_i}\end{aligned}$$

$$+\mathbf{s}_i^T\mathbf{H}_i(\mathbf{q}_i)\mathbf{e}_i(\mathbf{r}_i) - \mathbf{s}_i^T\mathbf{C}_i(\mathbf{q}_i,\dot{\mathbf{q}}_i)\mathbf{s}_i - \mathbf{s}_i^T\mathbf{K}_{DR_i}\mathbf{s}_i + \mathbf{r}_i^T\mathbf{C}_i(\mathbf{q}_i,\mathbf{s}_i + \dot{\mathbf{q}}_{r_i})\mathbf{r}_i - \mathbf{r}_i^T\mathbf{C}_i(\mathbf{q}_i,\dot{\mathbf{q}}_i)\mathbf{r}_i$$

$$-\mathbf{r}_i^T\mathbf{C}_i(\mathbf{q}_i,\mathbf{s}_i)\left[\mathbf{s}_i + 2\dot{\mathbf{q}}_{r_i}\right] - \mathbf{r}_i^T\mathbf{H}_{Rdi}\mathbf{r}_i - \mathbf{r}_i^T\mathbf{K}_{DR_i}\mathbf{s}_i + \Delta\mathbf{F}_i^T\xi_i\Delta\boldsymbol{\lambda}_i \tag{188}$$

Applying the property $\mathbf{C}(\mathbf{q},\mathbf{x})\mathbf{y} = \mathbf{C}(\mathbf{q},\mathbf{y})\mathbf{x}, \quad \forall \mathbf{x},\mathbf{y} \in \mathbb{R}^n$, we have:

$$\dot{V}_i(\mathbf{x}_i) = \frac{1}{2}\mathbf{s}_i^T\dot{\mathbf{H}}_i(\mathbf{q}_i)\mathbf{s}_i + \frac{1}{2}\mathbf{r}_i^T\dot{\mathbf{H}}_i(\mathbf{q}_i)\mathbf{r}_i +$$
$$\mathbf{s}_i^T\mathbf{K}_{R_i}\mathbf{r}_i + \mathbf{s}_i^T\mathbf{J}_{\varphi i}^T(\mathbf{q}_i)\left[\Delta\boldsymbol{\lambda}_i + k_{Fi}\Delta\mathbf{F}_i\right] - \mathbf{s}_i^T\mathbf{C}_i(\mathbf{q}_i,\dot{\mathbf{q}}_{r_i})\mathbf{s}_i$$

$$+\mathbf{s}_i^T\mathbf{H}_i(\mathbf{q}_i)\mathbf{e}_i(\mathbf{r}_i) - \mathbf{s}_i^T\mathbf{C}_i(\mathbf{q}_i,\dot{\mathbf{q}}_i)\mathbf{s}_i - \mathbf{s}_i^T\mathbf{K}_{DR_i}\mathbf{s}_i + \mathbf{r}_i^T\mathbf{C}_i(\mathbf{q}_i,\mathbf{s}_i + \dot{\mathbf{q}}_{r_i})\mathbf{r}_i - \mathbf{r}_i^T\mathbf{C}_i(\mathbf{q}_i,\dot{\mathbf{q}}_i)\mathbf{r}_i$$

$$-\mathbf{r}_i^T\mathbf{C}_i(\mathbf{q}_i,\mathbf{s}_i)\left[\mathbf{s}_i + 2\dot{\mathbf{q}}_{r_i}\right] - \mathbf{r}_i^T\mathbf{H}_{Rdi}\mathbf{r}_i - \mathbf{r}_i^T\mathbf{K}_{DR_i}\mathbf{s}_i + \Delta\mathbf{F}_i^T\xi_i\Delta\boldsymbol{\lambda}_i \tag{189}$$

In order to simplify (189), we recall that:

$$\mathbf{s}_i^T\mathbf{K}_{R_i}\mathbf{r}_i - \mathbf{r}_i^T\mathbf{K}_{DR_i}\mathbf{s}_i = \mathbf{s}_i^T\mathbf{K}_{R_i}\mathbf{r}_i - \mathbf{s}_i^T\mathbf{K}_{DR_i}\mathbf{r}_i$$

$$= \mathbf{s}_i^T\mathbf{K}_{R_i}\mathbf{r}_i - \mathbf{s}_i^T\mathbf{K}_{R_i}\mathbf{r}_i - \mathbf{s}_i^T\mathbf{D}_i\mathbf{r}_i = -\mathbf{s}_i^T\mathbf{D}_i\mathbf{r}_i \tag{190}$$

Note that:

$$\mathbf{Q}_i(\mathbf{q}_i)\mathbf{J}_{\varphi_i}^{\mathrm{T}}(\mathbf{q}_i) = \left[\mathbf{I} - \mathbf{J}_{\varphi_i}^{+}(\mathbf{q}_i)\mathbf{J}_{\varphi_i}(\mathbf{q}_i)\right]\mathbf{J}_{\varphi_i}^{\mathrm{T}}(\mathbf{q}_i)$$

$$= \left[\mathbf{J}_{\varphi_i}^{\mathrm{T}}(\mathbf{q}_i) - \mathbf{J}_{\varphi_i}^{+}(\mathbf{q}_i)\mathbf{J}_{\varphi_i}(\mathbf{q}_i)\mathbf{J}_{\varphi_i}^{\mathrm{T}}(\mathbf{q}_i)\right]$$

$$= \mathbf{J}_{\varphi_i}^{\mathrm{T}}(\mathbf{q}_i) - \mathbf{J}_{\varphi_i}^{\mathrm{T}}(\mathbf{q}_i)\left[\mathbf{J}_{\varphi_i}(\mathbf{q}_i)\mathbf{J}_{\varphi_i}^{\mathrm{T}}(\mathbf{q}_i)\right]^{-1}\left[\mathbf{J}_{\varphi_i}(\mathbf{q}_i)\mathbf{J}_{\varphi_i}^{\mathrm{T}}(\mathbf{q}_i)\right]$$

$$= \mathbf{J}_{\varphi_i}^{\mathrm{T}}(\mathbf{q}_i) - \mathbf{J}_{\varphi_i}^{\mathrm{T}}(\mathbf{q}_i) = 0 \tag{191}$$

Also we define

$$\dot{\mathbf{q}}_{r_i} = \mathbf{Q}_i(\mathbf{q}_i)\left[\dot{\mathbf{q}}_{di} - \Lambda_i(\hat{\mathbf{q}}_i - \dot{\mathbf{q}}_{di})\right] + \mathbf{J}_{\varphi_i}^{+}(\mathbf{q}_i)\left[\dot{\mathbf{p}}_{di} - \beta_i\Delta\mathbf{p}_i + \xi_i\Delta\mathbf{F}_i\right] \tag{192}$$

with $\beta_i > 0 \in \mathbb{R}$ and $\xi \in \mathbb{R}^{m_i \times m_i}, \Lambda_i \in \mathbb{R}^{n_i \times n_i}, \xi > 0$ and $\Lambda_i > 0$.

Consider:

$$\Delta \mathbf{F}_i = \int_0^t \Delta \boldsymbol{\lambda}_i(\zeta)\, d\zeta \tag{193}$$

Define also:

$$\begin{aligned} \mathbf{s}_i &= \dot{\mathbf{q}}_i - \dot{\mathbf{q}}_{r_i} \\ &= \mathbf{Q}_i(\mathbf{q}_i)\dot{\mathbf{q}}_i + \mathbf{J}_{\varphi_i}^{+}(\mathbf{q}_i)\dot{\mathbf{p}}_i - \dot{\mathbf{q}}_{r_i} \end{aligned} \tag{194}$$

where

$$\dot{\mathbf{q}}_{r_i} = \mathbf{Q}_i(\mathbf{q}_i)\left[\dot{\mathbf{q}}_{di} - \Lambda_i\left(\dot{\mathbf{q}}_i - \dot{\mathbf{q}}_{di} + \hat{\mathbf{q}}_i - \dot{\mathbf{q}}_i\right)\right] + \mathbf{J}_{\varphi_i}^{+}(\mathbf{q}_i)\left[\dot{\mathbf{p}}_{di} - \beta_i \Delta \mathbf{p}_i + \xi_i \Delta \mathbf{F}_i\right] \tag{195}$$

then, we have:

$$\begin{aligned} \mathbf{s}_i = {} & \mathbf{Q}_i(\mathbf{q}_i)\dot{\mathbf{q}}_i + \mathbf{J}_{\varphi_i}^{+}(\mathbf{q}_i)\dot{\mathbf{p}}_i - \\ & \mathbf{Q}_i(\mathbf{q}_i)\left[\dot{\mathbf{q}}_{di} - \Lambda_i\left(\dot{\mathbf{q}}_i - \dot{\mathbf{q}}_{di} + \hat{\mathbf{q}}_i - \dot{\mathbf{q}}_i\right)\right] - \mathbf{J}_{\varphi_i}^{+}(\mathbf{q}_i)\left[\dot{\mathbf{p}}_{di} - \beta_i \Delta \mathbf{p}_i + \xi_i \Delta \mathbf{F}_i\right] \end{aligned} \tag{196}$$

then,

$$\begin{aligned} \mathbf{s}_i &= \mathbf{Q}_i(\mathbf{q}_i)\left[\dot{\mathbf{q}}_i - \dot{\mathbf{q}}_{di} + \Lambda_i\left(\dot{\mathbf{q}}_i - \dot{\mathbf{q}}_{di} + \hat{\mathbf{q}}_i - \dot{\mathbf{q}}_i\right)\right] + \mathbf{J}_{\varphi_i}^{+}(\mathbf{q}_i)\left[\dot{\mathbf{p}}_i - \dot{\mathbf{p}}_{di} + \beta_i \Delta \mathbf{p}_i - \xi_i \Delta \mathbf{F}_i\right] \\ &= \mathbf{Q}_i(\mathbf{q}_i)\left[\dot{\tilde{\mathbf{q}}}_i + \Lambda_i \tilde{\mathbf{q}} - \Lambda_i \mathbf{z}_i\right] + \mathbf{J}_{\varphi_i}^{+}(\mathbf{q}_i)\left[\Delta \dot{\mathbf{p}}_i + \beta_i \Delta \mathbf{p}_i - \xi_i \Delta \mathbf{F}_i\right] \end{aligned} \tag{197}$$

Recall from (197) that we have:

$$\mathbf{s}_i = \underbrace{\mathbf{Q}_i(\mathbf{q}_i)\left[\dot{\tilde{\mathbf{q}}}_i + \Lambda_i \tilde{\mathbf{q}} - \Lambda_i \mathbf{z}_i\right]}_{\text{bounded}} + \underbrace{\mathbf{J}_{\varphi_i}^{+}(\mathbf{q}_i)\left[\Delta \dot{\mathbf{p}}_i + \beta_i \Delta \mathbf{p}_i - \xi_i \Delta \mathbf{F}_i\right]}_{\text{bounded}} \tag{198}$$

Now, substituting (191) in (198), and solving by $\mathbf{s}_i^T \mathbf{J}_{\varphi_i}^T(\mathbf{q}_i)$, we have:

$$\mathbf{s}_i^T\mathbf{J}_{\varphi_i}^T\left(\mathbf{q}_i\right)=\underbrace{\left(\cancel{\mathbf{Q}_i\left(\mathbf{q}_i\right)\left[\dot{\mathbf{q}}_i+\mathbf{A}_i\left(\hat{\mathbf{q}}_i-\mathbf{q}_{di}\right)\right]}+\mathbf{J}_{\varphi_i}^{+}\left(\mathbf{q}_i\right)\left[\Delta\dot{\mathbf{p}}_i+\beta_i\Delta\mathbf{p}_i-\xi_i\Delta\mathbf{F}_i\right]\right)^T}_{\mathbf{s}_i^T}\mathbf{J}_{\varphi_i}^T\left(\mathbf{q}_i\right)$$

$$=\left(\mathbf{J}_{\varphi_i}^{+}\left(\mathbf{q}_i\right)\left[\Delta\dot{\mathbf{p}}_i+\beta_i\Delta\mathbf{p}_i-\xi_i\Delta\mathbf{F}_i\right]\right)^T\mathbf{J}_{\varphi_i}^T\left(\mathbf{q}_i\right)$$

$$=\left[\Delta\dot{\mathbf{p}}_i+\beta_i\Delta\mathbf{p}_i-\xi_i\Delta\mathbf{F}_i\right]^T\left[\mathbf{J}_{\varphi_i}^{+}\left(\mathbf{q}_i\right)\right]^T\mathbf{J}_{\varphi_i}^T\left(\mathbf{q}_i\right) \tag{199}$$

Since $\Delta\dot{\mathbf{p}}_i+\beta_i\Delta\mathbf{p}_i=0$ in order to fulfill the constraints, we have:

$$\mathbf{s}_i^T\mathbf{J}_{\varphi_i}^T\left(\mathbf{q}_i\right)=-\left[\xi_i\Delta\mathbf{F}_i\right]^T\left[\mathbf{J}_{\varphi_i}^{+}\left(\mathbf{q}_i\right)\right]^T\mathbf{J}_{\varphi_i}^T\left(\mathbf{q}_i\right) \tag{200}$$

Substituting $\mathbf{J}_{\varphi_i}^{+}\left(\mathbf{q}_i\right)=\mathbf{J}_{\varphi_i}^T\left(\mathbf{q}_i\right)\left[\mathbf{J}_{\varphi_i}\left(\mathbf{q}_i\right)\mathbf{J}_{\varphi_i}^T\left(\mathbf{q}_i\right)\right]^{-1}$ in (200), we get:

$$\mathbf{s}_i^T\mathbf{J}_{\varphi_i}^T\left(\mathbf{q}_i\right)=-\xi_i\Delta\mathbf{F}_i\left[\mathbf{J}_{\varphi_i}^T\left(\mathbf{q}_i\right)\left[\mathbf{J}_{\varphi_i}\left(\mathbf{q}_i\right)\mathbf{J}_{\varphi_i}^T\left(\mathbf{q}_i\right)\right]^{-1}\right]^T\mathbf{J}_{\varphi_i}^T\left(\mathbf{q}_i\right)$$

$$=-\,\xi_i\Delta\mathbf{F}_i\left[\mathbf{J}_{\varphi_i}\left(\mathbf{q}_i\right)\mathbf{J}_{\varphi_i}^T\left(\mathbf{q}_i\right)\right]^{-T}\mathbf{J}_{\varphi_i}\left(\mathbf{q}_i\right)\mathbf{J}_{\varphi_i}^T\left(\mathbf{q}_i\right)$$

$$-\,-\xi_i\Delta\mathbf{F}_i\left[\mathbf{J}_{\varphi_i}\left(\mathbf{q}_i\right)\mathbf{J}_{\varphi_i}^T\left(\mathbf{q}_i\right)\right]^{-T}\mathbf{J}_{\varphi_i}\left(\mathbf{q}_i\right)\mathbf{J}_{\varphi_i}^T\left(\mathbf{q}_i\right)$$

$$=-\xi_i\Delta\mathbf{F}_i\underbrace{\left[\mathbf{J}_{\varphi_i}\left(\mathbf{q}_i\right)\mathbf{J}_{\varphi_i}^T\left(\mathbf{q}_i\right)\right]^{-T}\left[\mathbf{J}_{\varphi_i}\left(\mathbf{q}_i\right)\mathbf{J}_{\varphi_i}^T\left(\mathbf{q}_i\right)\right]^T}_{\mathbf{I}}$$

$$=-\xi_i\Delta\mathbf{F}_i \tag{201}$$

Thus, we have:

$$\mathbf{s}_i^T\mathbf{J}_{\varphi_i}^T\left(\mathbf{q}_i\right)\left[\Delta\boldsymbol{\lambda}_i+k_{Fi}\Delta\mathbf{F}_i\right]=-\Delta\mathbf{F}_i^T\xi_i\left[\Delta\boldsymbol{\lambda}_i+k_{Fi}\Delta\mathbf{F}_i\right]$$

$$=-\,\Delta\mathbf{F}_i^T\xi_i\Delta\boldsymbol{\lambda}_i-\xi_i\Delta\mathbf{F}_i^Tk_{Fi}\Delta\mathbf{F}_i$$

$$=-\,\Delta\mathbf{F}_i^T\xi_i\Delta\boldsymbol{\lambda}_i-k_{Fi}\Delta\mathbf{F}_i^T\xi_i\Delta\mathbf{F}_i \tag{202}$$

Substituting (190) and (202) in (189), we have:

$$\dot{V}_i(\mathbf{x}_i) = \underbrace{\frac{1}{2}\mathbf{s}_i^T\dot{\mathbf{H}}_i(\mathbf{q}_i)\mathbf{s}_i - \mathbf{s}_i^T\mathbf{C}_i(\mathbf{q}_i,\dot{\mathbf{q}}_i)\mathbf{s}_i}_{\text{skew-symmetric}} +$$

$$\underbrace{\frac{1}{2}\mathbf{r}_i^T\dot{\mathbf{H}}_i(\mathbf{q}_i)\mathbf{r}_i - \mathbf{r}_i^T\mathbf{C}_i(\mathbf{q}_i,\dot{\mathbf{q}}_i)\mathbf{r}_i}_{\text{skew-symmetric}} + \underbrace{\mathbf{s}_i^T\mathbf{K}_{R_i}\mathbf{r}_i - \mathbf{r}_i^T\mathbf{K}_{R_i}\mathbf{s}_i}_{(190)}$$

$$+\underbrace{\mathbf{s}_i^T\mathbf{J}_{\varphi_i}^T(\mathbf{q}_i)\left[\Delta\boldsymbol{\lambda}_i + k_{Fi}\Delta\mathbf{F}_i\right]}_{(202)} - \mathbf{s}_i^T\mathbf{C}_i\left(\mathbf{q}_i,\dot{\mathbf{q}}_{r_i}\right)\mathbf{s}_i + \mathbf{s}_i^T\mathbf{H}_i(\mathbf{q}_i)\mathbf{e}_i(\mathbf{r}_i) - \mathbf{s}_i^T\mathbf{K}_{R_i}\mathbf{s}_i$$

$$+\mathbf{r}_i^T\mathbf{C}_i\left(\mathbf{q}_i,\mathbf{s}_i + \dot{\mathbf{q}}_{r_i}\right)\mathbf{r}_i - \mathbf{r}_i^T\mathbf{C}_i(\mathbf{q}_i,\mathbf{s}_i)\left[\mathbf{s}_i + 2\dot{\mathbf{q}}_{r_i}\right] - \mathbf{r}_i^T\mathbf{H}_{Rdi}\mathbf{r}_i + \Delta\mathbf{F}_i^T\xi_i\Delta\boldsymbol{\lambda}_i$$

$$= -\mathbf{s}_i^T\mathbf{D}_i\mathbf{r}_i - \cancel{\Delta\mathbf{F}_i^T\xi_i\Delta\boldsymbol{\lambda}_i} - k_{Fi}\Delta\mathbf{F}_i^T\xi_i\Delta\mathbf{F}_i - \mathbf{s}_i^T\mathbf{C}_i\left(\mathbf{q}_i,\dot{\mathbf{q}}_{r_i}\right)\mathbf{s}_i + \mathbf{s}_i^T\mathbf{H}_i(\mathbf{q}_i)\mathbf{e}_i(\mathbf{r}_i) - \mathbf{s}_i^T\mathbf{K}_{R_i}\mathbf{s}_i$$

$$+\mathbf{r}_i^T\mathbf{C}_i\left(\mathbf{q}_i,\mathbf{s}_i + \dot{\mathbf{q}}_{r_i}\right)\mathbf{r}_i - \mathbf{r}_i^T\mathbf{C}_i(\mathbf{q}_i,\mathbf{s}_i)\left[\mathbf{s}_i + 2\dot{\mathbf{q}}_{r_i}\right] - \mathbf{r}_i^T\mathbf{H}_{Rdi}\mathbf{r}_i + \cancel{\Delta\mathbf{F}_i^T\xi_i\Delta\boldsymbol{\lambda}_i}$$

$$= -\mathbf{s}_i^T\mathbf{D}_i\mathbf{r}_i - k_{Fi}\Delta\mathbf{F}_i^T\xi_i\Delta\mathbf{F}_i - \mathbf{s}_i^T\mathbf{C}_i\left(\mathbf{q}_i,\dot{\mathbf{q}}_{r_i}\right)\mathbf{s}_i + \mathbf{s}_i^T\mathbf{H}_i(\mathbf{q}_i)\mathbf{e}_i(\mathbf{r}_i) - \mathbf{s}_i^T\mathbf{K}_{DR_i}\mathbf{s}_i$$

$$+\mathbf{r}_i^T\mathbf{C}_i\left(\mathbf{q}_i,\mathbf{s}_i + \dot{\mathbf{q}}_{r_i}\right)\mathbf{r}_i - \mathbf{r}_i^T\mathbf{C}_i(\mathbf{q}_i,\mathbf{s}_i)\left[\mathbf{s}_i + 2\dot{\mathbf{q}}_{r_i}\right] - \mathbf{r}_i^T\mathbf{H}_{Rdi}\mathbf{r}_i \tag{203}$$

Organizing (203), we have:

$$\dot{V}_i(\mathbf{x}_i) = \mathbf{r}_i^T\mathbf{D}_i\mathbf{s}_i - k_{Fi}\Delta\mathbf{F}_i^T\xi_i\Delta\mathbf{F}_i - \mathbf{s}_i^T\mathbf{C}_i\left(\mathbf{q}_i,\dot{\mathbf{q}}_{r_i}\right)\mathbf{s}_i - \mathbf{s}_i^T\mathbf{K}_{DR_i}\mathbf{s}_i - \mathbf{r}_i^T\mathbf{C}_i(\mathbf{q}_i,\mathbf{s}_i)\left[\mathbf{s}_i + 2\dot{\mathbf{q}}_{r_i}\right]$$

$$-\mathbf{r}_i^T\mathbf{H}_{Rdi}\mathbf{r}_i + \mathbf{s}_i^T\mathbf{H}_i(\mathbf{q}_i)\mathbf{e}_i(\mathbf{r}_i) + \mathbf{r}_i^T\mathbf{C}_i\left(\mathbf{q}_i,\mathbf{s}_i + \dot{\mathbf{q}}_{r_i}\right)\mathbf{r}_i \tag{204}$$

The expression is valid for all t. However, at $t = 0$ we necessarily have:

$$\left\|\mathbf{x}_i(0)\right\| \le a_i \le \mathbf{x}_{\max i} \tag{205}$$

Now we define the following relationships:

$$\mu_{1i} = \max_{\|\mathbf{x}\| \le \mathbf{x}_{max_i}} \left\| \mathbf{C}_i \left(\mathbf{q}_i, \dot{\mathbf{q}}_{r_i} \right) \right\| \tag{206}$$

$$\mu_{2i} = \max_{\|\mathbf{x}\| \le \mathbf{x}_{max_i}} \left\| \mathbf{C}_i \left(\mathbf{q}_i, \mathbf{s}_i + \dot{\mathbf{q}}_{r_i} \right) \right\| \tag{207}$$

$$\mu_{3i} = \max_{\|\mathbf{x}\| \le \mathbf{x}_{max_i}} \left\| \mathbf{C}_i \left(\mathbf{q}_i, \mathbf{s}_i \right) \left[\mathbf{s}_i + 2\dot{\mathbf{q}}_{r_i} \right] \right\| \tag{208}$$

$$\mu_{4i} = \mathbf{M}_{e_i} \left(\mathbf{x}_{max_i} \right) \boldsymbol{\lambda}_{Hi} \tag{209}$$

$$\boldsymbol{\lambda}_{Di} = \boldsymbol{\lambda}_{\max_i} \left(\mathbf{D}_i \right) \tag{210}$$

This means that, at least at $\mathrm{t} = 0$, we have:

$$\dot{V}_i \left(\mathbf{x}_i \right) \le -\lambda_{\min i} \left(\mathbf{K}_{DR_i} \right) \left\| \mathbf{s}_i \right\|^2 - \lambda_{\min i} \left(\mathbf{H}_{Rdi} \right) \left\| \mathbf{r}_i \right\|^2 - k_{Fi} \lambda_{\min i} \left(\xi_i \right) \left\| \Delta \mathbf{F} \right\|^2 + \mu_{1i} \left\| \mathbf{s}_i \right\|^2$$

$$+\mu_{2i} \left\| \mathbf{r}_i \right\|^2 + \mu_{3i} \left\| \mathbf{r}_i \right\| \left\| \mathbf{s}_i \right\| + \boldsymbol{\lambda}_{D_i} \left\| \mathbf{r}_i \right\| \left\| \mathbf{s}_i \right\| + \mathbf{s}_i^T \mathbf{H}_i \left(\mathbf{q}_i \right) \mathbf{e}_i \left(\mathbf{r}_i \right) \tag{211}$$

Since $\mathbf{H}_i \left(\mathbf{q}_i \right)$ satisfies

$$\lambda_{h_i} \left\| x \right\|^2 \le \mathbf{x}^T \mathbf{H}_i \mathbf{x} \le \lambda_{H_i} \left\| x \right\|^2 \ \forall \mathbf{q}_i, \ \mathbf{x} \in \mathbb{R}^{n_i}$$

where

$\lambda_{h_i} = \min_{\forall q_i \in \mathbb{R}^{n_i}} \lambda_{min} \left(\mathbf{H}_i \right)$, $\lambda_{H_i} = \max_{\forall q_i \in \mathbb{R}^{n_i}} \lambda_{\max} \left(\mathbf{H}_i \right)$, and $0 < \lambda_{h_i} \le \lambda_{H_i} < \infty$.

Considering (184) and (209), we obtain:

$$\mathbf{s}_i^T \mathbf{H}_i \left(\mathbf{q}_i \right) \mathbf{e}_i \left(\mathbf{r}_i \right) = \left\| \mathbf{s}_i \right\| \lambda_{H_i} \mathbf{M}_{e_i} \left\| \mathbf{r}_i \right\|$$

$$= \mu_{4i} \left\| \mathbf{r}_i \right\| \left\| \mathbf{s}_i \right\| \tag{212}$$

Substituting (212) in (211), we have:

$$\dot{V}_i\left(\mathbf{x}_i\right) \le -\left[\lambda_{\min i}\left(\mathbf{K}_{DR_i}\right) - \mu_{1i}\right]\left\|\mathbf{s}_i\right\|^2 - \left[\lambda_{\min i}\left(\mathbf{H}_{Rdi}\right) - \mu_{2i}\right]\left\|\mathbf{r}_i\right\|^2 - k_{Fi}\lambda_{\min i}\left(\xi_i\right)\left\|\Delta\mathbf{F}\right\|^2$$

$$+\left[\mu_{3i} + \boldsymbol{\lambda}_{D_i} + \mu_{4i}\right]\left\|\mathbf{r}_i\right\|\left\|\mathbf{s}_i\right\| \tag{213}$$

Clearly, we observe that in order to validate (213) for all t, we need a proper selection of $\lambda_{\min i}\left(\mathbf{K}_{DR_i}\right)$ and $\lambda_{\min i}\left(\mathbf{H}_{Rdi}\right)$. Note that:

$$\begin{aligned} &-\left[\left\|\mathbf{s}_i\right\| - \frac{1}{2}\left(\mu_{3i} + \boldsymbol{\lambda}_{D_i} + \mu_{4i}\right)\left\|\mathbf{r}_i\right\|\right]^2 - \left\|\mathbf{s}_i\right\|^2 + \\ &\left(\mu_{3i} + \lambda_{D_i} + \mu_{4i}\right)\left\|\mathbf{r}_i\right\|\left\|\mathbf{s}_i\right\| - \frac{1}{4}\left(\mu_{3i} + \lambda_{D_i} + \mu_{4i}\right)^2\left\|\mathbf{r}_i\right\|^2 \le 0 \end{aligned} \tag{214}$$

This suggests make the following definitions:

$$\lambda_{\min i}\left(\mathbf{K}_{DR_i}\right) = \mu_{1i} + 1 + \delta_{1i} \tag{215}$$

$$k_{d_i} = \frac{-\lambda_{\min i}\left(\mathbf{K}_{R_i}\right) + \mu_{2i} + \frac{1}{4}\left(\mu_{3i} + \lambda_{D_i} + \mu_{4i}\right)^2 + \delta_{2i}}{\lambda_{h_i}} \tag{216}$$

where $\delta_{1i} > 0$ and $\delta_{2i} > 0$. Substituting (215) in (213):

$$\dot{V}_i\left(x_i\right) \le -\left[\mu_{1i} + 1 + \delta_{1i} - \mu_{1i}\right]\left\|\mathbf{s}_i\right\|^2 - \left[\lambda_{\min i}\left(\mathbf{H}_{Rdi}\right) - \mu_{2i}\right]\left\|\mathbf{r}_i\right\|^2 - k_{Fi}\lambda_{\min i}\left(\xi_i\right)\left\|\Delta\mathbf{F}\right\|^2$$

$$+\left(\mu_{3i} + \lambda_{D_i} + \mu_{4i}\right)\left\|\mathbf{r}_i\right\|\left\|\mathbf{s}_i\right\|$$

$$\le -\left[1 + \delta_{1i}\right]\left\|\mathbf{s}_i\right\|^2 - \left[\lambda_{\min i}\left(\mathbf{H}_{Rdi}\right) - \mu_{2i}\right]\left\|\mathbf{r}_i\right\|^2 - k_{Fi}\lambda_{\min i}\left(\xi_i\right)\left\|\Delta\mathbf{F}\right\|^2$$

$$+\left(\mu_{3i} + \lambda_{D_i} + \mu_{4i}\right)\left\|\mathbf{r}_i\right\|\left\|\mathbf{s}_i\right\| \tag{217}$$

Substituting (167) in (213):

$$\dot{V}_i\left(x_i\right) \le -\left[1+\delta_{1i}\right]\left\|\mathbf{s}_i\right\|^2 - \left[\lambda_{h_i} k_{d_i} - \lambda_{\min i}\left(\mathbf{K}_{R_i}\right) - \mu_{2i}\right]\left\|\mathbf{r}_i\right\|^2 - k_{Fi}\lambda_{\min i}\left(\xi_i\right)\left\|\Delta\mathbf{F}\right\|^2$$

$$+\left(\mu_{3i}+\lambda_{D_i}+\mu_{4i}\right)\left\|\mathbf{r}_i\right\|\left\|\mathbf{s}_i\right\| \tag{218}$$

Ordering (218) we have:

$$\begin{aligned}&\dot{V}_i\left(\mathbf{x}_i\right) \le -k_{Fi}\lambda_{\min i}\left(\xi_i\right)\left\|\Delta\mathbf{F}\right\|^2 - \left\|\mathbf{s}_i\right\|^2 - \delta_{1i}\left\|\mathbf{s}_i\right\|^2 + \\ &\left(\mu_{3i}+\lambda_{D_i}+\mu_{4i}\right)\left\|\mathbf{r}_i\right\|\left\|\mathbf{s}_i\right\| - \left[\lambda_{h_i} k_{d_i} + \lambda_{\min i}\left(\mathbf{K}_{R_i}\right) - \mu_{2i}\right]\left\|\mathbf{r}_i\right\|^2\end{aligned} \tag{219}$$

Substituting (216) in (219), we get:

$$\dot{V}_i\left(\mathbf{x}_i\right) \le -k_{Fi}\lambda_{\min i}\left(\xi_i\right)\left\|\Delta\mathbf{F}\right\|^2 - \left\|\mathbf{s}_i\right\|^2 - \delta_{1i}\left\|\mathbf{s}_i\right\|^2 + \left(\mu_{3i}+\lambda_{D_i}+\mu_{4i}\right)\left\|\mathbf{r}_i\right\|\left\|\mathbf{s}_i\right\|$$

$$-\left[\cancel{\lambda_{h_i}}\frac{-\lambda_{\min i}\left(\mathbf{K}_{R_i}\right)+\mu_{2i}+\frac{1}{4}\left(\mu_{3i}+\lambda_{D_i}+\mu_{4i}\right)^2+\delta_{2i}}{\cancel{\lambda_{h_i}}} + \lambda_{\min i}\left(\mathbf{K}_{R_i}\right) - \mu_{2i}\right]\left\|\mathbf{r}_i\right\|^2$$

$$\le -k_{Fi}\lambda_{\min i}\left(\xi_i\right)\left\|\Delta\mathbf{F}\right\|^2 - \left\|\mathbf{s}_i\right\|^2 - \delta_{1i}\left\|\mathbf{s}_i\right\|^2 + \left(\mu_{3i}+\lambda_{D_i}+\mu_{4i}\right)\left\|\mathbf{r}_i\right\|\left\|\mathbf{s}_i\right\|$$

$$-\left[-\cancel{\lambda_{\min i}\left(\mathbf{K}_{R_i}\right)} + \cancel{\mu_{2i}} + \frac{1}{4}\left(\mu_{3i}+\lambda_{D_i}+\mu_{4i}\right)^2 + \delta_{2i} + \cancel{\lambda_{\min i}\left(\mathbf{K}_{R_i}\right)} - \cancel{\mu_{2i}}\right]\left\|\mathbf{r}_i\right\|^2$$

$$\le -k_{Fi}\lambda_{\min i}\left(\xi_i\right)\left\|\Delta\mathbf{F}\right\|^2 - \left\|\mathbf{s}_i\right\|^2 - \delta_{1i}\left\|\mathbf{s}_i\right\|^2 + \left(\mu_{3i}+\lambda_{D_i}+\mu_{4i}\right)\left\|\mathbf{r}_i\right\|\left\|\mathbf{s}_i\right\|$$

$$-\left[\frac{1}{4}\left(\mu_{3i}+\lambda_{D_i}+\mu_{4i}\right)^2 + \delta_{2i}\right]\left\|\mathbf{r}_i\right\|^2 \tag{220}$$

Replacing (214) into (220), we have:

$$\dot{V}_i\left(\mathbf{x}_i\right) \le -k_{Fi}\lambda_{\min i}\left(\xi_i\right)\left\|\Delta\mathbf{F}\right\|^2 - \cancel{\left\|\mathbf{s}_i\right\|^2} - \delta_{1i}\left\|\mathbf{s}_i\right\|^2 + \cancel{\left(\mu_{3i}+\lambda_{D_i}+\mu_{4i}\right)\left\|\mathbf{r}_i\right\|\left\|\mathbf{s}_i\right\|}$$

$$-\left[\frac{1}{4}\left(\mu_{3i}+\lambda_{D_i}+\mu_{4i}\right)^2+\delta_{2i}\right]\left\|\mathbf{r}_i\right\|^2-\left[\delta_{2i}\right]\left\|\mathbf{r}_i\right\|^2$$

$$\leq -k_{Fi}\lambda_{\min i}\left(\xi_i\right)\left\|\Delta\mathbf{F}\right\|^2-\delta_{1i}\left\|\mathbf{s}_i\right\|^2-\delta_{2i}\left\|\mathbf{r}_i\right\|^2 \tag{221}$$

Considering (171), suppose that it satisfies

$$\left\|\mathbf{x}_i\right\| \leq \mathbf{x}_{\max} \tag{222}$$

then,

$$\left\|\mathbf{x}_i\right\|^2=\left\|\mathbf{s}_i\right\|^2+\left\|\mathbf{r}_i\right\|^2+\left\|\Delta\mathbf{F}\right\|^2 \leq \mathbf{x}_{\max}^2 \tag{223}$$

Changing (223) into (220), we have:

$$\dot{V}_i\left(\mathbf{x}_i\right) \leq -\delta_i\left\|\mathbf{x}_i\right\|^2 \tag{224}$$

where $\delta_i=\min\left(\delta_{1i},\delta_{2i},k_{Fi}\lambda_{\min i}\left(\xi_i\right)\right)$. Note that (224) is independent of $\mu_{1i},\mu_{2i},\mu_{3i}$ and μ_{4i}. This makes $\dot{V}$ negative. Consequently, $\left\|\mathbf{x}_i\right\|$ cannot be larger than $\mathbf{x}_{\max_i}$. This means that (224) is valid for all $t \geq 0$.

Note that the inverse of the matrix always exists as $\xi_i > 0$. Moreover, the right side of (272) is bounded, hence $\Delta\lambda_i$ is bounded. It is additionally observed that $\dot{\mathbf{s}}_i$ is also bounded with respect to (270). Finally, in order to prove the stability of the whole system we may define:

$$V_i\left(\mathbf{x}_i\right)=\sum_{i=1}^{l} V_i\left(\mathbf{x}_i\right) \tag{225}$$

CONCLUSION

We presented the fundamentals of cooperative manipulation, starting on a brief historical overview. The kinematics and dynamics properties are considered. The workspace intersection was analyzed, and the object manipulation workspace was defined. Selecting the appropriate trajectory in order to bring the robot end-effector at the vicinity of the body and choosing properly the contact points between the object and the end-effectors. In order to ensure a right grip between the body and the end-effectors of the robot manipulators. We first must be planning a soft trajectory that carries the end-effectors of all robot manipulators at the vicinity of the body, then we must be planning the contact by determining the properly contact points, and finally we must hold the object ensuring the sufficient force so that no relative motion there exists. The tracking control problem for cooperative robots without velocity measurements is considered. The control law is a decentralized approachment which takes into account motion constraints rather than the held object dynamics. By assuming bounded that robot manipulators dynamics are well-known and contact force's measurements are available, an observer for each robot manipulator is proposed so that requires only the knowledge of the respective inertia matrix. Despite the fact, the stability analysis is complex, the controller and especially the observers are not. The controller and the observer are presented in the same way on (Gudiño-Lau *et al.*, 2004). The robots used to describe the cooperative systems are the CRS A255 and CRS A465. The assumption that only one robot owns a force sensor, results were better for this one.

REFERENCES

Arimoto, S., Miyazaki, F., & Kawamura, S. (2003). Cooperative Motion Control of Multiple Robot Arms or Fingers. *Proceeding of the IEEE International Conference on Robotics and Automation*, 4, 1407-1412. doi:10.1109/ROBOT.1987.1087820

Arimoto, S., Yun-Hui, L., & Tomohide, N. (1994). Principle of orthogonalization for hybrid control of robot arms. *Automatic Control-World Congress*; 3, 335-340.

Arteaga-Pérez, M. (1998). On the properties of a dynamic model of flexible robot manipulators. *Journal of Dynamic Systems, Measurement, and Control, 120*(1), 8–14. doi:10.1115/1.2801326

Bonitz, R. G., & Hsia, T. C. (1994). Force decomposition in cooperating manipulators using the theory of metric spaces and generalized inverses. *IEEE Transactions on Robotics and Automation, 2*, 1521–1527. doi:10.1109/ROBOT.1994.351372

Bonitz, R. G., & Hsia, T. C. (1996). Internal force-based impedance control for cooperating manipulators. *IEEE Transactions on Robotics and Automation, 12*(1), 78–89. doi:10.1109/70.481752

Caccavale, F., Chiacchio, P., & Chiaverini, S. (1999). Stability analysis of a joint space control law for a two-manipulator system. *Proceeding of the 35th IEEE Conference on Decision Control*, 3, 85-88. doi:10.1109/9.739077

Caccavale, F., Chiacchio, P., & Chiaverini, S. (2000). Task-Space regulation of cooperative manipulators. *Automatica, 36*(6), 879–887. doi:10.1016/S0005-1098(99)00215-0

Chang, K. S., Holmberg, R., & Khatib, O. (1995). The augmented object model: cooperative manipulation and parallel mechanisms dynamics. *Proceedings of the 2000 IEEE International Conference on Robotics and Automation*, 1, 470-475. doi:10.1109/ROBOT.2000.844099

Chiacchio, P., Chiaverini, S., & Siciliano, B. (1996). Direct and inverse kinematics for coordinated motion tasks of a two-manipulator system. *Journal of Dynamic Systems, Measurement, and Control, 118*(4), 691–697. doi:10.1115/1.2802344

Dauchez, P., & Zapata, R. (1985). Co-ordinated control of two cooperative manipulators: the use of a kinematic model. *Proc. 15th Int. Symp. Ind. Robots*. 641–648.

Dellinger, W. F., & Anderson, J. N. (1992). Interactive force dynamics of two robotic manipulators grasping a non-rigid object. *Proceeding of IEEE International Conference on Robotics and Automation*, 3, 2205-2210. doi:10.1109/ROBOT.1992.219930

Díaz Baca, C. S. (2007). *Manipulación cooperativa robot-robot y humano-robot. Aplicación a sistemas flexibles de desensamblado automático.* (Doctoral dissertation), Universidad de Alicante.

Dong, S., & Yun-Hui, L. (2001). Position and Force Tracking of a Two-Manipulator System Manipulating a Flexible Beam Payload. *Proceedings 2001 ICRA. IEEE International Conference on Robotics and Automation*, 4, 3484-3488. doi:10.1109/ROBOT.2001.933157

Fonseca Ferreira, N. M., & Tenreiro Machado, J. A. (2003). Fractional Order Position/Force Control of Two Cooperating Manipulators. *Proceedings of the ICCC 2003 IEEE International Conference on Computational Cybernetics*, 8(29), 29-31. doi:10.1155/2010/375858

Fujii, S., & Kurono, S. (1975). Coordinated computer control of a pair of manipulators. *Proceeding 4th IFToMM World Congress,* 411-417.

Gudiño-Lau, J., Arteaga, M. A., Muñoz, L. A., & Parra-Vega, V. (2004). On the Control of Cooperative Robots without Velocity Measurements. *IEEE Transactions on Control Systems Technology*, *12*(4), 600–608. doi:10.1109/TCST.2004.824965

Gueaieb, W., Karray, F., Al-Sharhan, S., & Basir, O. (2002). A hybrid adaptive fuzzy approach for control of cooperative manipulators. *Proceedings. ICRA '02. IEEE International Conference on Robotics and Automation*, 2, 2153-2158. doi:10.1109/ROBOT.2002.1014858

Hayati, S. (1988). Hybrid Position/force control of multi-arm cooperating robots. *Proceedings of the IEEE International Conference on Robotics and Automation*, 3, 82-89. doi:10.1109/ROBOT.1986.1087650

Hsu, P. (1986). Coordinated Control of Multiple Manipulators Systems. *IEEE Transactions on Robotics and Automation*, *9*(4), 400–410. doi:10.1109/70.246051

Hsu, P. (1989). Control of Multiple Manipulator system-trajectory tracking load distribution, internal force control and decentralized architecture. *Proceeding of the IEEE International Conference on Robotics and Automation*, 2, 1234-1239. doi:10.1109/ROBOT.1989.100149

Hu, Y. R., Goldenberg, A. A., & Zhou, C. (1995). Motion and force control of coordinated robots during constrained motion tasks. *The International Journal of Robotics Research*, *14*(4), 351–365. doi:10.1177/027836499501400404

Kelly, R., & Santibáñez, V. (2003). *Control de Movimiento de Robots Manipuladores.* Madrid: Pearson Educación.

Kousuge, K., Kogan, M., Furuta, K., & Nosaki, K. (1989). Coordinated Motion Control of Robot Arm Based on Virtual Internal Model. *Proceeding of IEEE International Conference on Robotics and Automation*, 2, 1097-1102. doi:10.1109/ROBOT.1989.100127

Liu, Y., Arimoto, S., Parra-Vega, V., & Kitagaki, K. (1997). Decentralized adaptive control of multiple manipulators in cooperation. *International Journal of Control*, *67*(5), 649–674. doi:10.1080/002071797223938

Luecke, G. R., & Lai, K. W. (1997). A joint error-feedback approach to internal force regulation in cooperating manipulator systems. *Journal of Robotic Systems*, *14*(9), 631–648. doi:10.1002/(SICI)1097-4563(199709)14:9<631::AID-ROB1>3.0.CO;2-M

Nakamura, Y., Naga, K., & Yoshikawa, T. (1989). Mechanics of coordinative manipulation by multiple robotics mechanism. *Proceeding of the IEEE International Conference on Robotics and Automation*, 4, 991-998. doi:10.1109/ROBOT.1987.1087941

Nakano, E., Ozaki, S., Ishida, T., & Kato, I. (1974). Cooperational control of the anthropomorphous manipulator MELAR. *Proceeding 4th International Symposium on Industrial Robots*. 251-260.

Naniwa, T., Arimoto, S., & Wada, K. (1997). Learning and adaptive controls for coordination of multiple manipulators without knowing physical parameters of an object. *Proceedings of IEEE International Conference on Robotics and Automation*, 2, 1496-1502. doi:10.1109/ROBOT.1997.614350

Parra-Vega, V., Rodríguez-Ángeles, A., Arimoto, S., & Hirzinger, G. (2001). High precision constrained grasping with cooperative adaptive control. *Journal of Intelligent & Robotic Systems*, *32*(3), 235–254. doi:10.1023/A:1013987209547

Perdereau, V., & Drouin, M. (1996). Hybrid external control for two robot coordinated motion. *Robotica*, *14*(2), 141–153. doi:10.1017/S0263574700019056

Pfeffer, L. E., & Cannon, R. H. Jr. (1993). Experiments with a Dual-Armed, Cooperative, Flexible Drive-train Robot System. *Proceedings of International Conference on Robotics and Automation*, 3, 601-608. doi:10.1109/ROBOT.1993.291839

Ramadorai, A., Tarn, T., & Bejczy, A. (1992). Task definition, decoupling and redundancy resolution by nonlinear feedback in multi-robot object handling. *Proceeding of the IEEE International Conference on Robotics and Automation*, 1, 467-474. doi:10.1109/ROBOT.1992.220296

Rivera Dueñas, J. C. (2012). *Control de posición/fuerza-torque para manipuladores con compensación de fricción*. (Doctoral dissertation), Available from UNAM Theses database. (tes.TES01000681507)

Sagredo Hernández, L. R. (2006). *Control de robots cooperativos con un observador*. (Unpublished master dissertation), Available from UNAM Theses database. (tes.TES01000650102)

Scheider, S. A., & Cannon, S. H. (1992). Object impedance control for cooperative manipulation: Theory and experimental result. *IEEE Transactions on Robotics and Automation*, *8*(3), 383–394. doi:10.1109/70.143355

Siciliano, B., & Villani, L. (1999). *Robot Force Control*. Klumer Academic Publishers. doi:10.1007/978-1-4615-4431-9

Szewczyk, J., Morel, G., Bidaud, P., & Basañez Villaluenga, L. (1997). Distributed Impedance Control of Multiple Robot Systems, *Proceeding of the IEEE International Conference on Robotics and Automation*. USA. 2, 1801-1806. doi:10.1109/ROBOT.1997.614414

Takase, K., Inoue, H., Sato, K., & Hagiwara, S. (1974). The design of an articulated manipulator with torque control ability. *Proceeding 4th. International Symposium on Industrial Robots*. 261-270.

Tao Jian, M., & Luh, J. Y. (1991). Position and Force Control for Two Coordinating Robots. *Proceeding of the IEEE International Conference on Robotics and Automation*, 1, 176-181. doi:10.1109/ROBOT.1991.131575

Tarn, T. J., Bejczy, A. K., & Yun, X. (1987). Design of dynamics control of two cooperating robot arm. *Proceeding IEEE International Conference on Robotics and Automation*, 4, 7-13. doi:10.1109/ROBOT.1987.1088028

Tarn, T. J., Bejczy, A. K., & Yun, X. (1988). New nonlinear control algorithms for multiple robot arms. *IEEE Transactions on Aerospace and Electronic Systems*, *24*(5), 571–583. doi:10.1109/7.9685

Tinós, R., Terra, M., & Bergrman, M. (2002). *Dynamic Load-Carrying Capacity of Cooperative Manipulators with Passive Joints*. Congreso Brasileiro de Automática.

Uchiyama, M., & Dauchez, P. (1988). A symmetric hybrid position/force control scheme for the coordination of two robots. *Proc. 1988 IEEE Int. Conf. on Robotics and Automation*, 1, 350-356. doi:10.1109/ROBOT.1988.12073

Uchiyama, M., & Dauchez, P. (1993). Symmetric kinematic formulation and non-master/slave coordinated control of two-arm robots. *Advanced Robotics*, *7*(4), 361–383. doi:10.1163/156855393X00221

Uchiyama, M., & Dauchez, P. (1998). A symmetric hybrid position/force control scheme for the coordination of two robots. *Proceeding of the IEEE International Conference on Robotics and Automation*, 350-356. doi:10.1109/ROBOT.1988.12073

Uchiyama, M., Iwasawa, N., & Hakomori, K. (1987). Hybrid position/force control for coordination of a two-arm robot. *Proc. 1987 IEEE Int. Conf. on Robotics and Automation*, 4, 1242-1247. doi:10.1109/ROBOT.1987.1087766

Walker, I. D., Freeman, R., & Marcus, S. (1991). Analysis of motion and internal grasped by multiple cooperating manipulators. *The International Journal of Robotics Research*, *10*(4), 396–409. doi:10.1177/027836499101000408

Walker, I. D., Freeman, R., & Marcus, S. (1991). Analysis of motion and internal force loading of objects grasped by multiple cooperating manipulators. *The International Journal of Robotics Research*, *10*(4), 396–409. doi:10.1177/027836499101000408

Walker, M., Kim, D., & Dionise, J. (1989). Adaptive Coordinated motion control of two manipulators arms. *Proceeding of the IEEE International Conference on Robotics and Automation*, 2, 1084-1090. doi:10.1109/ROBOT.1989.100125

Wen, J. T., & Kreutz-Delgado, K. (1992). Motion and force control of multiple robotic manipulators. *Automatica*, *28*(4), 729–743. doi:10.1016/0005-1098(92)90033-C

Williams, D., & Khatib, O. (1993). The virtual linkage: a model for internal forces in multi-grasp manipulation. *Proc. 1993 IEEE Int. Conf. on Robotics and Automation*. 1025-1030. doi:10.1109/ROBOT.1993.292110

Woong, K., & Beom, H. L. (1999). General Redundancy Optimization Method for Cooperating Manipulators Using Quadratic Inequality Constraints. *IEEE Transactions on Systems, Man, and Cybernetics. Part A, Systems and Humans*, *29*(1), 41–51. doi:10.1109/3468.736359

Yoshida, K., Kurazame, R., & Umentani, Y. (1991). Dual-arm coordination in space free-flying robot. *Proceeding of the IEEE International Conference on Robotics and Automation*, 3, 2516-2521. doi:10.1109/ROBOT.1991.132004

Yoshikawa, T., & Zheng, X. Z. (1990). Coordinated dynamic hybrid position/force control for multiple robot manipulators handling one constrained object. *Proceedings of 1990 IEEE International Conference on Robotics and Automation*, 2, 1178-1183. doi:10.1109/ROBOT.1990.126156

Yun, X., & Kumar, V. (1991). An approach to simultaneous control of trajectory and interaction forces in dual-arm configuration. *IEEE Transactions on Robotics and Automation*, *5*(5), 618–625. doi:10.1109/70.97873

Chapter 3
Mobile Robot Path Planning using Voronoi Diagram and Fast Marching

S. Garrido
Carlos III University of Madrid, Spain

L. Moreno
Carlos III University of Madrid, Spain

ABSTRACT

This chapter presents a new sensor-based path planner, which gives a fast local or global motion plan capable to incorporate new obstacles data. Within the first step, the safest areas in the environment are extracted by means of a Voronoi Diagram. Within the second step, the fast marching method is applied to the Voronoi extracted areas so as to get the trail. This strategy combines map-based and sensor-based designing operations to supply a reliable motion plan, whereas it operates at the frequency of the sensor. The most interesting characteristics are high speed and reliability, as the map dimensions are reduced to a virtually one-dimensional map and this map represents the safest areas within the environment.

INTRODUCTION

Mobile robot motion planning has become a very important research topic since the 1980's. Many researchers have worked extensively to get efficient ways to solve issues related to this topic. Two different ways can be followed to solve this problem. The first approach considers that the global map (surroundings and obstacle data) and the robot characteristics are known. The second one uses local sensor data and robot characteristics.

In recent years, sensor-based path planning has emerged as a compromise between efficient optimized trajectories and ability to react to unexpected environmental events. Different variations of the classical methods have been implemented to attain an operative sensor-based path planning. One among these techniques is the Voronoi-based path planning.

DOI: 10.4018/978-1-4666-8693-9.ch003

Many researchers have studied the Voronoi Diagrams. An advantage of Voronoi-based path planning is that it diminishes the problem to one dimension while the trajectories follow the most clearance map. This implies that the trajectories are the safest ones. Moreover, the Voronoi-based paths can be viewed as a type of topological navigation and, as a result of that, it is like the human one in some aspects. However, Voronoi-based strategies have the typical difficulties of drawing the Voronoi Diagram: finding out lines and polygons, finding the vertices and nodes, and making a tree to find the trail. As an alternative to unravel these issues, we propose the employment of image processing techniques.

In order to calculate the trajectory, the new motion planning technique relies on the mixture of a Voronoi Diagram and the expansion of a wave from the beginning point to the goal following the Voronoi Diagram. It uses local sensor data and also the known map of the surroundings.

The Voronoi Diagram is constructed using image processing methods (skeletonization) based on the Breu technique in 2D (Breu, 1995) and Gagvani (1997) in 3D. The Voronoi Diagram (skeleton) is dilated and inverted to get a road map. The following step is to calculate the trajectory within the image generated by the thick Voronoi Diagram employing a wave expansion. The wave expansion is calculated solving the Eikonal equation. The Fast Marching method has been implemented to solve this equation. There are other similar techniques that can also be used for the same purpose (Tsiksitlis, 1995), (Mauch, 2003). After that, the obtained path verifies the smoothness and safety considerations required for mobile robot path planning.

The approach implemented in this work relies on the potential field technique of the wave expansion, which repels a robot far away from obstacles and towards the goal employing a rigorously designed artificial potential function. Since the potential field is like a Lyapunov function, the stability of the system is assured.

Finally, this strategy has been extended to be used with non-holonomic robots. It associates a vector field that permits to use the strategy with non-holonomic robots (the Voronoi Diagram thread has become a road). This is done by employing a Voronoi Diagram thicker than the robot. In this thick diagram, a repulsive potential is made using the propagation of a wave that starts at the edges. This potential field is employed as a slowness or refraction index in a second wave propagation from the goal point up to the actual position of the robot. This potential field is attractive to the goal and repulsive from the perimeters of the thick Voronoi Diagram. There are no different local minima than the destination. This potential and its associated vector field permit the motion planning of the non-holonomic robot.

The main advantages of this technique are:

- It is straightforward to implement.
- The speed is very high.
- The trajectories are smooth.
- The trajectories are the safest ones.
- It works on-line.
- It works with non-holonomic robots.
- The technique is complete, i.e., the strategy is capable of finding a path if it exists.

RELATED WORK

Voronoi Diagram

The Voronoi idea has been used for four centuries. In his 'Traite de la Lumiere', printed in 1644, Rene Descartes uses diagrams just like Voronoi to indicate the disposition of matter within the solar system and its surroundings. Algorithms to build Voronoi Diagrams have appeared since the Seventies. See the surveys by (Aurenhammer, 1991, 2000), (de Berg, 2000), and (Okabe, 1992) on numerous algorithms, applications, and generalizations of Voronoi Diagrams.

The Generalized Voronoi Diagram is the set of points where the distance to the two nearest objects is the same (Choset, 2005). Several algorithms are proposed to construct Generalized Voronoi Diagrams (GVD) using the distance data provided by diverse external sensors like sonar sensors, laser scanners, and stereo cameras. An interesting strategy for construction of a Generalized Local Voronoi Diagram (GLVD) can be found in (Mahkovic, 1998). They use measurements from a laser scanner. First, the points that belong to the same object are clustered, then the Voronoi Diagram is generated, and finally, the perimeters outside the visible region and those that both side generator points belonging to the same object are deleted. In (Sudha, 1999), they construct a Voronoi Diagram from a binary image of the workspace obtained employing a digital camera. This category of algorithms involves progressive construction procedures based on the exploration of unknown surroundings (Nagatani, 1998), (Choset, 2000) and (Choset, 2005). These algorithms do not take into account the encoder errors and they cannot be utilized in a large environment. (Bhattacharya, 2008) provides an algorithm based on the Voronoi Diagram to compute an optimal path between source and destination in the presence of simple disjoint polygonal obstacles. They evaluate the quality of the path based on clearance from obstacles, overall length, and smoothness.

There are some path planning algorithms inspired by the Voronoi Diagram idea. An example is the MAPRM methodology (Wilmarth, 1999), which retracts sampled configurations onto the medial axis (or GVD). The EquiDistance Diagram (EDD) (Keerthi, 1999) is based on the Voronoi roadmap idea. It consists of a roadmap shaped by connecting the local maxima of a clearance function. The main disadvantage of these strategies is that the roadmap is built offline and the environment information should be provided earlier. (Bortoff, 2000) proposes a two step path-planning algorithm for UAVs. The algorithm generates a stealthy path through a set of enemy radar sites of known location, and provides an intuitive way to trade-off stealth versus path length. In the first step, a suboptimal rough-cut path is generated through the radar sites by constructing and searching a graph based on Voronoi polygons. In the second step, a set of nonlinear Ordinary Differential Equations (ODE) is simulated, using the graph solution as an initial condition. The ODE describe the dynamics of a set of virtual masses located in a virtual force field. The virtual forces push the masses away from the radars and toward one another.

(Breu, 1995) presents a linear time (and therefore asymptotically optimal) algorithmic rule for computing the Euclidean distance transform of a two-dimensional binary image. The algorithm relies on the construction and regular sampling of the Voronoi Diagram whose sites consist of the unit (feature) pixels within the image. Their algorithmic rule constructs the Voronoi Diagram wherever it intersects the horizontal lines passing through the image constituent centers. They additionally discuss the extensions to higher dimensional images and to different distance functions. The strategy proposed in this paper relies on this algorithmic rule.

Introduction to Voronoi Diagram and Skeleton

The Voronoi Diagram concept can be explained in an intuitive way. Given a finite set of objects in an area, all locations within the area are related to the nearest member of the set. The result is a partition of the area into a group of regions, Voronoi regions. The GVD is formed by the frontier between these regions. Given its widespread use, it is not stunning that this idea has been discovered persistently in many alternative places.

The close relationship between the Voronoi Diagram and the Medial Axis has already been reviewed in literature (Ogniewicz,1995). Within the field of computer vision, the skeleton of an object or image is called the Medial Axis transform. It coincides with the definition of the GVD utilized in AI. This is often to not be confused with the Voronoi Diagram of a distinct set of purposes: for every point x within the set P, there exists a boundary polygonal enclosing all the intermediate points lying nearer to x than to alternative points within the set P, and also the set of those polygons for a given purpose set is named the Voronoi Diagram.

A very illustrative definition of the skeleton is given by the prairie-fire analogy: the boundary of an object is ready on fire and also the skeleton is formed from the loci wherever the fire fronts meet and quench one another.

Based on Blum's definition of a skeleton (Blum, 1967), the skeleton S of a group D is the locus of the centers of maximal disks. A maximal disk of D could be a closed disk contained in D that is interiorly tangent to the boundary δD, which is not contained in the other disk in D. Every maximal disc should be tangent to the boundary in a minimum of two completely different points. With each skeleton point $s \in S$ we have a tendency to conjointly store the radius r(s) of its maximal disk. The skeleton $\sigma(D)$ is the set of centers of maximal disks in D. The skeleton is desired to be a 'graph-like' retraction of the first set. For a good overview of skeletonization techniques, see (Smith, 1987).

A class of skeletonization approaches relies on the Distance Transform (DT). In this paper, algorithms based the DT are used to compute the skeleton (or Voronoi Diagram) of the binary image that represents the environment. This associates each pixel in the image with its distance to the closest obstacle (See an example in Figure 3). The DT is vital for natural neighbor interpolation of a function and for morphological operations on images, such as Medial Axis and skeleton computation, and polygonal shape morphing. Several efficient algorithms exist for the computation of the DT, or approximations of this unit (meaning that the distances are not precisely the true distances), for images. The Medial Surface/ Axis can be defined as the locus of greatest circles (2D) or ball (3D). A very efficient algorithm that has been used for the calculation of the Voronoi Diagram (2D) in this work is thanks to (Breu, 1995). For its approach, it is not necessary to calculate vertices and nodes of the Voronoi Diagram and it may be used for all types of walls and obstacles, even curved ones. It runs in O(m) time, wherever m is that the number of image pixels. This algorithmic rule relies on the construction and regular sampling of the Voronoi Diagram whose sites consist of the unit (feature) pixels within the image. The algorithm constructs the Voronoi Diagram wherever it intersects the horizontal lines passing through the image pixel centers. This algorithm has been used to calculate the Voronoi Diagram in 2D environments.

An important disadvantage of the skeletonization strategies is the existence of spurious branches on the generated skeletons. This unwanted fact is caused by the irregularities of the boundary. Boundary noise would cause the Voronoi Diagram to be very dense, which might result in the necessity to prune it to get the medial axis. (Ogniewicz, 1995) reduced skeletons shaped from raster boundary points to a straightforward form. This was achieved by pruning the hair nodes of the skeleton till a given minimum circumcircle was reached. On the other hand, it has been demonstrated that hair nodes correspond to a location of minimum curvature on the boundary. This implies that for a sampled boundary curve, three adjacent points are co-circular, with their center at the skeleton hair. For the simplification of the skeleton, the hairs ought to be retracted to their parent node location. The central point of the three is moved towards the parent node till it coincides with the parent node circumcircle, leading to smoothing outward-pointing salients of the boundary.

The process is iterated from the other side of the boundary, additionally retracting these salients. This could displace points involved within the first smoothing step. However, the method is convergent and a small number of iterations suffices to supply a smoothened curve having a similar number of points as the original one, but with a simplified skeleton.

In order to extract the skeleton in 3D, the algorithm presented in (Gagvani,1997) has been implemented. The skeletonization methodology for 3D objects relies on a quasi-Euclidean DT. Once the DT value at each object point is known, a linked list of all the object points is made. One pass is completed over the list to check if a local maximum of it satisfies the condition:

$$MNTp < DTp - TP$$

where TP is a thinness parameter, DTp is the DT of the voxel p, and $MNTp$ is the mean of the neighbors' DT .

Selecting a thinness parameter that removes most of the 'hairs' within the skeleton reduces the impact of boundary noise.

INTRODUCTION TO FAST MARCHING

The Fast Marching Method (FMM) is a computational algorithm to find the time of arrival to every point of the area of an expanding wave that starts in a point or a set of points.

Conceptually, it may be thought as a continuous version of the Dijkstra's algorithm, (Dijkstra, 1959). It is based on the assumption that the information solely flows outward from the seeding area (wave source).

The FMM was proposed by J. A. Sethian (Sethian, 1996) to approximate the solution of the Eikonal equation, a non-linear partial equation encountered in the propagation of an electromagnetic wave, such as light.

Let assume a 2D map, where $x = (x, y)$ is a point on the map with its coordinates defined in a Cartesian reference system, $T(x)$ is a front wave arrival time function for every point of the map, and $F(x)$ is the velocity of the wave propagation in each point x. Let also assume that a wave starts propagating at $x_0 = (x_0, y_0)$ at time with velocity $F(x) \geq 0$. The Eikonal equation defines the time of arrival of the propagating wavefront $T(x)$ at each point x of the map, in which the propagation speed depends on the point F(x), according to:

$$|\nabla T(x)|\, F(x) = 1 .$$

With the discretization of the gradient $\nabla T(x)$ (Chiang, 1992), it is possible to solve the Eikonal equation at each point ... Simplifying the notation, the last equation becomes:

$$T_1 = min\left(T\left(i-1,\ j\right); T\left(i+1;\ j\right)\right)$$
$$T_2 = min(T\left(i;\ j-1; T\left(i;\ j+1\right)\right)$$

$$\left(\frac{T(i,j)-T_1}{\Delta x}\right)^2 + \left(\frac{T(i,j)-T_2}{\Delta y}\right)^2 = \frac{1}{F(i,j)^2}$$

The FMM consists on the solution the last equation, when all the parameters are known except $T(i, j)$. This method is iterative, beginning at the source point of the wave (or waves) where $T(i_0, j_0)$. The next iteration solves the value $T(i, j)$ for the neighbors of the points solved in the previous iteration. Using as an input a binary grid map, in which velocity $F(i, j) = 0$ (black) means obstacle and $F(i, j) = 1$ (white) means free space, the output of the algorithm is a map of distances. These distances are equivalent to the time or arrival of the expanding wave at each point of the map, as shown in Figure 1.

The FMM is often directly used as a path planner algorithm. By applying gradient descent from any point of the distance map, a path can be obtained with the source of the wave as a goal point. The paths provided by the FMM are optimal in distance terms, but they do not accomplish the smoothness and safety constraints that most robotic applications need. These paths run too near to obstacles and walls and have sharp curves.

Figure 1. The function $T(x, y)$ gives a cone-shaped surface. The height T gives the set of points reached at time T. The surface on the right below is called the arrival time surface, because it gives the arrival time.

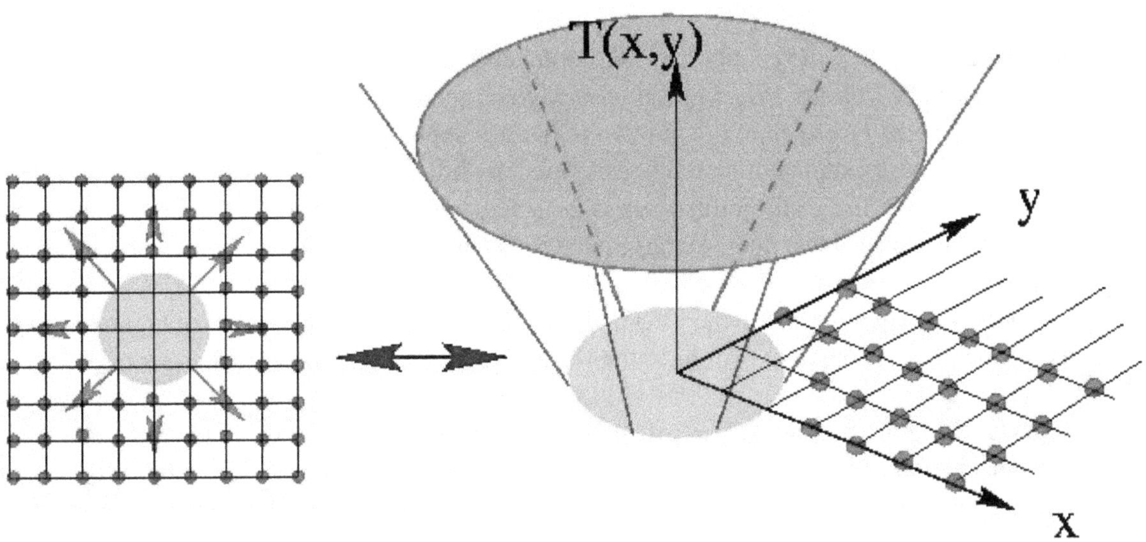

IMPLEMENTATION OF THE METHOD

This methodology operates in two main steps. The first step consists on the calculation of the Voronoi Diagram of the 2D or 3D map of the environment (that is, the cells located equidistant to the obstacles). This method is completed by means of morphological operations on the image of the environment map. In the second step, the FMM is applied to make a potential $T(x)$ that represents the propagation of an electromagnetic wave. This potential is employed to calculate the trajectories based on the potential surface defined by the slowness map.

To be more precise, the planner follows the subsequent steps:

- **Modeling.** A grid-based map of the environment is the model to be used. The a priori known map (if it is known) and, therefore, the sensor data, are integrated in an updated image of the environment with black and white cells. The sole criteria to select the grid size is the computation time. The grid does not need to be uniform, being attainable to use a triangular mesh.
- **Object enlarging.** To avoid impractical trajectories, the objects and walls are enlarged by the radius of the mobile robot before the calculation of the Voronoi Diagram to make sure it neither collides with obstacles or walls nor accepts passages narrower than the robot size. This permits robot to be considered as as a point from now on.
- **Filter.** In order to get rid of the tiny 'hairs' of skeletons, an averaging filter is applied. This filter is applied over the binary image to eliminate the corners and avoid the appearance of hairs.
- **Voronoi Diagram.** Morphological image process techniques are applied to the map to create the Voronoi Diagram. To be precise, a skeletonization of the image based mostly in the (Breu, 1995) methodology (2D) and (Gagvani, 1997) (3D) is applied to get the Voronoi Diagram as shown in Figure 3(left). This skeleton is an image in which the points of the Voronoi diagram are black and the rest of the points are white. There are no vertices/nodes. It is possible to calculate the nodes, however it's more sensible to search the path with the image with the FMM, because the calculation time is smaller and it works even with curved obstacles and walls.
- **Dilatation.** Afterwards, a dilatation is completed to have a thick Voronoi Diagram (see Figure 3(right)) wherein to calculate the propagation of the wave front of the FMM. This is often done in order to get 2 advantages. First, the sharp angles between segments of the diagram are smoothened, which improves the continuity of the generated trajectories and frees them from sharp angles. Second, the Voronoi Diagram is thickened to eliminate excessive narrowing, and thereby to permit a far better propagation of the wave front after applying the FMM. In this way, the path is calculated. This thickening of all the lines of the Voronoi Diagram is completed within the same amount. However, the grade of dilatation is the solely design parameter of the method. It is small enough to have no isolated points or large enough to have a tube in which non-holonomic vehicles maneuver.
- **Viscosity Map**. In order to use the FMM, it is necessary to invert the image (see Figure 4(left)) to get a difficulty (or slowness) map since the wave travels faster within the clearer zones and slower within the darker ones.
- **Fast Marching methodology.** The next step is to calculate the flight in the image generated by the Voronoi Diagram employing a wave expansion. The wave equation is approximated using the paraxial approximation utilized in optics. This way, the Eikonal equation is obtained. To solve the

Eikonal equation, the FMM has been used. The FMM is employed to make a potential $T(x)$ that represents the propagation of an electromagnetic wave in a viscosity map. The time is added as a third axis in 2D or the fourth axis in 3D. The origin of the wave is the goal point, which continues propagating till it reaches the present position of the robot. The FMM produces a funnel-shaped surface in which the time is the last axis, i.e., $T(x)$.

- **Path Calculation.** This potential field is employed to calculate the trajectories based on the potential surface defined by the slowness map using the gradient technique with the present position of the robot as the starting point and the goal position as the final point (see Figure 4(right)). The gradient determines the direction of travel once the FMM has calculated $T(x)$.

The algorithm steps are summarized in the flowchart of Figure 2. The path is obtained inside the safest areas provided by the thickened Voronoi Diagram and is correctly smooth, particularly at the angles, since

Figure 2. Flowchart of the proposed planning algorithm

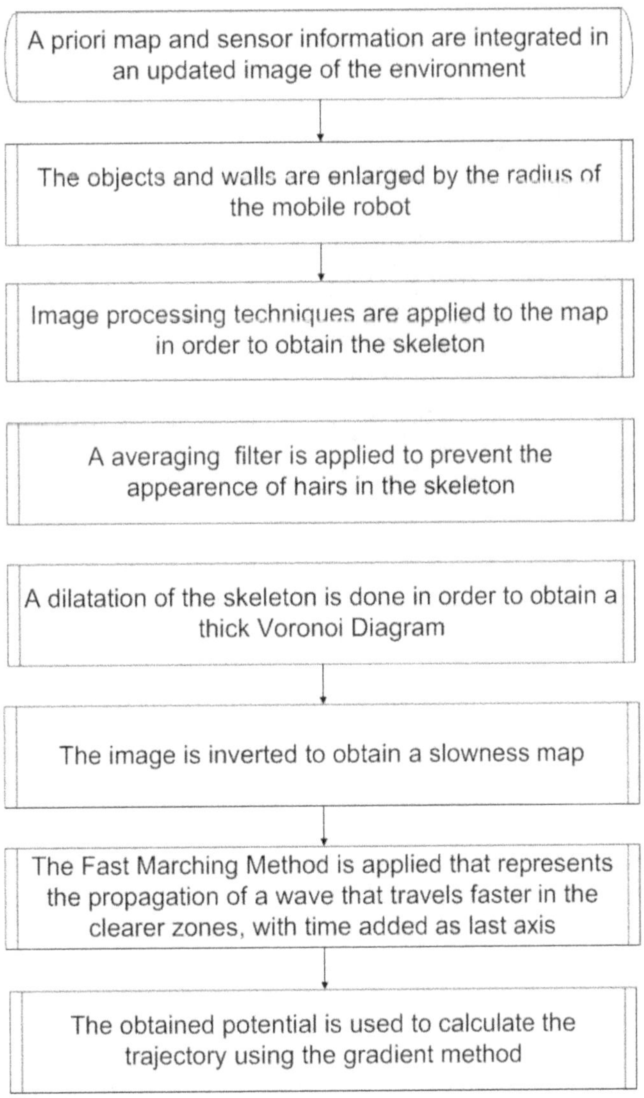

Figure 3. Map of the room used in the first experiment (left) Voronoi Diagram (right) Thickened Voronoi Diagram

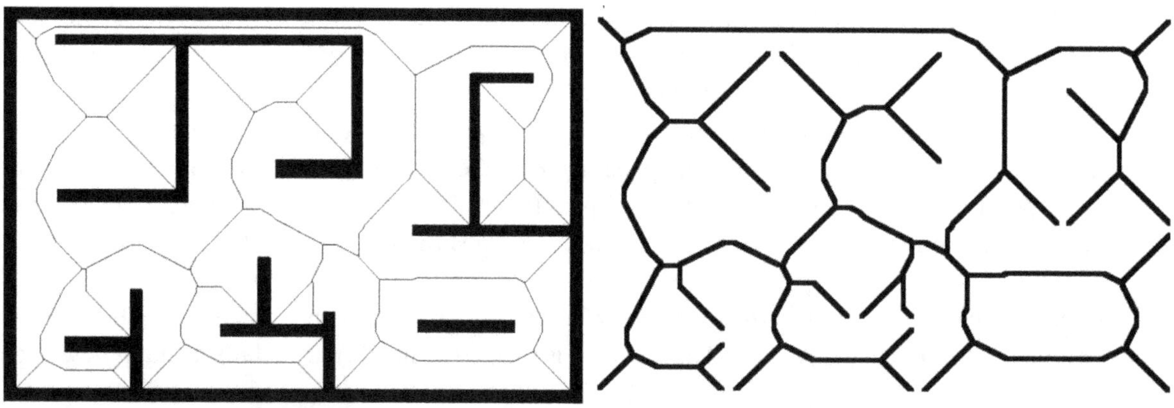

Figure 4. Trajectory calculated by Fast Marching Method in the inverted image of the thickened Voronoi Diagram of the room (left) Inverted image of the thickened Voronoi Diagram (right) Trajectory calculated by Fast Marching Method

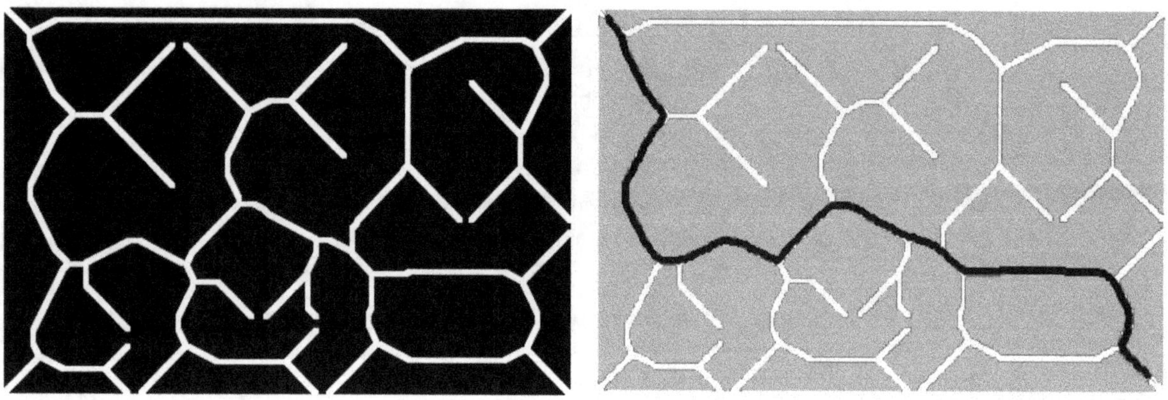

the fast marching technique selects a continuous path in gradient terms. Moreover, trajectory extraction is extremely fast as a result of the fast marching technique propagates in a nearly one-dimensional curve (although this is often an approximation since the Voronoi Diagram is thickened within the perpendicular direction to the diagram curves). The proposed method can also be used for sensor-based planning, working directly on a raw laser sensor image of the environment, as shown in Figures 5 and 6. The data shown corresponds to the corner between two perpendicular aisles of the Carlos III University to the left of the robot, which is located in the center of the bottom of the scene. The image obtained from raw data may contain scattered pixels (see Figure 5(left)), and the dilatation of the image achieves the continuity and thickening of the lines representing the walls and obstacles (see Figure 5(right)). There are similar spurious lines behind the walls because the Voronoi Diagram is made with an open image, but the path followed by the robot is correct. At this point the path planning steps described above are

Figure 5. Data read by the robot (Local map) (left) Raw laser data (right) Thickened laser data

Figure 6. Voronoi Diagram and Trajectory of the Local map (left) Voronoi Diagram (VD) (right) Trajectory calculated with FM over the VD

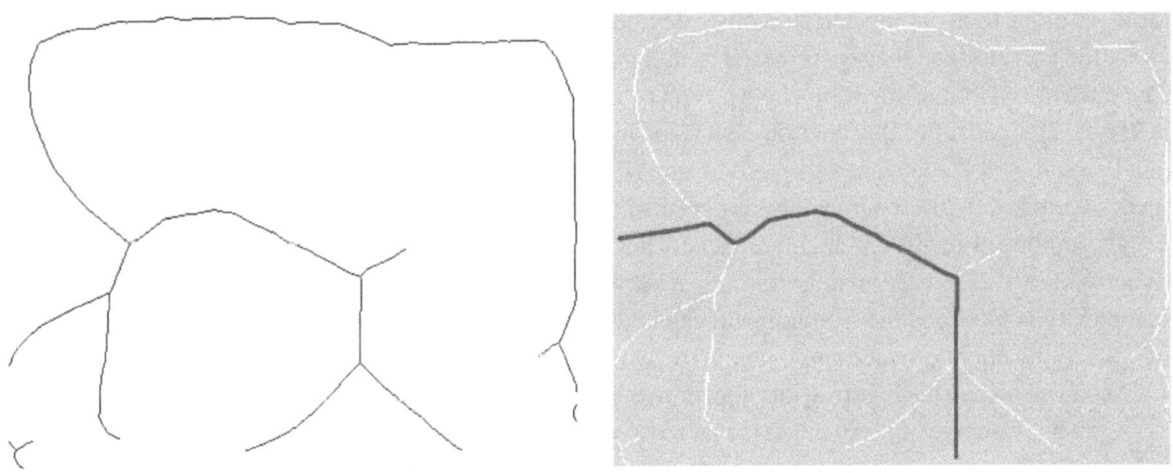

applied in the same sequence. The capability of using raw sensor data combined with the speed of the calculation allows this methodology to be used online to recalculate the trajectory at any time, avoiding this way dynamic obstacles and objects not present in the original map.

RESULTS

The proposed method has been tested using the manipulator robot MANFRED. It has a coordinated control of all the Degrees Of Freedom (DOF) in the system (the mobile base has 2 DOF and the manipulator has 6 DOF) to achieve smooth movement. This mobile manipulator uses a sensorial system based on vision and 3D laser telemetry to perceive and model 3D environments. The mobile manipulator will include all the capabilities needed to localize and avoid obstacles and to navigate safely through the environment.

To illustrate the capabilities of the proposed method, different tests are presented in this section. In the first test (room test with a resolution of 628 ×412 pixels), a difficult test room environment and the floor of the laboratory environment have been used (Figure 7).

In the second test (laser test), the method is applied to a local environment path planning task where the laser scanner measurement data are used to generate the trajectories, as it was explained in the previous section. Figures 5 and 6 illustrate the achieved good trade-off between trajectory length, distances to obstacles and smooth changes in the trajectory.

The method has capabilities to generate adequate paths on a global scale as shown in Figures 7 and 8. The results for the complete lab floor are shown in Figure 7 top (the environment map of the Robotics Lab of the Carlos III University) and Figure 8 bottom (the path obtained after applying the FMM to the previous Voronoi Diagram image). The method provides smooth trajectories that can be used at low control levels without any additional smooth interpolation process.

The last test (lab test) is dedicated to illustrate the capability of the proposed method of adapting to a changing environment taking into account possible dynamic features of the environment such as moving obstacles and persons in the vicinity of the robot, as shown in Figure 8. During motion, the robot observes the environment with its laser scanner, introducing new information in the map and planning a new trajectory. Local observations (obstacle in the middle of the corridor) generate modified trajectories in order to avoid the detected obstacles. In the bottom map of the figure, the obstacles detected block the corridor and the sensor-based global planner generates a completely different safe trajectory. The dimensions of the environment are 116x14 m (the cell resolution is 12 cm). The image resolution is 966 ×120 pixels. For this environment (lab floor) the first step (the Voronoi extraction) takes 0.05 s in a Pentium 4 at 2.2 Ghz, and the second step (Fast Marching) takes 0.15 s for a long trajectory, which makes a total of 0.20 s to apply the Voronoi FM method.

The proposed method is highly efficient from a computational point of view because it operates directly over a 2D image map (without extracting adjacent maps), and due to the fact that Marching complexity is O(n) and the Voronoi path calculation is also of complexity O(n), where n is the number of cells in the environment map.

As can be seen in the Figure 8, the algorithm is perfectly able to make a correct planning of trajectories.

A great advantage of this method is that it can work with any kind of shaped objects, as demonstrated in Figure 9. The curvature in the shape of the obstacle is not a restriction for the application of the method. In Figure 10, the proposed method has been applied to an open environment not limited totally by walls,

Figure 7. Trajectory calculated with Fast Marching over the Voronoi Diagram (Global map)

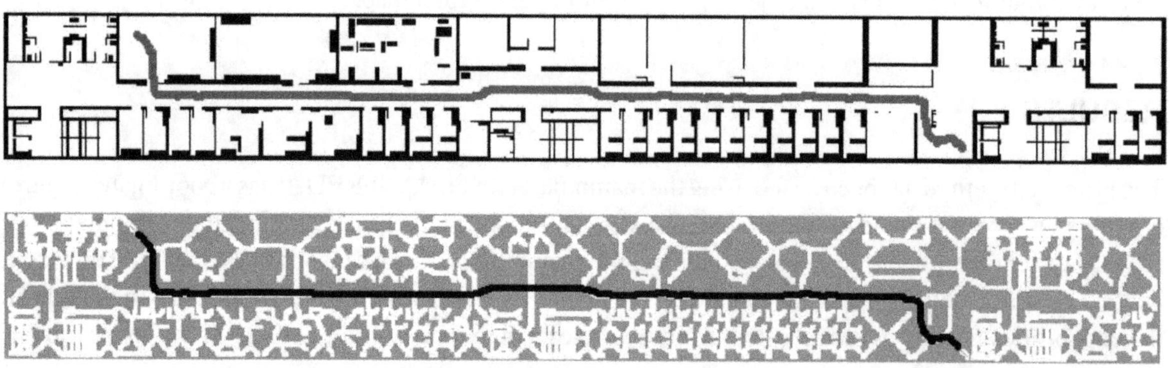

Figure 8. Evolution of the path when the robot reads information about the new obstacles that are not in the previous map and the robot can no pass through the corridor

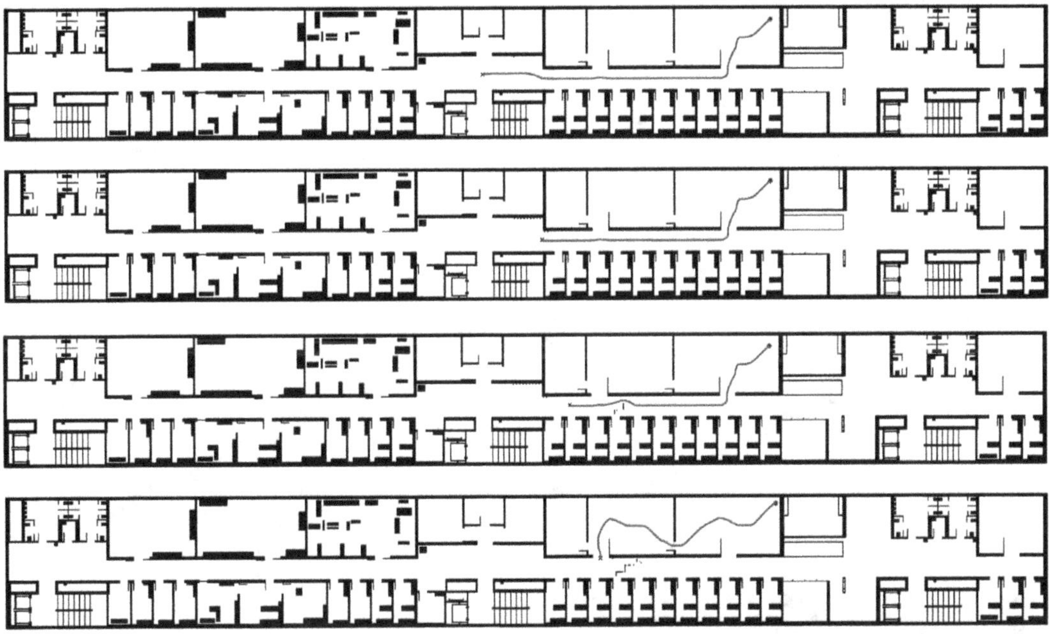

Figure 9. Voronoi Diagram and trajectory in an environment with curved shapes

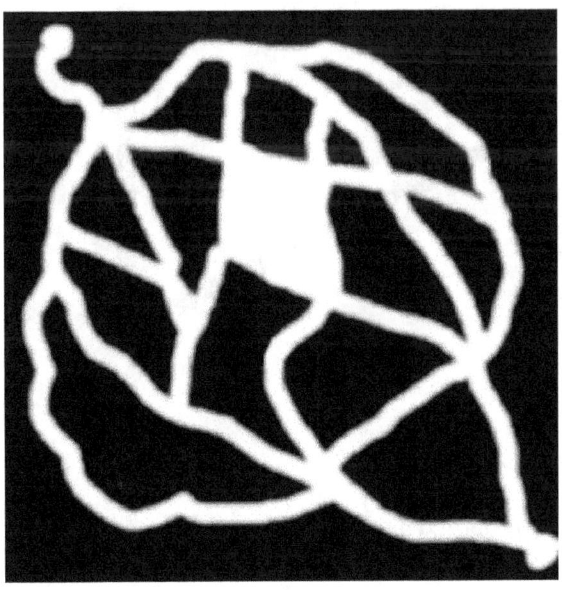

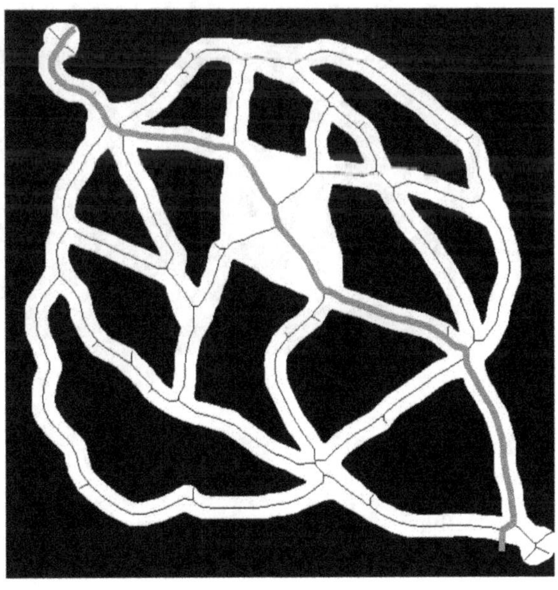

in which there are circular obstacles. This map represents a corner with columns. As can be seen in the Figure, the algorithm is perfectly able to make a correct planning of trajectories.

Figure 11 shows the application of the proposed method to find a path in 3D. The FMM can even work in more dimensions, but the method must undergo the problem of the dimensionality.

Figure 10. Trajectory calculated by Fast Marching Method in the Voronoi Diagram of an open aisle with columns. (left) Used environment (Not totally enclosed by walls), (right) Thickened Voronoi Diagram, (c) Inverted image of the thickened Voronoi Diagram and (d) Trajectory calculated by Fast Marching Method

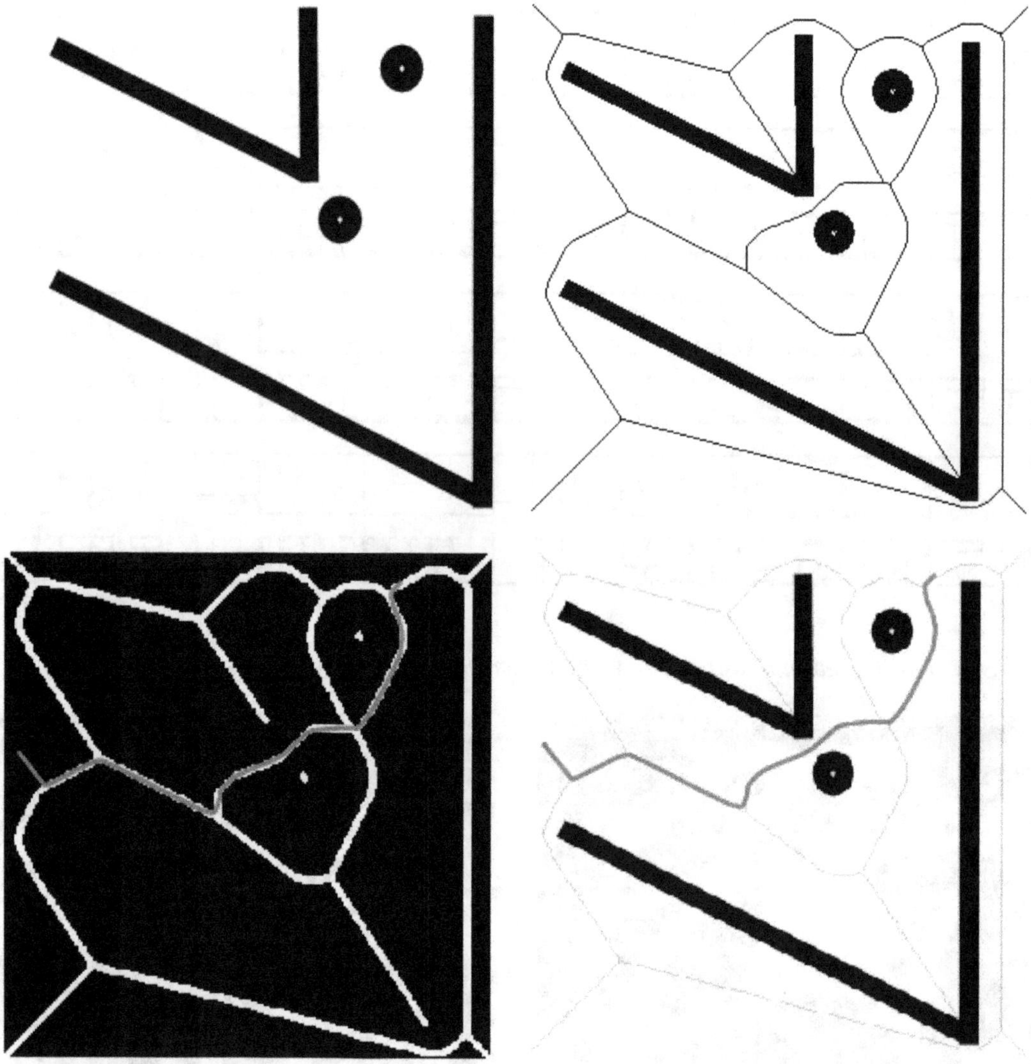

CONCLUSION

A sensor-based path planner is presented in this paper. The proposed method is able to deal simultaneously with both global and local planning requirements. The advantages of the approach can be summarized by the fact that the trajectories obtained are smooth and safe, and at the same time, free of local traps due to the integration of the real-time sensor information in the recalculation of the path.

The method is easy to implement, the algorithm is very fast and it can work online. It works in cluttered and changing environments with moving obstacles. The method is complete, i.e., the method is capable of finding a trajectory if it exists.

Figure 11. Trajectory calculated with the proposed method in 3D

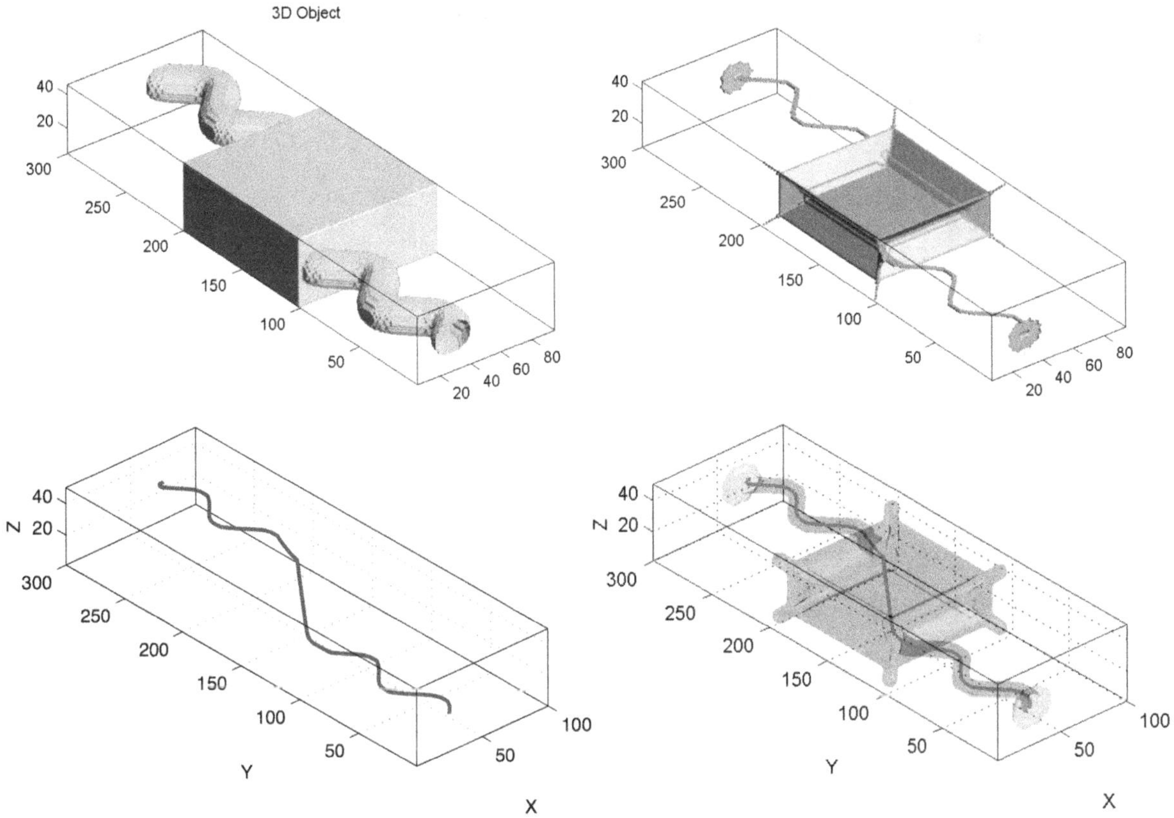

As demonstrated along this work, the method can perform in all types of environments without restrictions in the form of the obstacles. The planner works with curved forms, open environments (not totally enclosed by walls), and concavities.

The other main advantage of this method is that non-holonomic restrictions, such as steering angle limits, can be easily incorporated into the algorithm and it still generates smooth trajectories.

The algorithm complexity is O(n), where n is the number of cells in the environment map, which let us use the algorithm on line.

REFERENCES

Adalsteinsson, D., & Sethian, J. (1995). A Fast Level Set method for propagating interfaces. *Journal of Computational Physics*, *118*(2), 269–277. doi:10.1006/jcph.1995.1098

Aurenhammer, F. (1991). Voronoi Diagrams: A survey of a fundamental geometric data structure. *ACM Computing Surveys*, *23*(3), 345–405. doi:10.1145/116873.116880

Aurenhammer, F., & Klein, R. (2000). In J. Sack & J. Urrutia (Eds.), *Handbook of computational geometry* (pp. 201–290). doi:10.1016/B978-044482537-7/50006-1

Bhattacharya, P., & Gavrilova, M. L. (2008). Roadmap-based path planning-Using the Voronoi diagram for a clearance-based shortest path. *Robotics & Automation Magazine, IEEE, 15*(2), 58–66. doi:10.1109/MRA.2008.921540

Blum, H. (1967). In M. I. T. Press (Ed.), *"A transformation for extracting new descriptors of shape," in Models for Perception of Speech and Visual Form, W. W. Dunn* (pp. 153–171). Cambridge, Mass.

Bortoff, S. A. (2000). Path planning for UAVs. *Proceedings of the 2000 American Control Conference*, Vol. 1. *No. 6. IEEE.*

Breu, H., Gil, J., Kirkpatrick, D., & Werman, M. (1995). Linear time Euclidean Distance Transform algorithms. *IEEE Transactions on Pattern Analysis and Machine Intelligence, 17*(5), 529–533. doi:10.1109/34.391389

Chiang, C. S. (1992). *The Euclidean Distance Transform.* [Ph. D. Thesis]. Dept. Comp. Sci. Purdue University.

Choset, H. (1996). *Sensor based motion planning: The hierarchical generalized Voronoi graph.* Ph. D. Thesis, California Institute of Technology, Pasadena, California.

Choset, H. (2005). *Principles of Robot Motion: Theory, Algorithms, and Implementations*. The MIT Press.

Davis, J. L. (1988). *Wave Propagation in Solids and Fluids*. Springer. doi:10.1007/978-1-4612-3886-7

de Berg, M., van Krefeld, M., Overmars, M., & Schwarzkopf, O. (2000). *Computational geometry: Algorithms and applications, (2nd Ed).* Springer. doi:10.1007/978-3-662-04245-8

Dijkstra, E. (1959). A note on two problems in conexion with graphs. *Numerische Mathematik, 1*(1), 269–271. doi:10.1007/BF01386390

Gagvani, N., & Silver, D. (1997). *Parameter controlled skeletonization of three dimensional objects (Tech. Rep.). CAIP-TR-216.* Rutgers State University of New Jersey.

Garrido, S., Moreno, L., Abderrahim, M., & Martin, F. (2006). Path planning for mobile robot navigation using Voronoi Diagram and Fast Marching. In *Proc of IROS'06. Beijing. China.* (pp. 2376-2381). doi:10.1109/IROS.2006.282649

Garrido, S., Moreno, L., & Blanco, D. (2006). Voronoi diagram and Fast Marching applied to path planning. In *Proc. of ICRA* (pp. 3049-3054). doi:10.1109/ROBOT.2006.1642165

Garrido, S., Moreno, L., Blanco, D., & Munoz, M. L. (2007). Sensor-based global planning for mobile robot navigation. *Robotica, 25*(02), 189–199. doi:10.1017/S0263574707003384

Keerthi, S., Huang, C., & Gilbert, E. (1999). Equidistance diagram- a new roadmap method for path planning. In *Proc. ieee int. conf. on Robotics and Automation* (pp. 682-687).

Mahkovic, R., & Slivnik, T. (1998). Generalized Local Voronoi Diagram of visible region. In *In proc. IEEE Int. Conf. on Robotics and Automation, (pp.* 349-355*), Leuven, Belgium*. doi:10.1109/ROBOT.1998.676424

Mauch, S. (2003). *Efficient algorithms for solving static Hamilton-Jacobi equations*. Doctoral dissertation, California Inst. of Technology.

Nagatani, K., Choset, H., & Thrun, S. (1998). Towards exact localization without explicit localization with the generalized Voronoi graph. In *Proc. IEEE int. conf. On Robotics and Automation* (pp. 342-348). Leuven, Belgium. doi:10.1109/ROBOT.1998.676421

Ogniewicz, R., & Kubler, O. (1995). Hierarchic Voronoi Skeletons. *Pattern Recognition*, *28*(3), 343–359. doi:10.1016/0031-3203(94)00105-U

Okabe, A., Boots, B., & Sugihara, K. (1992). Spatial Tessellations: Concepts and Aplications of Voronoi Diagrams"., 1992. (J. Wiley & Sons, Eds.). Chichester, UK.

Sethian, J. (1996). *Level Set Methods*. Cambridge University Press.

Smith, R. W. (1987). Computer processing of line images: A survey. *Pattern Recognition*, *20*(1), 7–15. doi:10.1016/0031-3203(87)90013-6

Sudha, N., Nandi, S., & Sridharan, K. (1999). A parallel algorithm to construct Voronoi diagram and its vlsi architecture. In *Proc. IEEE Int. Conf. on Robotics and Automation*, *(pp.* 1683-1688*)*. doi:10.1109/ROBOT.1999.770351

Tsitsiklis, J. N. (1995). Efficient algorithms for globally optimal trajectories. *IEEE Transactions on Automatic Control*, *40*(9), 1528–1538. doi:10.1109/9.412624

Wilmarth, S. A., Amato, N. M., & Stiller, P. F. (1999). MAPRM: A probabilistic roadmap planner with sampling on the medial axis of the free space. In *Proc. IEEE Int. Conf. on Robotics and Automation* (p. 1024-1031). doi:10.1109/ROBOT.1999.772448

Yatziv, L., Bartesaghi, A., & Sapiro, G. (2005). A fast O(n) implementation of the Fast Marching algorithm. *Journal of Computational Physics*, *212*(2), 393–399. doi:10.1016/j.jcp.2005.08.005

KEY TERMS AND DEFINITIONS

Dijkstra's Algorithm: Also called graph's shortest path algorithm is an algorithm for determining the shortest given a source vertex to the other vertices in a weighted graph in each edge path. The graph may represent, for example, a road network. Its name refers to Edsger Dijkstra who first described it in 1959.

Fast Marching: The Fast Marching Method (FMM) is a computational algorithm to find the time of arrival to every point of the area of an expanding wave that starts in a point or a set of points. Conceptually, it may be thought as a continuous version of the Dijkstra's algorithm, (Dijkstra, 1959). It is based on the assumption that the information solely flows outward from the seeding area (wave source).

Path Planning: A valid path planning is that sequence of points in the open space that connect the points of origin and destination and that the robot doesn't invade the obstacles zones and satisfy movement constraints and possibly optimize some aspect of the movement.

Robot Motion Planning: This concept is similar to Path Planning, but a Motion Planning considers also the speed and turning commands sent to the robot's wheels.

Voronoi Diagram: The Voronoi diagram is a decomposition of a metric space regions associated with the presence of objects or obstacles, so that in this decomposition, each object is assigned a metric space region formed by points that are closest to it than to any of the other objects. I.e., we divide the space so that to each point or object we assign the region formed by everything that is closer to it than to any other points or objects.

Section 2
Introduction of Automation and Control

Chapter 4
Interactive and Collaborative Virus-Evolutionary CNC Machining Optimization Environment

N. A. Fountas
School of Pedagogical and Technological Education, Greece

N. M. Vaxevanidis
School of Pedagogical and Technological Education, Greece

C. I. Stergiou
Piraeus University of Applied Sciences, Greece

R. Benhadj-Djilali
Kingston University, UK.

ABSTRACT

Research on the area of sculptured surface machining optimization is currently directed towards the implementation of artificial intelligence techniques. This chapter aims at presenting a novel approach of optimizing machining strategies applied to manufacture complex part geometries. Towards this direction a new genetic-evolutionary algorithm based on the virus theory of evolution is developed as a hosted module to a commercial and widely known CAM system. The new genetic algorithm automatically evaluates pairs of candidate solutions among machining parameters for roughing and finishing operations so as to optimize their values for obtaining optimum machining programs for sculptured parts in terms of productivity and quality. This is achieved by introducing new directions of manipulating manufacturing software tools through programming and customization. The environment was tested for its efficiency and has been proven capable of providing applicable results for the machining of sculptured surfaces.

1 INTRODUCTION

Sculptured (or free-form) surfaces (Choi & Jerard, 1998) constitute the fundamental geometrical features of specialized products found in aerospace, automobile, shipbuilding and consumer electronics industries. Such products are aircraft fuselage, turbine blades, compressors and impellers, automobile body panels, mobile phone castings and moulds/dies. Manufacturing processes to produce free-form products

DOI: 10.4018/978-1-4666-8693-9.ch004

include casting, injection molding, stamping and metal cutting operations (Kalpakjian & Schmid, 2008). In the case of metal cutting processes, 3- and/or 5-axis machine tools are employed to remove the extra material from the raw stock as it occurs in prismatic parts comprising regular surfaces. The difference between sculptured and regularly-surfaced products lies on the fact that an inherent freedom exists when designing sculptured surfaces whilst regular ones contain features that can be easily designed and thus characterized in terms of their machining strategies and tools to be applied for their manufacturing. Moreover, morphological features of sculptured surfaces involve combinatorial machining operations as opposed to common surfaces that may be normally manufactured by individual stages.

The machine tools utilized for sculptured surface machining are computer numerical control (CNC) machining centers that realize material removal by repetitively moving a rotating milling cutter along pre-determined trajectories known as tool-paths (Choi & Jerard, 1998). The quality of tool-paths is of paramount importance and characterizes not only the efficiency but final product quality as well. Tool-path generation is a direct output of computer-aided manufacturing (CAM) environment, and is a result from end-users' selections made for strategies and process parameter sets in terms of machining modeling. This implies that tool-path creation lies thoroughly on human expertise and experimentations towards the selection of optimal parameter values satisfying pre-determined quality objectives. Since manufacturing software has been characterized as being the center of gravity for modeling complex machining operations (Lopez de Lacalle et al., 2007), it is essential that research efforts ought to be directed and carried out utilizing CAD/CAM technology.

State of-the-art research studies and cutting-edge approaches concerning tool-path generation involve CAM systems either to integrate them with new tool positioning algorithms introduced as novel machining strategies (Warkentin et al., 2000; Wang et al., 2006) or optimize mathematical expressions/ predictive models including discrete quality targets (dependent variables) and process parameters (independent variables) from which targets are greatly influenced (Karagiannis et al., 2013; Vaxevanidis et al., 2013). Even though both research directions have already provided remarkable results to academia and industry, the latter seems to be the practically viable one against tool positioning methodologies. Unless other mechanisms are implemented to evaluate given quality objectives rather than mathematical relations used as "objective functions", machining optimization with the use of CAM systems can be a quite promising research field, directly applicable to industry.

Noticeable research contributions falling to both aforementioned categories include the work of Quinsat and Sabourin (2006) in which optimum machining direction is investigated when applying 3-axis machining for sculptured parts. In their work machining direction is considered to be key objective on optimizing both machining time and surface quality. On the basis that machining strategies applied to machine complex geometries are of major importance to objectives such as part quality, machining time and cutting forces, Ozturk et al., (2007) studied the proper selection of related machining parameters via simulations mainly conducted employing multi-axis machining. In order to utilize the potential of the five-axis milling process, Kersting and Zabel (2009) presented an optimization approach based on recent multi-objective evolutionary algorithms. Their optimization module was integrated in a simulation environment for 5-axis milling whilst provides a link in the process chain between CAM environment and actual machining operation. Zeroudi et al., (2012) took advantage of tool positioning points obtained by CAM software so as to calculate local inclination angle of the generated surface and then the tool engagement to the work piece material, hence; predicting cutting forces. Fountas et al., (2014) conducted simulation experiments by implementing both 3- and 5-axis sculptured surface machining tool-paths and their related machining parameters so as to study their influence on productivity and quality.

The concept of optimization modules embedded to CAM or NC verification systems has already been captured by manufacturing software vendors worldwide. Optimization algorithms currently implemented to virtual machining environments utilize tool positioning strategies so as to minimize the distance between the swept envelope of the tool and the ideally designed surface as stored in a parametric form in CAD/CAM systems, thus; achieving more efficient tool contact and leaving a machined surface as close to nominal one as possible. Moreover existing optimization algorithms take into account exact cutting depths (radial and axial) and angles of cutting tool trajectories to calculate material removal and optimize feeds-speeds for each cutting condition. Thereby, improved tool-paths identical to initial ones; are computed. Optimization under that essence refers to topological aspects only and does not concern technological or economical criteria often treated as constraints actually found in industry. In addition, optimization procedure is performed after end-user's initial parameter settings and decision making regarding machining strategies and tool-path styles. To avoid partial problem solving, oversimplifications and shortcomings concerning local scale optimization, this work proposes a novel optimization methodology based on artificial intelligence with the use of a new virus-evolutionary genetic algorithm and its integration to a CAM system. This approach not only is independent by human expertise since parameter value ranges and constraints are considered but eliminates common drawbacks reported in the literature about conventional genetic-evolutionary algorithms as well giving to the approach even more robustness, consistency and applicability.

The main contribution of this approach is the automatic value determination for crucial process parameters by an intelligent algorithm, towards its convergence to optimal solutions concerning discrete quality objectives referring to both productivity and product quality. By automating the collaborative environment, process planning time and cost are dramatically reduced whilst final values for machining parameters will lead to optimally formulated tool paths for sculptured surfaces. The methodology investigates both roughing and finishing stages applied to machine parts since series of processes are extensively performed in actual industrial manufacturing operations.

The study starts with an overview on the basics of genetic and evolutionary algorithms (GAs-EAs). The new virus-evolutionary genetic algorithm is presented next along with its basic features and operators as a solution to common shortcomings found in conventional GAs-EAs proposed so far in the literature. The system's integration to the CAM environment through software automation techniques is then presented and obtained outputs from its implementation are discussed. The study ends up with conclusions drawn by the application of the proposed methodology and future perspectives are given as new directions to further enhance the developed optimization approach.

2 GENETIC AND EVOLUTIONARY ALGORITHMS: AN OVERVIEW

Genetic algorithms (GAs) are successful heuristic search systems initially developed and proposed by John Holland in 1975 (Holland, 1975). GAs mimic natural selection procedure according to Darwin's theory. The key concept in GAs is keeping good solutions in their populations longer so that they have more opportunities to reproduce better offspring by combining or adjusting their chromosomes to "environmental" changes. A typical GA is consisted of genetic operators like selection, crossover and mutation whilst an initialization scheme is applied prior to them so as to formulate the initial seed of candidate solutions. Selection operators are applied for choosing parent individuals to be reproduced. Good solutions with better fitness values are more likely to be selected. Better fitness is referred to the type of optimization problem. In the case of maximization, candidate solutions with higher fitness are considered to be better than the rest, whereas in minimization problems candidate solutions that minimize fitness are favored. Selected solutions have higher possibility to be mated or stay in the next generation. Evolutionary algorithms (EAs) are in fact generalized GAs (Holland, 1975). They include more sophisticated operators than the ones in GAs and possibly implement other artificial intelligence, deterministic or heuristic algorithms, in order to reduce evaluation cost or increase the overall algorithm performance in terms of solution quality and exploitation. Yet again, the criterion by which candidate solutions in EA populations are evaluated is the fitness function. If the fitness function is scalar, then the method is called single-objective optimization, regardless of the number of physical objectives included. In the same sense, if the fitness function is a vector, EAs perform multi-objective optimization and the solutions are most commonly ranked based on "Pareto" optimality. Following sub-sections give further information concerning the basics of GAs-EAs.

2.1 Basic Structure of Genetic Algorithms

Basic Genetic Algorithms (GAs) have a simple architecture, but they have to be suitably extended for a given optimization problem to be solved. A number of genetic operators have been proposed to apply various optimization problems. The fundamental operations a standard genetic algorithm may perform under its basic architecture are the following:

2.1.1 Initialization and Encoding

When an optimization problem is to be tackled by implementing a GA, the solution space should be encoded into a string space. That is, to map variables from the solution space into a finite-length string space. Therefore, initialization through proper encoding schemes is required, to efficiently solve an optimization problem. Several encoding methods have been proposed so far (Tamaki et al., 1994; Fogel 1995). The most important aspect during encoding is to cover all the solution space with the mapped string space without redundancy. As a consequence, the phenotype of the string space should be equal to the problem solution space, in order to simplify the problem. Moreover, string space should be generated as a set of feasible candidate solutions in order to avoid unnecessary search from the algorithm. However, genetic operators work directly on the genotype (or "chromosome") of GAs. Thus, the performance of the genetic search depends highly on symbolic operations for the genotype. GAs performance increases when substrings have consistent benefit throughout the string space. This is based on the concept that an encoding method is deeply involved in crossover operators. The neighbourhood of a candidate solution in the solution space should be similar to the neighbourhood of the respective string in the string space. If the offspring generated by a crossover operator is not similar to its parents, the population would proceed towards a different evolutionary direction and genetic search would probably result in failure. On the contrary, if good offspring is generated, genetic search results in success. When relationships between encoding methods and genetic operators are taken into account, the smallest character length that represents a solution space should be preferred and an encoding method able to create consistent strings through genetic operators should be applied (Goldberg, 1989).

There are three types of variable encoding: (a) *real value*, (b) *binary* and (c) *gray binary*. The binary types are usually preferred, since they are said to offer no bias during the evolution being strings of binary digits. In some cases, encoding with real values (between 0 and 1) tends to make the GA converge faster, but this way the GA solutions are susceptible to be trapped to local optima. GAs perform best when elite substrings tend to appear more frequently in a population without having their genetic material altered by genetic operators. Although the search depends on genetic operators, genetic information inherited from parents to offspring is dependent on the encoding.

2.1.2 Selection

Selection simulates the process of natural selection and GAs need a similar mechanism to make a population evolve toward a better direction of optimal solutions. The scope of selection scheme is to reproduce a population by pre-selecting an individual with a selection probability proportional to its fitness value. In this kind of selection, an individual with a higher fitness can reproduce more offspring. Since selection is performed stochastically, it is possible that a number of identical (or the same) individuals may be chosen by chance. As a result, some types of strings tend to occupy ("take over") a population. This is called ''*genetic drift*'' (Holland, 1992) or ''*random drift*'' which often occurs in the case where the GA population size is relatively small and has strong elitist behaviour. Several selection schemes have been proposed in order to prevent a population from genetic drift. Selection schemes are classified into two main categories; the *proportional selection scheme* and the *competition selection scheme*. In the first category, selection is based on the fitness value of an individual compared to the total fitness value of the overall population. In the second category, selection is based on the fitness values of some other individuals. Some of the most commonly used selection schemes are presented next (Goldberg, 1989):

- "Roulette wheel" selection scheme: This scheme selects an individual with probability in proportion to its fitness values.
- "Elitist" selection scheme: This scheme preserves the fittest individual through all generation t, that is, the fittest individual is certainly selected prior to others into the next generation.
- "Tournament" selection scheme: the individual with the highest fitness value between m randomly pre-selected individuals is selected. Note that m is the number of competing members.
- "Ranking" selection scheme: This scheme is based on the rank of the individual's fitness value. Additionally, the individuals are selected by the number of their reproduction into the next generation based on the ranking table predefined earlier.
- "Expected value" selection scheme: This scheme is based on the expected value of the individual's fitness. According to a respective probability, the expected value of an individual is calculated. Then, the probability of the selected individual is decreased by 0.5. Thus, this selection scheme relatively prevents an individual from being selected more than twice in the population.

2.1.3 Genetic Operators

The genotype of an individual is changed by genetic operators, such as combination, inversion and duplication, while evolution is controlled under natural selection. GAs need from genetic operators to make the evolution feasible to all successive populations. Recombination generates new individuals with crossover operators. Mutation generates new individuals with some perturbation operators. Each operator contributes to evolution by performing specialized actions.

2.1.3.1 Crossover (Recombination)

The Crossover operator generates new individuals as solution candidates in GAs (Goldberg, 1989). GAs can search the solution space mainly by using one of the crossover operators. With the absence of crossover operators, GAs would be random search algorithms. The crossover operator exchanges each substring between two individuals and replaces old individuals with others of a new genotype. The recombination between two strings is performed according to the type of crossover operator. Depending on the number of break-points among individuals, the crossover mechanism recombines the strings of the genotype. Some of the crossover operators are the following:

- One-point crossover: One-point crossover is the most common from the rest of crossover operators. By choosing a break-point as a cross site, the one-point crossover recombines substrings between two individuals.
  Example: 100|11110 → 100|10010
  101|10010 → 101|11110
- Multi-point crossover: Multi-point crossover has more than one break-points on a string. Multi-point crossover recombines some substrings which cut by some break-points between individuals.
  Example: 1|11|1|1 → 10010
  0|00|0|0 → 01101
- Uniform crossover. In the uniform crossover operator, a mask pattern is randomly generated including "0" and "1" digits. Crossover between two individuals is then achieved by exchanging the characters in the genotype string according to the mask pattern.
  Example: 11|11|1 → 11001
  00|00|0 → 00110
  (Mask Pattern: 00110)
- Cycle crossover: Cycle crossover operator first selects a starting point and formulates a closed round of substrings between individuals. The following example describes the workflow of Cycle crossover:
  Example: Consider the Parents *P1* and *P2* as follows: *P1*: 31245, *P2*: 24351. Firstly the starting point on the string is chosen and then the selection of locus 2 of *P1* parent is done. A closed round of substring begins from this starting point. The characters 1 and 4 in the locus 2 of *P1*, *P2* to the same locus of offspring are copied, respectively:
  O1: *1*** and *O2*: *4***
  Next, the character of the locus existed the character 4 in the locus 2 of *P2* is copied as follows:
  O1: *1*4* and *O2*: *4***
  The process is repeated until the first chosen character of *P1* is reached again. Consequently, the closed round of substring is achieved as follows:
  O1: *1*45 and *O2*: *4*51
  Finally, characters are filled from the former parents to the latter offspring. The complete offspring becomes: *O1*: 21345 and *O2*: 34251
- Partially matched (PMX) crossover: PMX crossover is performed by executing two procedures. Like the cycle crossover operator, PMX crossover is a specialized mechanism which may be applied in order to overcome permutation problems (Goldberg, 1989). The workflow of PMX crossover is described below in the example:

Example: Consider the Parents *P1* and *P2* as follows: *P1*: 31|24|5, *P2*: 24|35|1. The exchange of substrings is performed between two break-points to both parents without the overlapping of characters *O1*: **35** and *O2*: **24**. Character 2 in locus 3 of *P1* is exchanged with character 3 in the same locus of *P2*, that is, the positions of the characters 2 and 3 in *P1*and the characters 3 and 2 in *P2*. The characters on the substring rounded by two break-points are also operated by this exchange, respectively. Consequently, partially matching substrings can be obtained as follows: *O1*: 2*354 and *O2*: 3524*. The remaining character is copied from *P1* to *O1* and from *P2* to *O2*. As a result, the complete offspring is generated as follows: *O1*: 21354 and *O2*: 35241.

In this way, genetic algorithms can generate new individuals of feasible candidate solutions satisfying the constraint of permutation problems (Bui and Moon, 1994). It can be clearly seen that an independent relation exists between coding and crossover. When it comes to the design of crossover operation, one should take into account the inheritance of genetic information from parents to offspring. The offspring should inherit adequate genetic information from parents.

2.1.3.2 Mutation

Mutation occurs as the replicating error in nature. In GAs, the mutation operator replaces a randomly selected character on the string with the other one (Goldberg, 1989). Mutation is performed regardless of individual fitness values. A classic mutation operation is the one-point changing per individual. Several mutation types are occurred in nature such as inversion, translocation and duplication. These are also the mutating mechanisms applied to GAs to simulate these phenomena. The types of mutation applied to GAs-EAs are as follows:

- Inversion: Inversion partially changes a character sequence from one direction to the opposite. At first, two points are chosen randomly and the string is cut at these points. Then the substring is linked in the reverse direction into the remained one. Assume a string included 5 characters. Two points on the loci 2 and 4 are selected: 1|234|5. By linking the substring, inversion results in the following state: 14325.
- Translocation (or Shift): This operator changes the character sequence with moving to the different position. In translocation, a substring is chosen randomly as occurs in Inversion. The substring is then translocated to a randomly chosen break point (or locus). Assume a string includes 5 characters. The segment from locus 1 to 2 is chosen as a substring; i.e., |12|345. The substring is inserted from locus 4 to 5 whilst the remaining substring of the string shifts to the leftmost point of the string as follows: 34512.
- Duplication: This mechanism overlaps the same substring on a string. The substring is randomly chosen as it occurs in inversion. The substring is then overwritten to a randomly selected locus. Assume a string included 5 characters. The segment from locus 1 to 2 is chosen as a substring, i.e: |12|345. The substring is overwritten over loci 4 and 5 and the new string is 12312.

Mutation operators have great influence on the GA performance because they seriously affect populations. GAs are able to search a global solution space, since the mutation mechanism randomly changes the strings of individuals. If the probability of mutation is high, the mutation operator often breaks important substrings on individuals. Therefore, it is essential that a very low mutation probability should be used to overcome this drawback.

2.2 Basics of Evolutionary Algorithms

An evolutionary algorithm's optimization philosophy is more or less the same to a GA's. In fact, there is not a discrete distinction between the definition of EAs and GAs. It can be claimed that GAs are a subset of the general EA family. EAs utilize a population of solutions/individuals that successively evolves in generations and constantly produces better solutions until the algorithm converges and the global optima is obtained. Operators help EA population find potentially better solutions by recombining, mutating and other operators that change the original solutions, while they substitute bad/old individuals with better/younger ones in the population. EAs, however, can be different to GAs in a number of manners. One common differentiation is that operators can be functions of time, in other words probabilities changing with respect to the current generation number or have completely different schemes that are chosen according to a specific probability or to the current generation number. A major trend that minimizes calculation cost is to divide EA population appropriately in subpopulations. Each subpopulation optimizes a certain fitness function, thus the EA subpopulations are competitive to one another and they exchange optimal values at the end of each generation (Figure 1). This notion comes from Game Theory, at which players in a game behave competitively to each other and try to make maximum profit, until a certain point is reached at which no player makes more profit by changing his playing strategy -"Nash equilibrium" point- (Nash, 1951). Alternatively, all subpopulations can optimize the same objective function, but calculations are carried out on different computers or on different CPUs. This is called parallelism of computers (or parallel systems) and algorithms can lead to serious cut-down of calculation cost. In either case, subpopulation are parts of a single population, thus there is need for intercommunication among them. This is performed introducing migration operators that control the way that individuals move to other subpopulations carrying all genetic information along with them. In some optimization problems, such as machining or flow optimization, the evaluators are very costly and take up too many resources, in order to calculate objective function values. This means that the EA must "pay" high calculation cost for all solutions, even for the bad ones. However, if there is a way to classify solutions as "bad" or "good" prior to evaluation, then only the "good" would be evaluated and cost associated with "bad" solutions/individuals would almost be eliminated. During the last years, classification problems are commonly solved using Artificial Neural Networks (ANNs). Both supervised and unsupervised ANNs can perform decently in solution classification, but it has been shown that the unsupervised ones work best at such problems in very small response times.

Including ANN operators in the EA optimizes its performance in terms of calculation time. Nevertheless, solutions/individuals classified as "good" should always be exactly evaluated by the EA. There is an inherent drawback in ANN implementation; there is always the possibility that a "good" solution is classified as "bad" and vice versa. This is overcome by the nature of EA, since "good" genetic material is never lost or ignored through the use of the other EA operators, such as mutation, crossover etc., that force it again back in the population either in the current or a following generation. As it is obvious, many variations and differentiations can be met in the relative literature. Apart from Game Theory and ANNs, heuristic methods are popular in the field of EAs. Heuristics are based on experience; they help

Figure 1. A Nash game with two players: information exchange after search tasks held in isolation

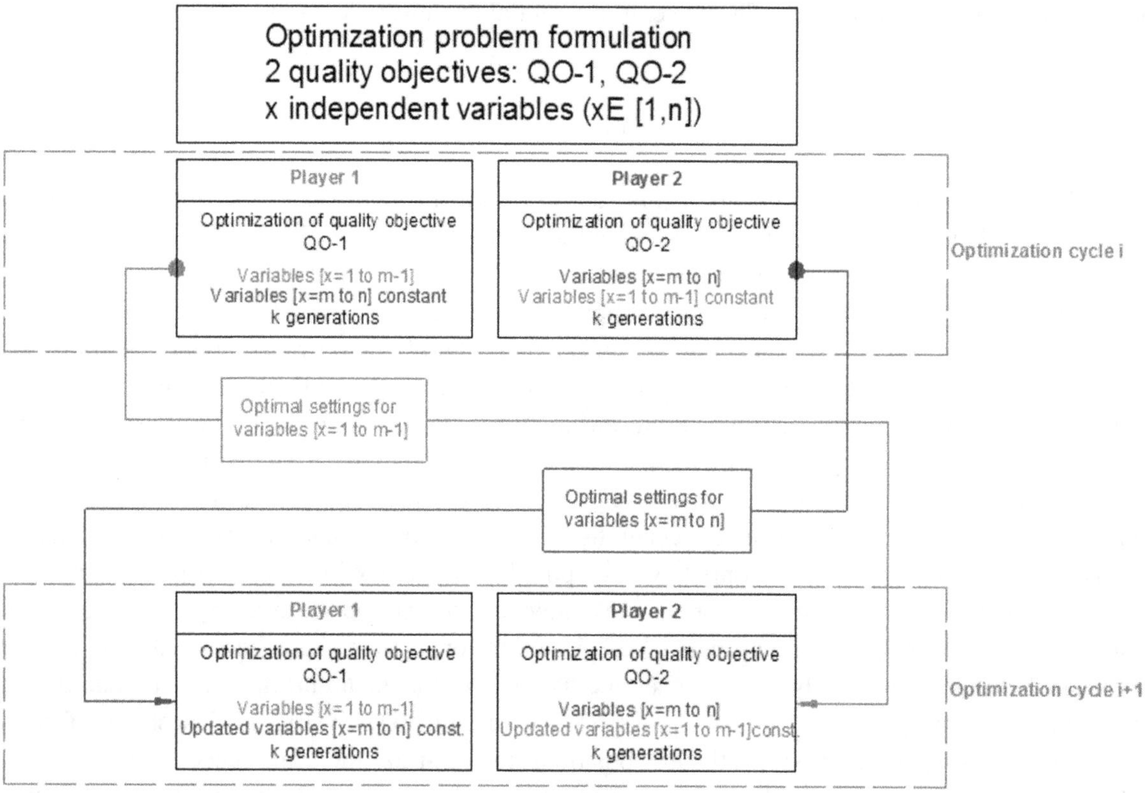

in minimizing calculation cost and facilitate the algorithm in finding global optima rapidly, in other words augmenting both exploration and exploitation abilities of the algorithm. In the same sense, other knowledge-based systems can be implemented in EAs, such as expert systems.

2.3 GA-EA Performance: Exploration and Exploitation

Exploration and exploitation assess the performance of GA-EA applications to optimization problems and are opposite to each other. Major target to all intelligent optimization algorithms is to balance these two contradictory criteria thus effectively performing both global and local search of the solution space.

2.3.1 Exploration and Exploitation Definitions

GAs-EAs need to satisfy the fundamental criteria of exploration and exploitation in order to successfully perform series of evaluations thus converging to an optimal solution for a given optimization problem. Exploration is the procedure of visiting thoroughly new regions of the solution domain. In other words, exploration is the process of rapidly searching local regions towards a solution. Exploitation is the procedure of visiting these regions of a search domain within the neighborhood of previously visited regions. That is; exploitation represents the ability of a GA or EA to rapidly search the entire solution space. Consecutively, a successfully developed optimization algorithm should maintain a good ratio between exploration and exploitation thus balancing the two conflicting criteria. Several discussions have already been provided pointing out that research should be continued on establishing well-structured genetic operators to satisfy the needs of exploration and exploitation since crossover and mutation play key role to exploration whereas selection greatly affects exploitation according to Eiben and Schippers (1998). Even though this assumption has found some merit, further research conducted by several other authors revealed that this problem still remains and no general agreement has been reached on this subject. Figure 2 depicts exploration and exploitation regions for a given function.

2.4 Drawbacks of Genetic and Evolutionary Algorithms

Despite their great success, GAs-EAs have the known problem of premature convergence. One reason causing this problem is that elite of outstanding solutions may dominate the population or proportional selection may favour all schemata (effective and ineffective) thus; slowing down efficiency and robustness of optimization algorithms. As a result the lack of genetic spread (also called diversity) might trap the search process to a local optimum. Solving this problem without modifying a GA or an EA is almost impossible because feature configurations of a GA-EA determine that converge towards some points during the search is needed. In order to ensure that a GA-EA reaches global optimum, trial-and-error experimentations are usually conducted, hoping that the GA-EA converges to different solutions per evaluation cycles and one of them is the global optimum. A more efficient approach is to tackle premature convergence problem by storing patterns during a search. Thus, effective patterns won't be withdrawn from the search domain and population diversity will be maintained as the algorithm evolves towards its convergence.

3 VIRUS EVOLUTIONARY GENETIC ALGORITHM

The Virus-evolutionary genetic algorithm is based on the virus theory of evolution and uses two populations. One is the host population, which works basically the same as the population of the GA. The other

Figure 2. Graphical depiction of a function's exploration and exploitation regions

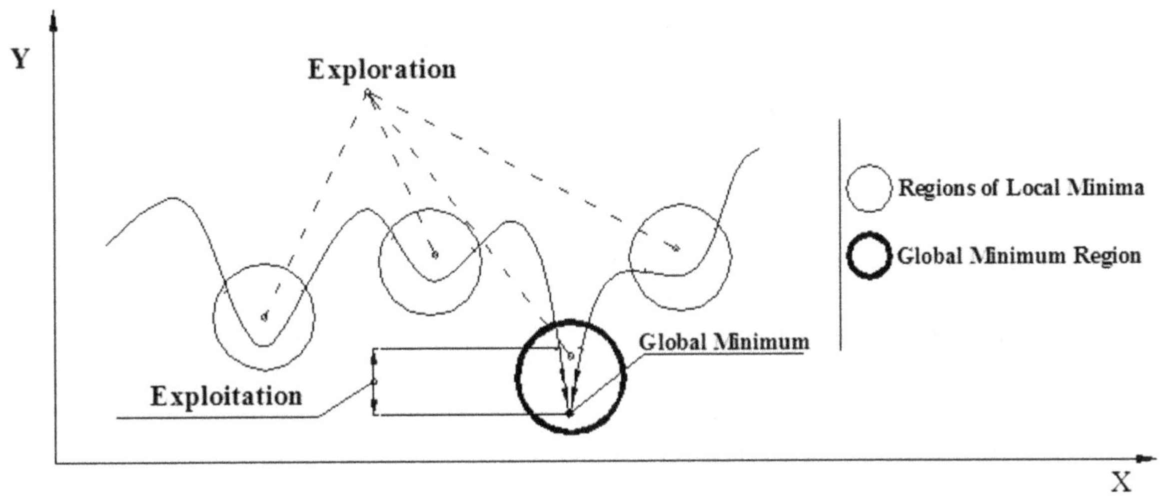

is the virus population. The virus population is for saving effective patterns during a search. When a new generation of the host population is created, these effective patterns offer information to make the new hosts better than their parents.

3.1 Virus Theory of Evolution: Brief Overview

In general, the virus theory of evolution is relied on *transduction* operation. Transduction is the process of transporting DNA segments across species. Hence, genetic changes occurred to a bacterium's DNA chain when a becteriophage carries DNA segments from another bacterium and locates them to its DNA chain. Viruses found in nature can be such bacteriophages. Viruses have the ability to penetrate to species' genetic material (DNA chain) and being transmitted directly from individuals of a phylum to another. This special ability of viruses to be transmitted directly from one kind to another is known as *horizontal propagation*. The incorporation of a host's DNA segments into effective viruses and subsequent transfer to other cells is widely known. Besides, entire virus genomes can be incorporated into germ cells and transmitted from generation to generation as *vertical inheritance*. (Anderson, 1970; Nakahara & Sagawa, 1989). The natural mechanism of "*vertical inheritance*" among genomes was successfully simulated by Kubota et al., (1996) through *reverse transcription* operator. The proposed algorithm's special features and operators involve functions presenting host and virus populations, reverse transcription and transduction operators. The rest of the functions for selection, crossover and mutation operators are as in conventional GAs - EAs.

3.2 Virus-Evolutionary Genetic Algorithm Architecture

Virus-evolutionary genetic algorithm has exactly the same structure as a conventional GA-EA with the difference that incorporates the two aforementioned virus operators (reverse transcription and transduction) so as to achieve both the generation of virus individuals and infection process applied to host individuals. Since a general overview for the basic architecture of GAs-EAs has already been given to section 2, only the special features for this algorithm will be discussed.

3.2.1 Initialization

Initialization process involves the random generation of host population. As a parameter representation scheme, binary encoding is selected since it facilitates the pattern formulation process. Initialization is achieved by creating the individuals' chromosomes according to the parameter value range and accuracy (number of binary digits to describe the phenotypic values for variables).

To realize the capabilities of both horizontal propagation and vertical inheritance the algorithm utilizes two populations for its benefit; the population of hosts and the population of viruses. The former reflects the set of candidate solutions; whilst the latter is developed as sets of sub-strings of host individuals. Virus individuals are formulated as substrings of host individuals through transduction operation. In the proposed algorithm transduction operator is initially applied to the first best fitted individuals after their fitness function calculation and ranking. Note that if the problem is of maximization the best fitted individuals are considered to be those that achieve high values; whilst in the case of minimization problems best fitted values for individuals are those with low values. Transduction operator is then performed to a couple of randomly selected host individuals from the population. This is done to remove the inherent bias owing to best fitted individuals by simultaneously favoring some randomly selected hosts that may produce high fitted offspring (Figure 3).

Transduction operator generates viruses from hosts by cutting sub-strings from the strings of host individuals as Figure 4 illustrates.

3.2.2 Virus Infection

Once the two populations have been created through the transduction operator the population of viruses infects host individuals with reverse transcription operator (Figure 5). Reverse transcription operator overwrites a virus' string on a host's string. Programmatically this is achieved by sequentially typing each binary digit or the virus' string to the host's string according to the determined locus for replacement. To do so, the indices of both the virus and the host together with their replacement loci for the strings are declared. As a result a new host, (infected host), is generated.

Reverse transcription and transduction operators are dynamically collaborated under the major goal of increasing more effective individual representation schemata that will rapidly lead to the global optimum solution for a given problem. Figure 6 depicts the operational sequence for reverse transcription and transduction operators towards their virus infection result.

3.2.3 Parameter Settings

In addition to common parameter settings that conventional GAs-EAs suggest (crossover probability, mutation probability and rate, etc), virus evolutionary genetic algorithm involves parameters to determine virus fitness, infection rate, the "life" of viruses, etc.

3.2.3.1 Fitness Evaluation

A virus individual (*virus "i"*) has a fitness value (*fitvirus (i)*) which is calculated to indicate its efficiency. The fitness of each virus is determined as the sum of the fitness of each infection caused by the current virus to the host population.

$$fitvirus(i) = \sum_{j \in S} fitvirus(i, j) \tag{1}$$

Moreover, the fitness of each infection is computer which is determined as the difference between the fitness value of the original host and the one calculated after its infection and is described as follows:

$$fitvirus(i, j) = fithost(j') - fithost(j) \tag{2}$$

Thereby, each virus has a measuring parameter for its infection strength; that is *fitvirus (i)*. *Fithost (j)* and *fithost (j')* are the fitness values of a host "*j*" before and after its infection, respectively. The indicator *fitvirus (i,j)* denotes the difference between the fitness values *fithost (j)* and *fithost (j')* which is equal to the improvement value obtained through the infection process. To the equations presented above, "*i*" denotes the virus number and "*S*" represents the sum of host individuals infected by the virus "*i*".

Figure 3. Virus population creation mechanism

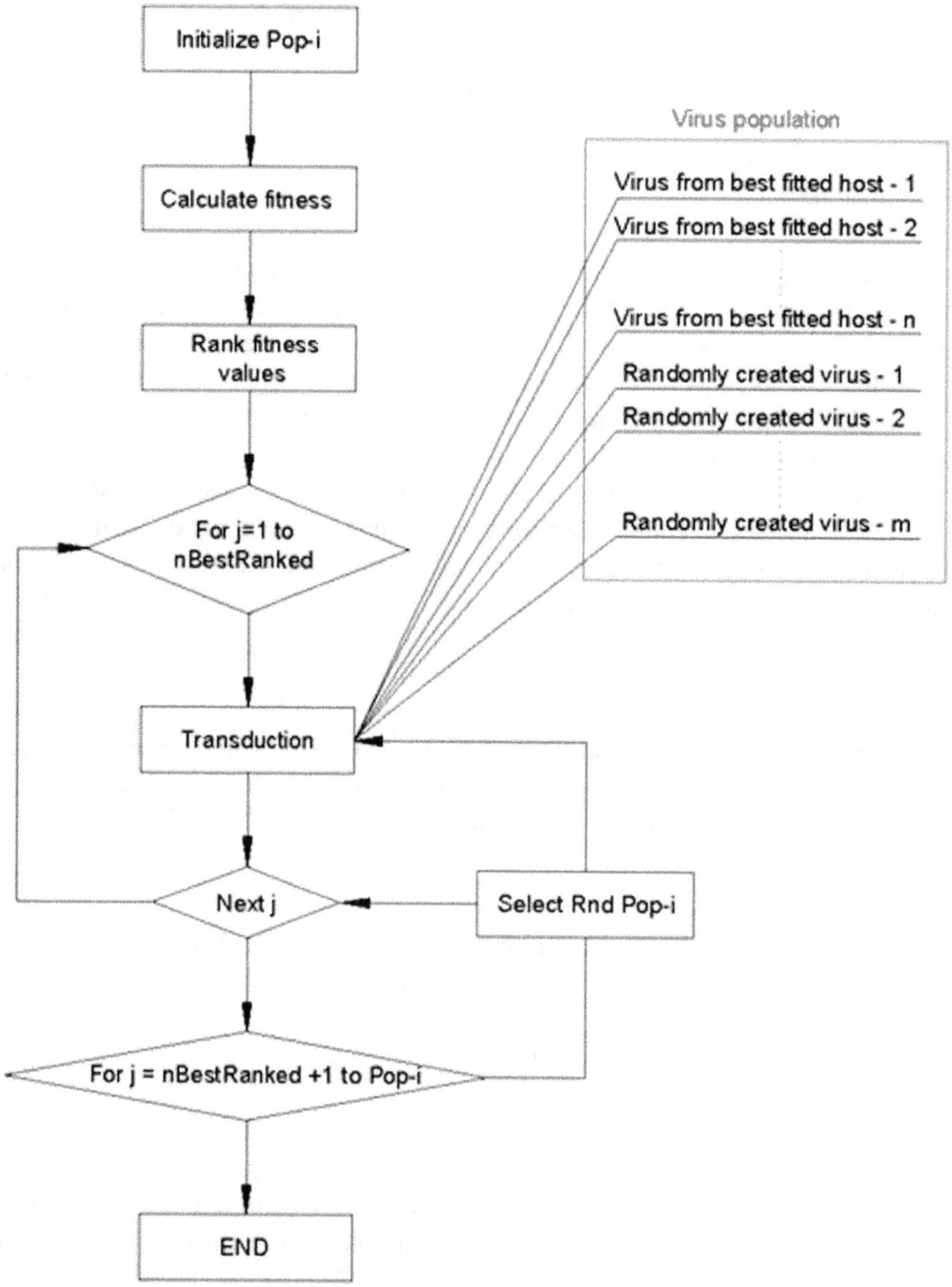

3.2.3.2 Infection Rate

The number of infections caused by each virus is controlled through infection rate *(infrate (i))*. If a virus has a positive fitness, the infection rate is increased according to a constant parameter "α". In contrast, if the virus has a negative value the infection rate is reduced. Kubota et al., (1996) suggested that maximum infection rate to be 0.1 and initial infection rate to be 0.05. The infection rate of each virus should

Figure 4. Transduction operation to create virus individuals

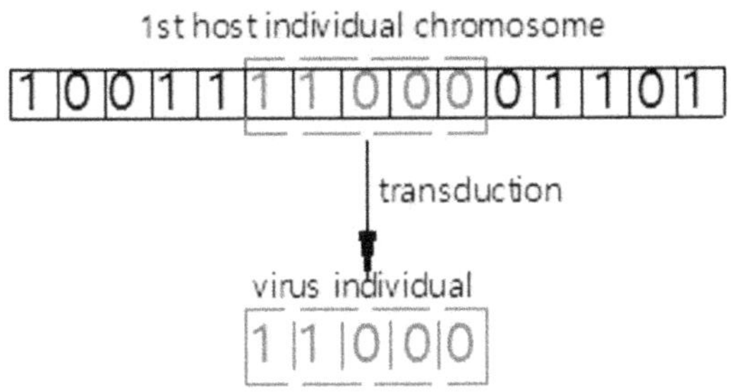

Figure 5. Reverse transcription operation to infect host individuals

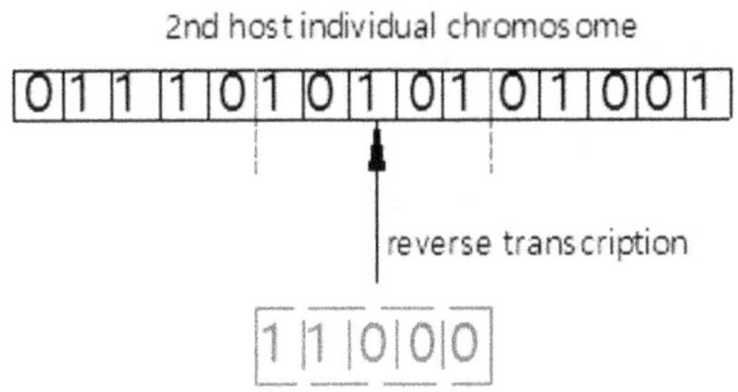

Figure 6. Generation of a new host individual through virus infection operators

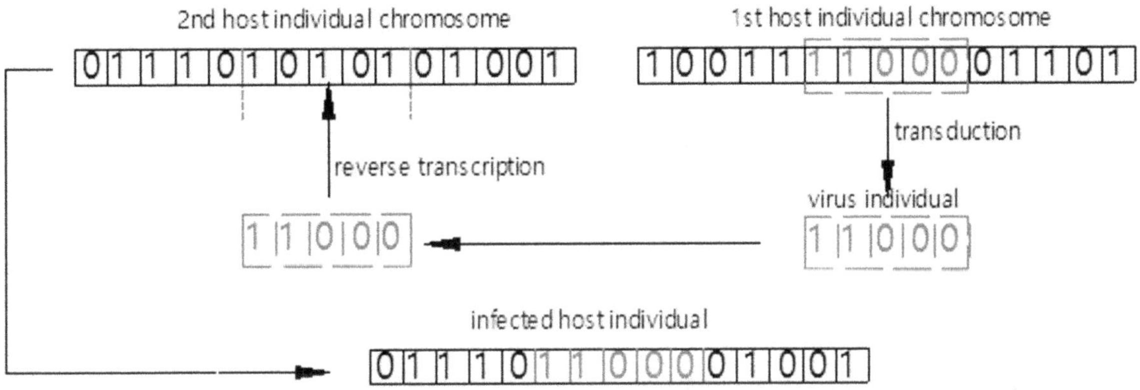

satisfy $0 \leq infrate\ (i) \leq 1.0$ in order to perform a reverse transcription operation to a given host population. Equation 3 represents the formulation of these settings for the determination of infection rate by Kubota et al., and is depicted below:

$$\inf rate_{i,t+1} = \{(1+a) \times \inf rate_{i,t}\} \text{ if } fitvirus\ (i) \geq 0 \quad (3)$$

$$\inf rate_{i,t+1} = \{(1-a) \times \inf rate_{i,t}\} \text{ if } fitvirus\ (i) < 0$$

Consecutively, the higher the infection rate of a virus "*i*" is, the higher its infection capability becomes and the virus infection rate is equal to the frequency increasing a schema in a population.

3.2.3.3 Virus Life

The operational duration for virus individuals is controlled by virus life. The virus life indicates the contribution and positive impact of virus infection to the evolution process. The virus life is programmatically computed as follows:

$$life_{i,t+1} = r \times life_{i,t} + fitvirus_i \tag{4}$$

Where "r" is a life reduction rate and "t" represents the generation being evaluated. If life is negative then the virus individual transduces a new substring from a random host to change its scheme and become efficient. If life is positive the virus individual transduces a partially new substring from one of the infected host individuals to manipulate it for its benefit (further increase its efficiency). The positive calculus of a virus' life implies that its fitness is quite high and the schema of the successfully infected host by this virus is strong. Hence partial transduction is applied to this host to generate another good sub-string pattern which is another virus; and that virus proceeds to further infections. In general efficient schemata in virus evolutionary genetic algorithm are increased exponentially through this mechanism and local regions of a given solution space are rapidly searched. What virus evolutionary genetic algorithm looses from calculating fitness for two populations in terms of the slight increase of computational cost; it is gained back by its effectiveness and fast response. To cope with the issue of computational time and cost, a steady-state scheme is utilized for this algorithm and replaces the weakest hosts by new fitted ones. Partial transduction in the case of a positive virus life is illustrated in Figure 7.

3.2.4 Optimization Outcome and Discussion

It has been shown so far that trying to satisfy both exploration and exploitation is a difficult task due to their trade-off. The on-going evolution referring to this particular research field indicates that no results have been found to thoroughly suggest an ideal strategy to tackle with this problem. Yet, new systems proposed tend to minimize negative effects when treating both exploration and exploitation as quality targets in terms of genetic and evolutionary algorithms efficiency.

One of the major benefits by utilizing virus theory of evolution in a GA-EA is the ability of creating a "hybrid" optimization system but without different modules. Virus individuals are parts of hosts

Figure 7. Partial transduction mechanism

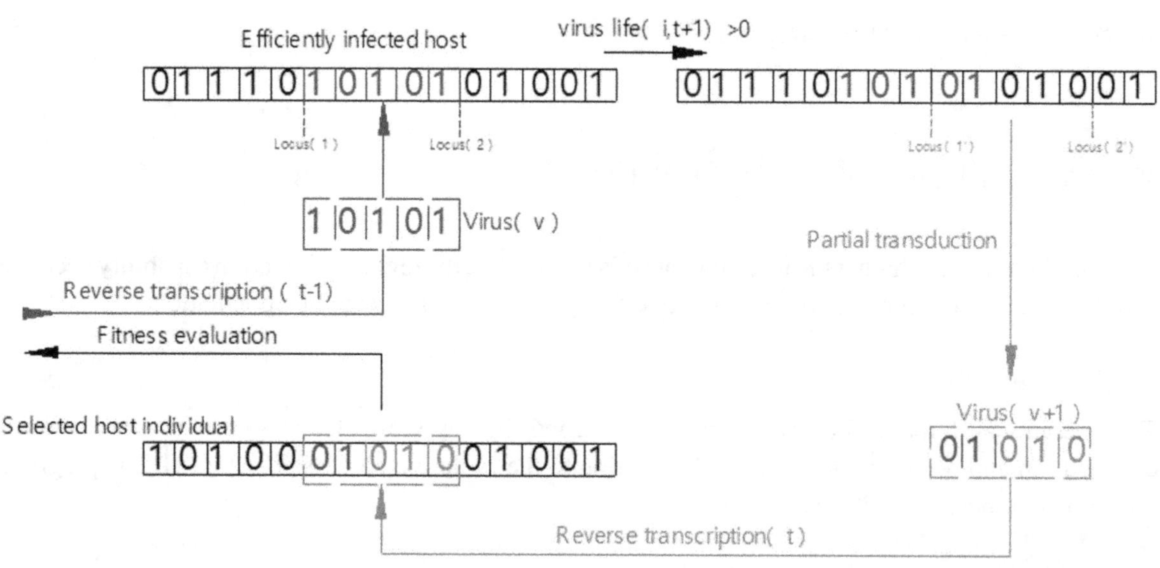

thus, properties characterizing the two populations are more or less common. Consecutively, additional computational cost is avoided whilst searching ratio towards the optimal solution is high. The proposed methodology directly increases effective schemata in a host population by virus infection operators (reverse transcription and transduction). The particular mechanism of reverse transcription actually acts as both crossover and mutation simultaneously, since reverse transcription generates new individuals overwriting on host individuals according to the virus infection rate. That is, the virus infection rate is equal to the frequency increasing a schema in a population. In the meanwhile transduction operator generates new viruses (virus individuals) building this way a virus population co-existed and co-evolved with the population of host individuals. The most important characteristics obtained by virus infection are summarized as follows:

- Reverse transcription operator directly increases effective schemata
- Reverse transcription operator directly generates new host individuals
- Transduction mechanism changes virus individuals in every generation

Co-evolution of viruses and hosts, allows the rapid solution of optimization problems whilst significantly reducing computational time and cost. The higher the infection strength of a virus "i" is, the higher its infection capability will be. As a result, the virus GA mainly performs local search by virus infection operators. To the contrary, when virus infections tend to be minimized, the virus GA utilizes the classic genetic operators (crossover and mutation) hence; performing a global search within the search space. As a matter of fact this kind of GA can self-adaptively change its searching ratio between global and local search regarding the optimization problem and the current state of viruses and hosts.

Due to GAs stochastic nature, determining convergence criteria is usually an arbitrary process. A common ground in many studies is that the algorithm terminates after a predetermined number of generations or when no further improvement is indicated to the solutions for a given number of generations. Thereby, virus evolutionary genetic algorithm can be adjusted to terminate its evaluation process after specifying a constant number of generations.

4 MANUFACTURING SOFTWARE AUTOMATION

Software automation deals with the ability of performing computer-based tasks without the user interaction, expanding software capabilities and developing independent functions for data exchange among two or more software modules. Through this particular technology it is possible to capture "know-how", add knowledge to procedures, increase productivity and add functionality to systems. In the case of manufacturing software, automatic interaction is essential when a large number of alternative scenarios need to be examined. These scenarios are formulated by selecting machining strategies, cutting tools and related machining parameters. These entities are extensively discussed in (Fountas et al., 2012, 2014).

4.1 Software Development for Automated Machining Modeling

To facilitate evaluation process between virus evolutionary genetic algorithm and CAM system, additional code was developed to customize the collaborative environment. The code embedded to the environment as stand-alone automation routines and was written in Visual Basic for Applications (VBA) utilizing the hosted programming development platform of *Dassault Systèmes* CATIA® V5R18 CAD/CAM/CAE package. The code developed undertakes to:

- Scan the machining modeling history tree, retrieve the machining operations with their related process parameters and pass the values generated by the genetic algorithm;
- Compute the tool path according the specified settings and run the machining simulation;
- Extract information and statistics corresponded to quality objectives depending on the machining phase (i.e: remaining volume and rough machining time for roughing; surface deviation and finish machining time for finishing).

Details about the code developed for the aforementioned procedures are described in the following sub-sections.

4.1.1 Automated Machining Setup Scanning and Parameter Value Assignment

Aiming towards automating process parameters found in most commercial CAM systems, one should identify the fundamental entities from which the parameters under interest are to be retrieved. In a machining setup document such entities are the *processes*, the *products* and the *resources* abbreviated as (PPR); and are related together and categorized in the document's PPR tree. Processes contain description of part operations, manufacturing programs and machining operations (i.e. tool changes and/or operation sequence). Products give a full design description of the NC setup including the geometry of the design part, its stock and support devices (fixtures, etc.). Resources describe the physical entities needed to produce the part including machine tools, cutting tools and tool assemblies. These three components are attached to the PPR tree depicted in Figure 8.

Manufacturing software vendors provide structures for automation objects to facilitate programming efforts of software developers (CATIA API, 2005). By utilizing the structure of automation objects for such documents, PPR tree is accessed so as to retrieve the machining operations containing the strategies and their related parameters for automation. Common functions of visual basic were applied to declare the parameters according their nature whereas special programming was conducted to pass the values from the code to CAM interface. The input file was a *.txt (or *.dat) file containing the values that are supposed to be generated form the intelligent algorithm. According the values of the input file, the parameters are updated and a new simulation scenario is executed. The values of parameters were retrieved from their associated process.

4.1.2 Automated Tool Path Computation

Major scope for machining optimization with the use of CAM software is to generate the optimum trajectory pattern that a cutting tool will follow to remove the material from a raw stock towards the production of the final part by metal cutting operations. The trajectory pattern is determined by the cutting strategy and the values for its related parameters. Figure 9 illustrates such cutting strategies.

So far, this operation is manually executed by experienced CAM software users worldwide in academic research and industry. To do so, the functions hosted to user forms and system windows need to be manually accessed and the commands responsible for tool path generation should be prompted.

Figure 8. Entities for processes, products and resources in a machining setup document (PPR tree)

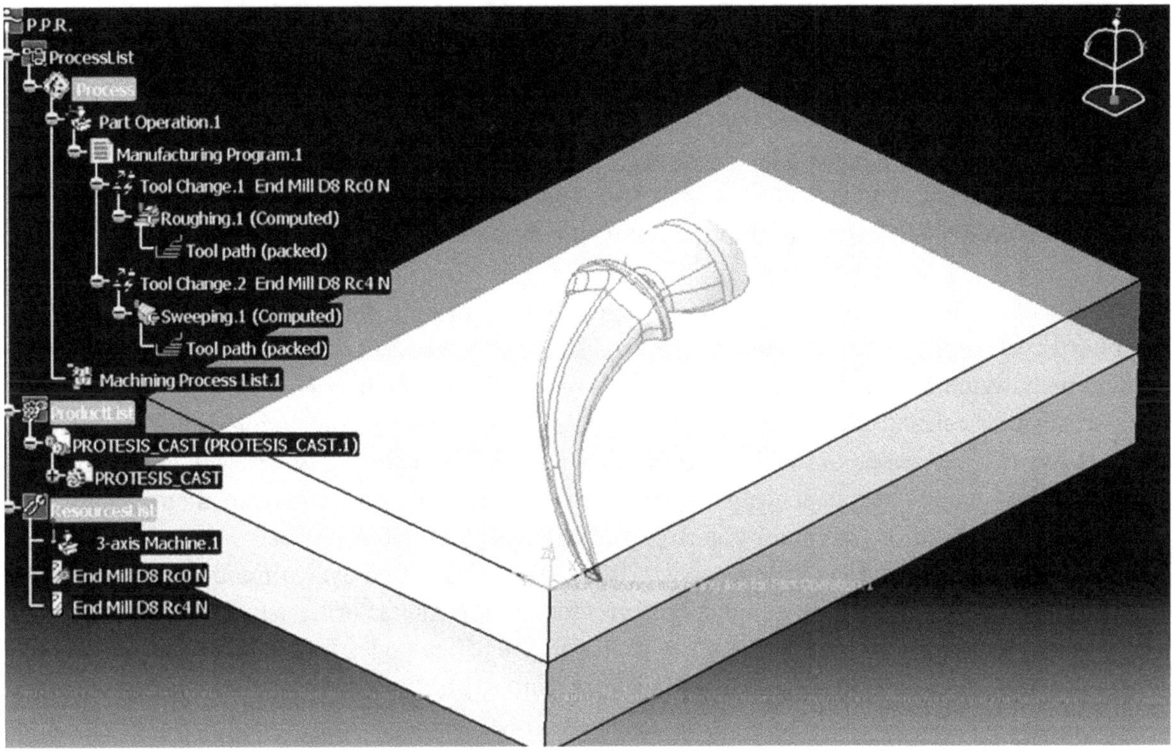

Figure 9. Tool path strategies for the machining of a sculptured part

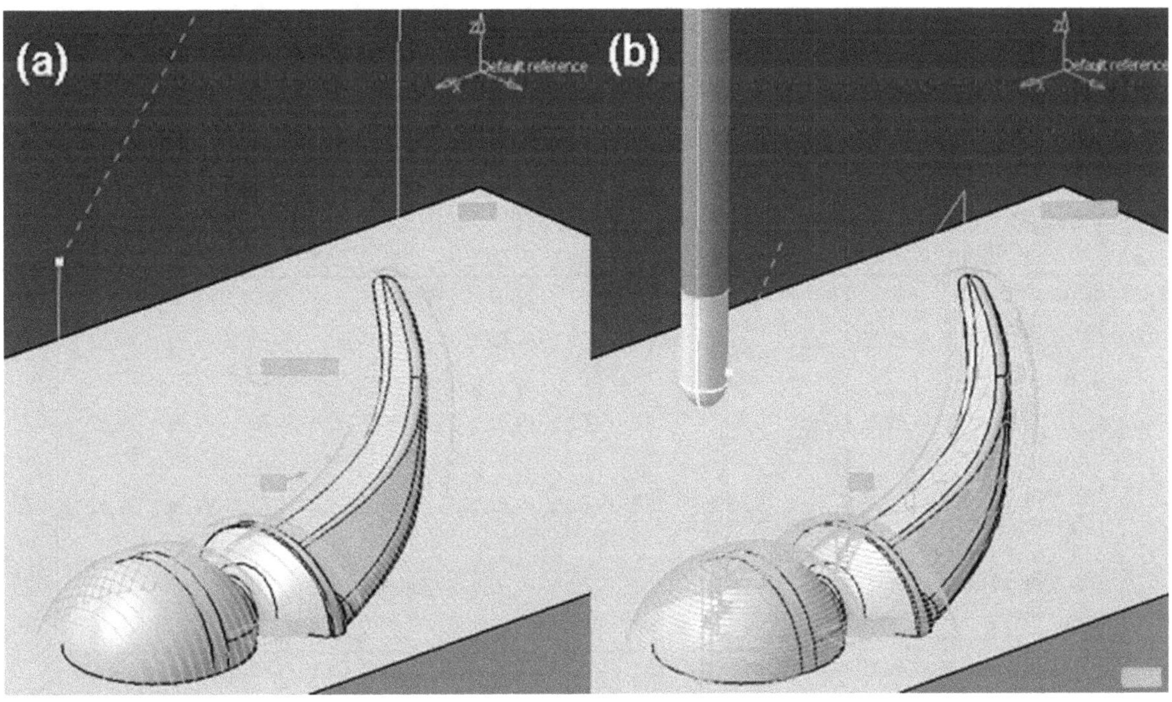

The majority of tool path automation cases suggest that the responsible objects ought to be retrieved and customized; yet; it depends on the software whether this approach is applicable. To cope with the problems experienced in this work, special properties of Microsoft® MSDN WinAPI were adopted and implemented. Thus; the visual basic code developed to automatically call the proper functions for tool path computation involves modules that search through the menus and dialog controls in order to trace the commands and manipulate them accordingly.

4.1.3 Automated Statistics and Data Extraction

Evaluation of machining scenarios involves different objectives regarding the operation. For roughing, calculated cycle time (CCT) and remained volume have been considered to formulate a common expression properly weighted as the quality objective to assess it. For finishing operations evaluation involves the assessment of calculated cycle time (CCT) for finishing; whilst surface deviation yielded and existing between the swept envelope of the tool and the ideal design model surface was treated as the objective indicating final quality. Yet again, the two quality objectives formulate a common mathematical expression as a weighted sum to turn multi-objective problem to a single-objective one. Further details about this approach can be found in (Fountas et al., 2012, 2014). Calculated cycle time from the operations retrieved by calling the respective automation object found in "Manufacturing activity" class (see Figure 10). The rest of the objectives automated by developing separate codes for technological objects like inertia properties (Figure 11a) and volume objects (Figure 11b).

Figure 10. The "Calculated Cycle Time-CCT" object retrieved to automatically compute operation cycles

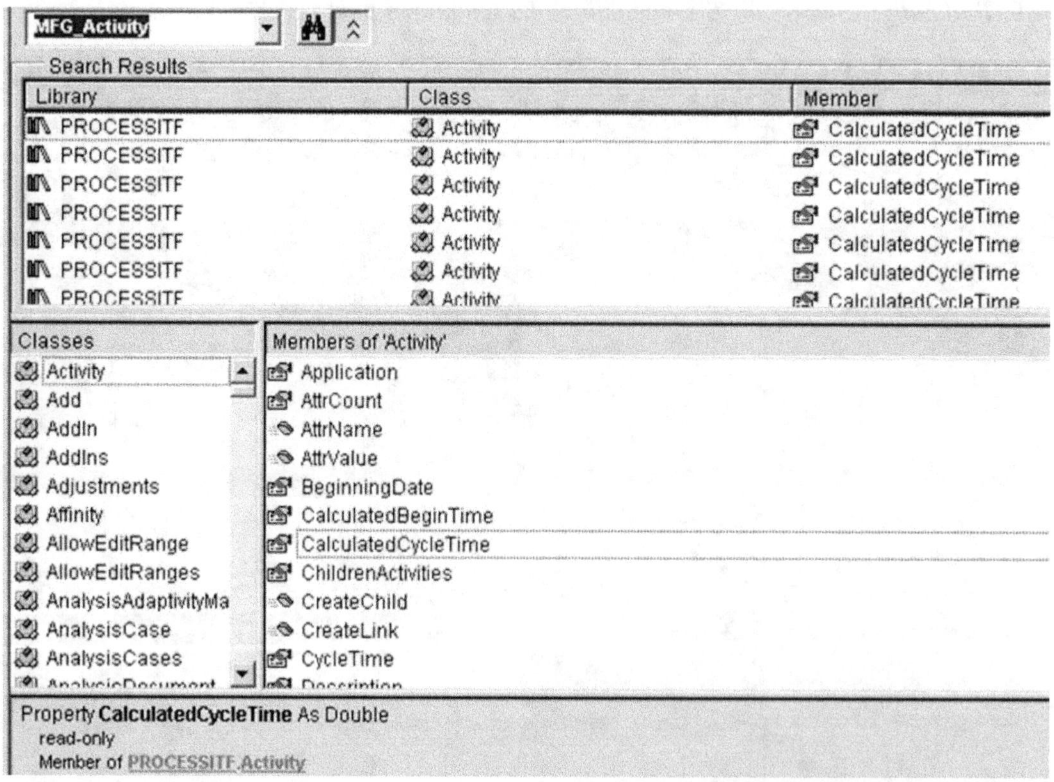

Figure 11. (a): "GetTechnologicalObject" and (b): "GetVolume" automation objects to compute inertia model properties

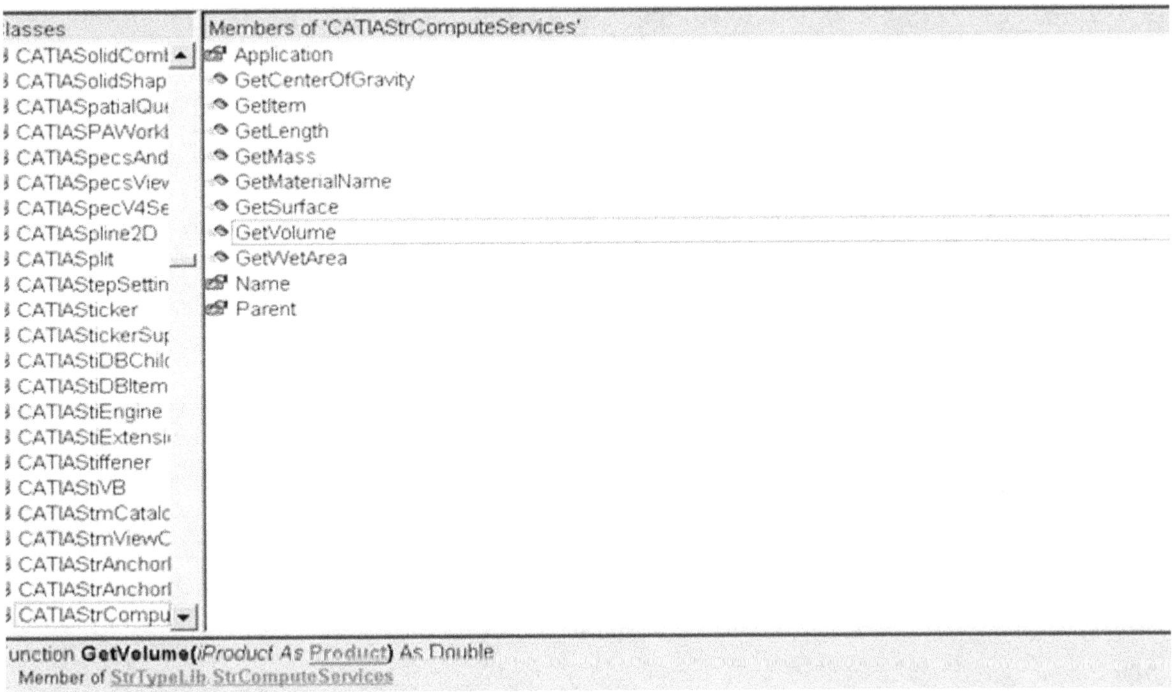

5 SOFTWARE COLLABORATION FOR MACHINING OPTIMIZATION

The environment optimizes machining operations for roughing and 3-/5-axis finishing processes. Optimization is achieved through the collaboration and data exchange between the virus evolutionary genetic algorithm presented in section 3 and CAM system via its customization routines presented in section 4. This section describes the entire operational workflow between the systems from initializing the environment and evaluating machining scenarios until reaching global optimum and obtaining the results.

5.1 Infrastructure and Optimization Workflow

The infrastructure of the proposed optimization methodology is consisted of two major components. The first component is CAM system and the second is the virus evolutionary genetic algorithm. Given the quality objectives to be optimized according the machining phase (i.e. roughing and/or finishing) the algorithm utilizes software automation routines to repetitively manipulate specification for parameter settings and tool path simulation process. Figure 12 illustrates the overall workflow of the entire automated environment between the GA and the CAM system's automated framework and Figure 13 gives an overview of the optimization workflow relating virus evolutionary genetic algorithm and CAM system.

When the framework is activated by the user a selection is made it terms of the manufacturing program type according the machining setup document (3 or 5 axis). Through this selection the programming functions corresponded to roughing and finishing stages are loaded. Further on the user is prompted to specify the importance of quality objectives through the determination of weights. The values for

Figure 12. The infrastructure for machining optimization using CAM environment and genetic algorithm

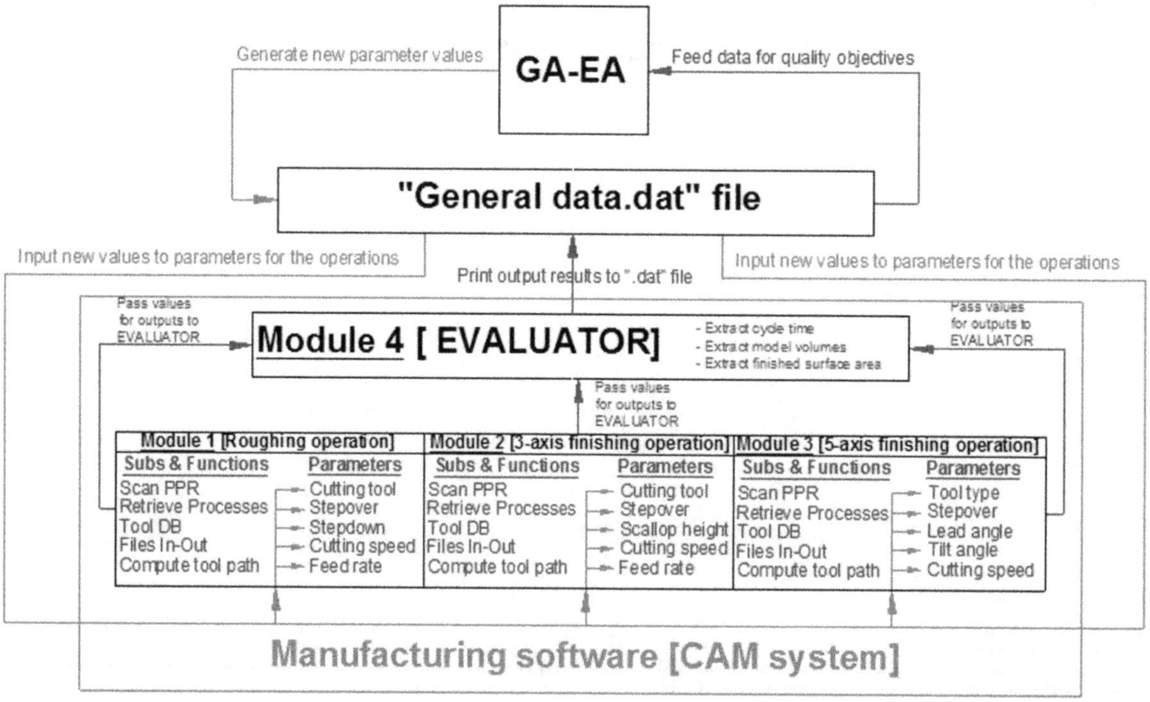

the weights are based on industrial demands and special needs; thus, they ought to vary depending on the case. Optimization settings involve the specification for process parameter values which are loaded as initial inputs to the system (initial seeding for the genetic algorithm). The document's "products-processes-resources" (PPR) tree (see Figure 8 sub-section 4.1.1) is scanned to retrieve operations. The procedure is initially executed for roughing whilst one of the finishing operations follows next in the sequence. If a previous run with the same variable values is detected, evaluator will propose the results of that run. To do so, Visual Basic's flag states were programmed in the code reducing this way overall performance time of the algorithm. The phenotype values for process parameters are loaded as soon as the machining operation is retrieved and a check whether they satisfy technological constraints, is performed. The involvement of technological constraints to machining optimization simulate actual status for physical operations since they prevent populations' following generations from taking values in vicinity that offers penalized objective values. Technological constraints deal with maximum motor power of the machine tool, feeds and speeds, maximum production time allowed, etc. Should the values violate the constraints; the system returns a penalized value for the corresponded objective function.

5.2 Implementation and Algorithm Settings

The proposed methodology was employed to prepare the machining program for the milling of a test sculptured model representing a prototype hip joint orthopedic part (Figure 14). The part was virtually machined preparing a 3-axis roughing / 3-axis finishing machining program. The algorithm run for 75

Figure 13. Overview of the optimization workflow relating virus evolutionary genetic algorithm and CAM environment

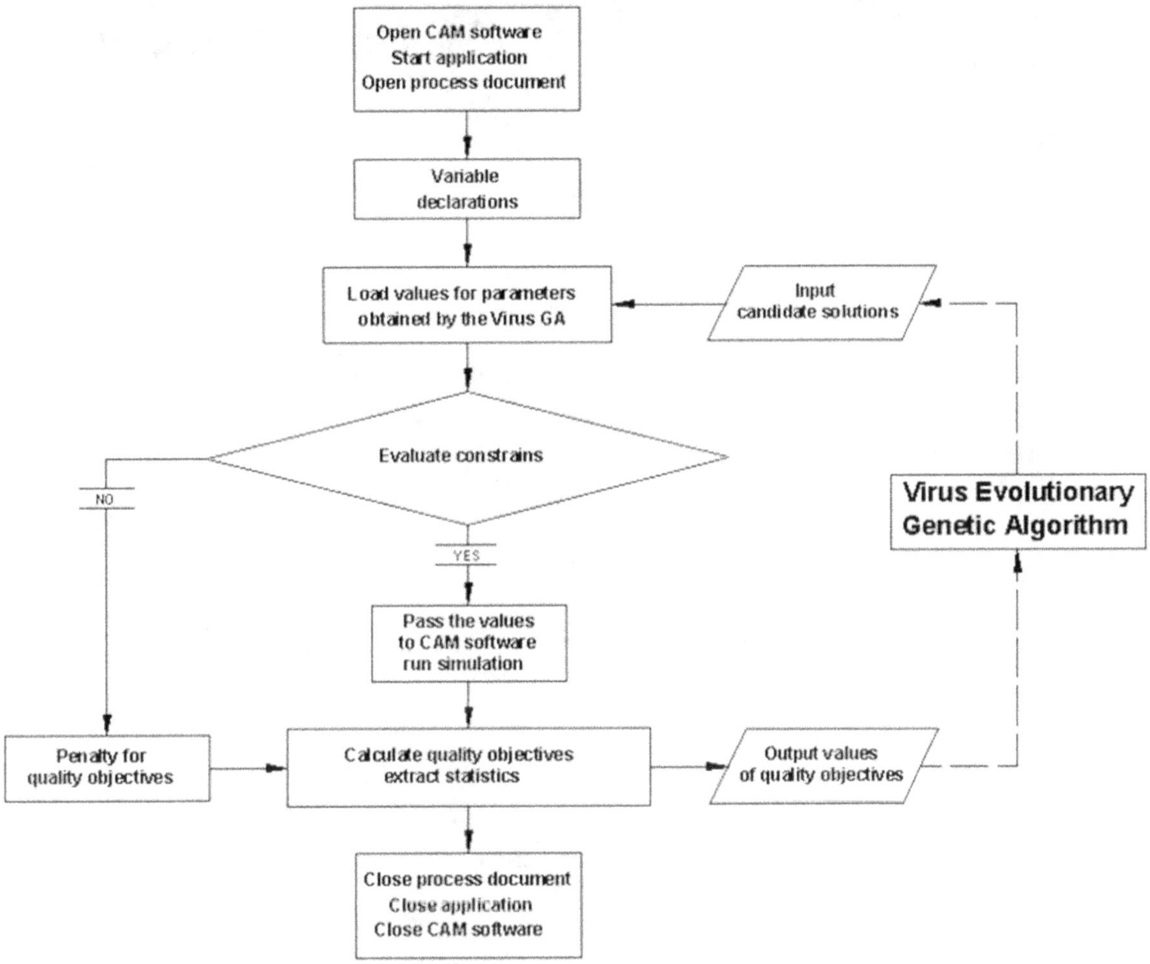

generations (stopping criterion) on its way to drive CAM software to perform the corresponded simulations and find optimal machining parameter sets for both roughing and finishing operations. Machining parameters and technological constraints specified, are presented below:

- Stock material: Ti-6Al-4V
- Machining strategies: Sweep roughing and sweeping for finishing
- Tool path style: Zig-zag
- Pmax = 10kW
- η = 0.7 (motor performance coefficient)
- Spindle speed range: 10-8000 rpm
- Feed percentage range: 10-95%
- Depth of cut range (Stepdown): 0.2-3.5mm

Figure 14. Test sculptured part: (a) roughing tool path, (b) finishing tool path, (c) roughed model, (d) finished model, (e) machining simulation in a virtual machine tool setup

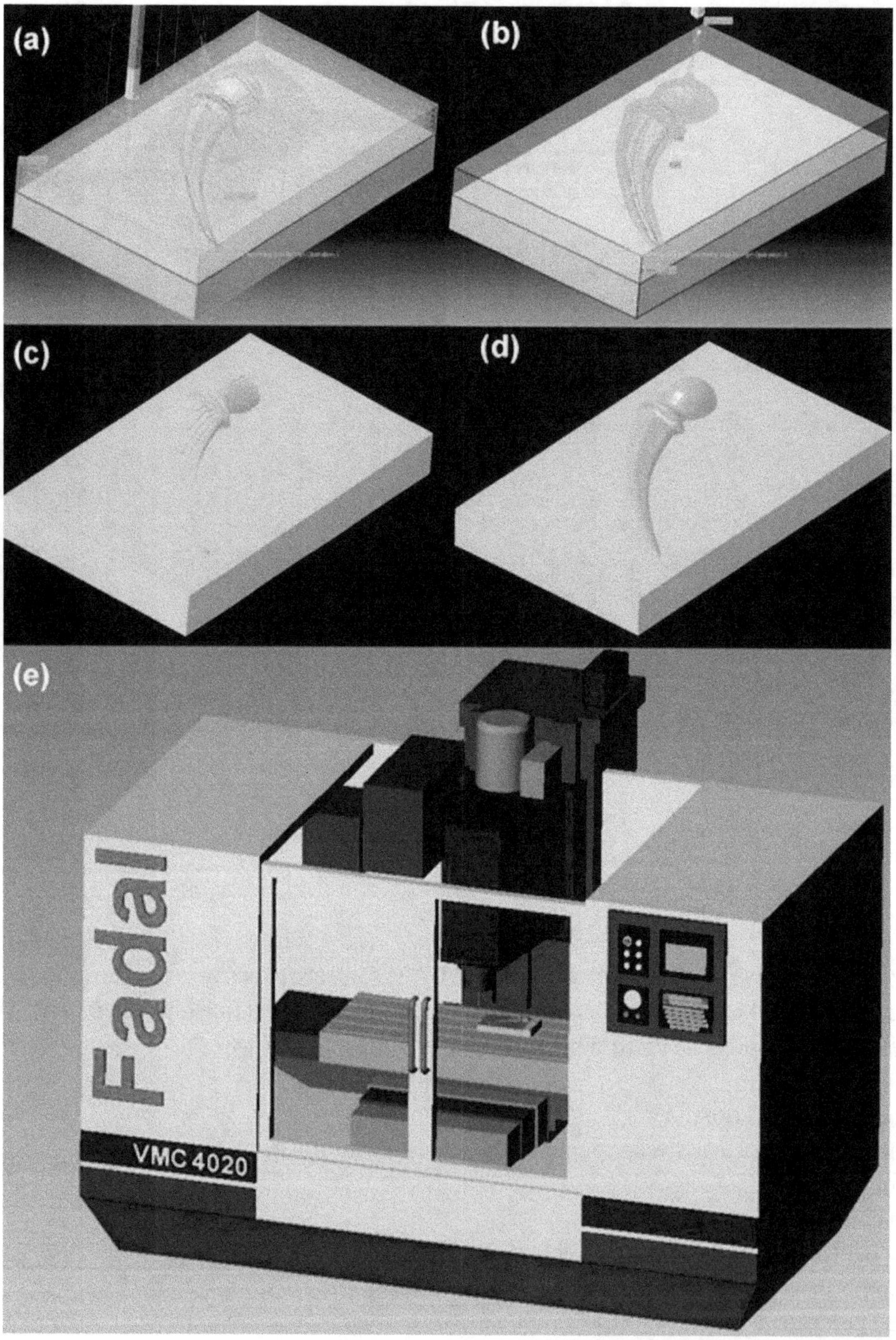

- Stepover percentage range = 5-100% of the tool diameter
- Available tools: TiAlN coated solid carbide end-mills (treated coating thickness 3μm, surface hardness 2800 HV, friction coefficient 0.3).

The developed optimization algorithm run on a personal computer with AMD Phenom ™ II X4 X965 Processor 3.42 GHz, 3.25 GB RAM, ATI Radeon HD 5800 Series graphics card on Windows XP Professional Version 2002 Service Pack 3. The algorithm run for 75 generations on a population that consisted of 50 individuals (about 50x75=3750 evaluations). Single objective optimization was conducted to both first series of evaluations.

In the first series of runs for roughing operation, single objective optimization was conducted with machining time (t_{mr}) and remaining volume (V_{rem}) as a common linear expression for the quality characteristic to be minimized. In the second series of GA runs for 3-axis finishing, the objective function was a linear combination of finish machining Time (t_{mf}) and surface deviation (S_{dev}) after proper normalization of their values, owing to different value sizes between them. t_m and S_{dev} were added together after weighing. The weights of each quality characteristic were of equal importance for the present study ($w_1 = w_2 = 0.5$).

Typical settings for algorithm parameters were adopted according DeJong (1975). These settings have been tested for their validity and efficiency in past research studies and found to produce good results. For virus operators preliminary experiments were conducted to decide the settings for the parameters involved. The algorithm's parameter settings are bulleted below:

- Host population size (n_h): 50
- Virus population size (n_v): 5 (1/10 of the host population)
- Number of generations: 75
- Initialization scheme: Binary
- Selection method:
- Crossover: One-point, (crossover rate - p_c) = 0.6
- Mutation: One-point, (mutation rate - p_m = 0.001)
- Selection scheme: Steady-state (deleting the weakest individual)
- Elitism: true
- Termination criterion = 75 generations
- Virus infection rate: Min 0.05, Max 0.1
- Virus life reduction coefficient: 0.2
- Virus string: variable with minimum number of digits, 10.

5.3 Optimization Results

The algorithm found to generate better results for a machining simulation with average settings in terms of process parameters which may empirically determined by to CAM environment. Nevertheless, it can generate satisfactory results when the initial seeding is created by specifying values for parameters close to minimum and maximum limit ranges as well. Table 1 presents the results obtained from the algorithm for both machining phases whilst Table 2 enumerates the cutting tools from which the algorithm selects

Table 1. Optimum machining parameters generated by the algorithm for process optimization

Roughing operation						
Tool ID	**Stepover (%Ø)**	**Stepdown** (*mm*)	**n** (*rpm*)	**f** (*% fmax*)	t_{mr} (*min*)	V_{rem} (*mm³*)
4	37	3.8	5124	92	7.38	33794.458
3-axis finishing operation						
Tool ID	Stepover (%Ø)	scallop height (*mm*)	n (*rpm*)	f (*% fmax*)	t_{mf} (*min*)	S_{dev} (*mm*)
8	87	0.017	6430	98	66.3	0.001247

Table 2. Cutting tools for machining simulations in CAM environment

Tool ID	**Cutting diameter** (*mm*)	**Shank diameter** (*mm*)	**Flute length** (*mm*)	**Overall length** (*mm*)
1	6	6	25	60
2	8	8	30	60
3	10	10	45	80
4	12	12	50	80
5	14	14	50	100
6	6	6	25	60
7	8	8	25	60
8	10	10	30	80
9	12	12	30	100
10	14	14	35	100

to perform the machining simulations. Note that the first 5 cutting tools are flat end mills suitable for roughing whereas the rest of the tools are ball end mills suitable for finishing when applying 3-axis machining technology.

Figure 15 illustrates the convergence diagrams according the evaluations performed by the virus evolutionary genetic algorithm. Figure 15a depicts the evaluations performed for roughing operation. Convergence was achieved at 54th generation. Figure 15b depicts the evaluations performed for the 3-axis finishing operation. Convergence was achieved at 49th generation.

7 CONTRIBUTION TO FLEXIBLE MANUFACTURING SYSTEMS (FMS)

Simulation of the dynamic production process can stress out the critical points in the production line, of a flexible manufacturing system, as well as the positions where excess material is accumulated or where idle times are higher than expected. Testing alternative cell layout scenarios is a low-cost procedure that is performed on computers. Thereby, engineers can decide on the optimal layout scenario without in fact having the manufacturing units moved around the plant. In this way, error and accident possibilities are significantly reduced. Moreover, only the necessary manufacturing units are incorporated in the cell, thus exploiting units that do not add to a specific production philosophy in terms of delivery dates or part quality. Material can thus follow the shortest possible path so as to become a final ready-to-ship

Figure 15. Convergence diagrams for the optimization process: (a) roughing operation, (b) 3-axis finishing operation

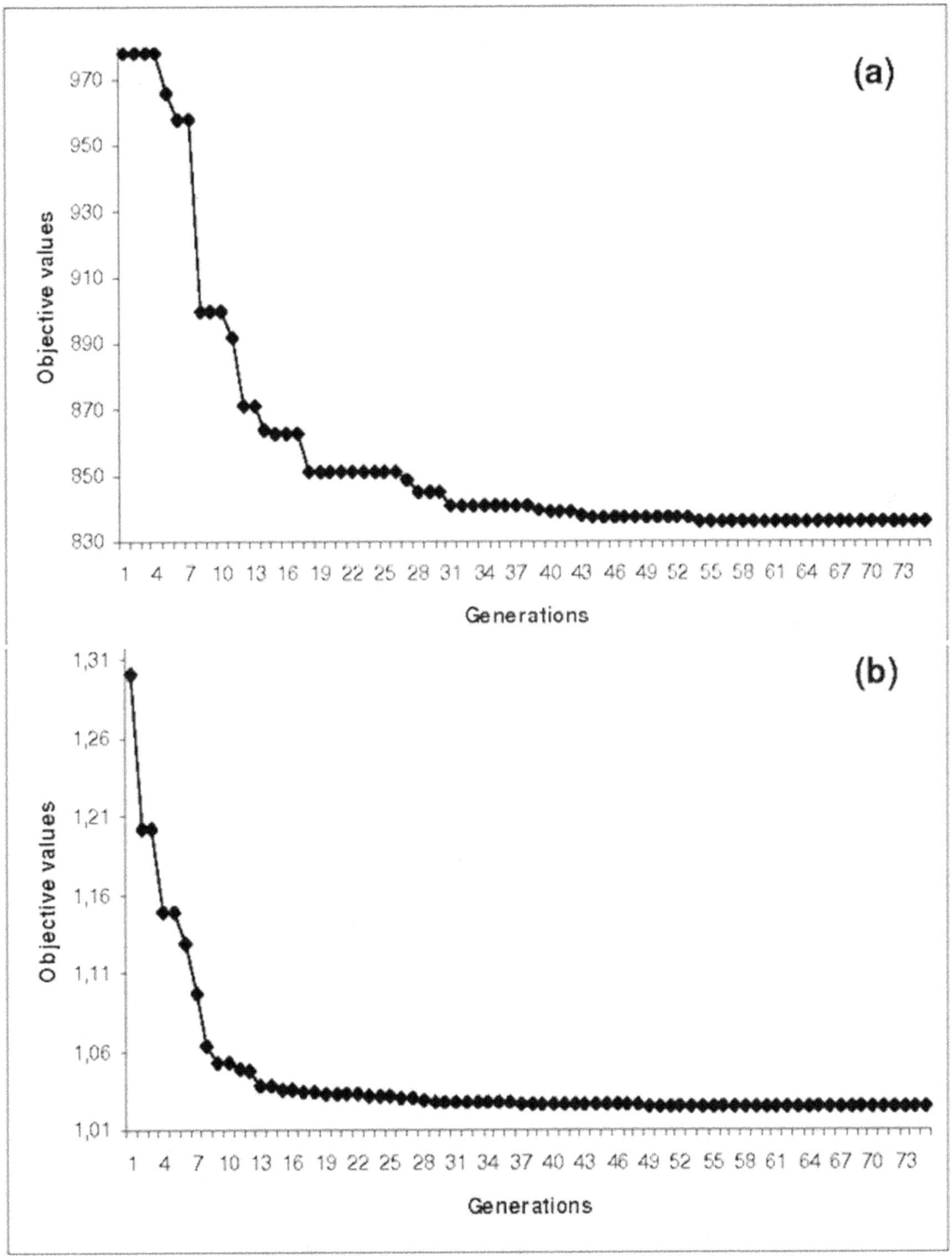

part. Using a virtual configuration composed by CNC machine tool models, robotic arms and conveyor belts (Figure 16); all prepared and tested through CAD/CAM, it is possible to define scheduling times for all industrial operations. Robot movement analysis offers the time needed to load and unload the machine tool. Conveyor belt velocity may be also known, hence the time a part resides or moves on it, can be easily computed. Should a manufacturing cell is simple a process planning study is more than enough to maximize productivity while reducing idle times. On the contrary, should a manufacturing cell is more complicated, special strategies may be implemented to tackle with optimum productivity cycles and beneficial performance. The process of flexible manufacturing systems planning optimization can be handled by appropriate artificial intelligence algorithms (Kubota et al., 1996) such as the one presented in this study.

Resources cooperation can be described with the following steps (Krimpenis et al., 2013):

Step 1. A conveyor belt forwards the raw part to a certain position and pushes the previously completed part to storage.
Step 2. A robotic takes up the raw part and carries it to the CNC machine tool vice.
Step 3. The vice automatically grips the raw part and robot moves away to safe position.
Step 4. Machining processes begin. Robot waits until all machining phases are completed.
Step 5. Robot moves towards the vice. Vice jaws open and robot picks up the completed part.
Step 6. Robot places the completed part to a specific position on the conveyor belt.
Step 7. Steps 1 to 6 are repeated until batch production is complete.

Based on the aforementioned data a production problem can be formulated towards the re-configuration and self-organization of a flexible manufacturing system on its way to determine optimum pallet location for placing fixturing devices for raw materials considering new part setups. Thereby, the virus-evolutionary genetic algorithm can be applied to an NC machining line and make it capable of reorganizing itself so that can process several types of machining jobs. Since the NC machining line is composed of arranged CNC machining centers the direction of the job's flow is "one-way" determined. A material

Figure 16. A flexible manufacturing cell involving a CNC machine tool, a robotic arm and a conveyor

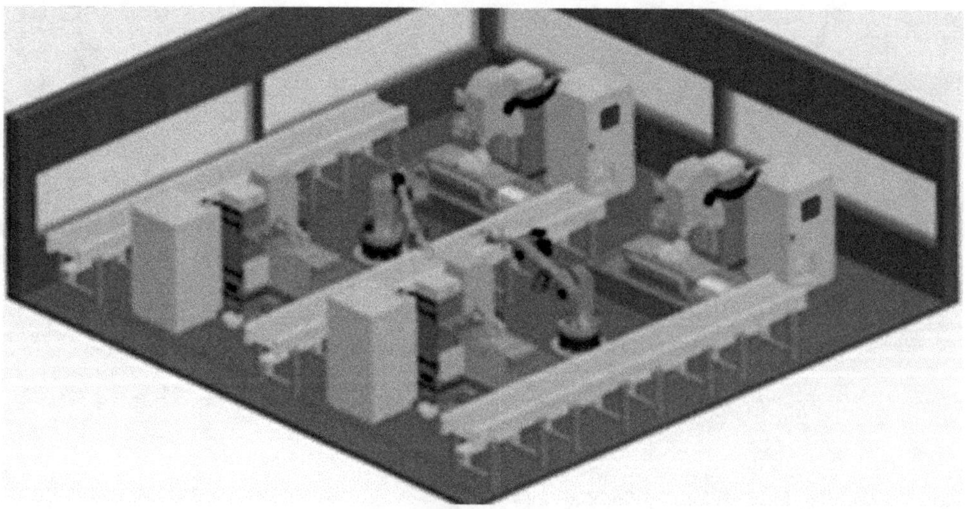

is transported from the storage place to the NC machine by a robotic system. After machining operation the final part is transported to the place for products by the same robotic system. Thus, the overall procedure is performed according to a flow shop scheduling. Assumptions of a problem formulation to optimize with artificial intelligent algorithms are as follows:

- A part operation is composed by *n* activities whilst the activity sequence is kept fixed.
- An activity requires machining operations by *m* cutting strategies and their number is fixed.
- A pallet on a CNC machining center may be equipped with three different fixturing devices regarding the work holding type that a cutting strategy requires to access the part and machine it.
- A CNC machining center is fed with a raw material from the entrance and the final part is placed on the exit after machining.
- The fixturing device a part setup requires is automatically chosen on the pallet and the machining process is executed.

Assumptions for the robotic system may be bulleted as follows:

- If a machining operation is concluded and a final part is on the exit, the robot prepares the next setup for a new part operation. Should the job setup is favored by the current fixture, a new raw material is supported on it; whereas should the job setup requires another work holding the robot removes the fixture device, places it back to the pallet whilst proceeds with a new fixture selection.
- If the CNC machining center in front of the robotic system is in process, the robot stands by until the machining operation is finished and then loads a new part setup (fixture and raw material).
- If the fixture a machining setup requires is not in the current pallet, then that pallet passes by the CNC machining center and the pallet next in line approaches to select one from the available fixture devices.

A location optimization problem suggests the selection of fixtures from pallets to facilitate a given setup. Thus, the objective is to reorganize fixtures on the pallets for minimizing the function determined as consuming time until finishing all part operations. Note also that transportation time of the robotic system is ignored compared to the NC machining time on a machining center and the loading/unloading time of raw materials and final parts respectively is constant. Let t_{mf} be a finishing time of a part operation OP_i. The function F is then defined as follows:

$$F = max(t_{mf\_1}, t_{mf\_2}, \ldots\ldots\ldots, t_{mf\_n}) \quad (5)$$

In this problem an individual would have a specific number of strings whose length is equal to the number of CNC machining centers. A gene would stand for the fixture type and the gene's locus "l" denotes the position of the CNC machine. Thus, the set of i-th genes on each string in the individual is the pallet of the fixtures on the i-th CNC machining center. Constraints can be introduced to the optimization problem such that overlapping of the same type of fixture on the same pallet may not be allowed. Operators (common and virus) can be normally specified to cope with optimization loops until convergence. Crossover is applied between the same number of strings of each individual and mutation replaces the gene with one of the rest of the genes excluding those of the pallet. The virus operators would replace string positions in an individual. Specifically, the virus operators replace individuals with the highest fitness values with those generated by crossover operator. Population size for host and virus individuals can be specified according the settings proposed in section 5.2. The string length of each part operation is the number of CNC machining centers in each part operation respectively. Through the applied intelligence with virus-evolutionary genetic algorithm, re-organization of fixtures could shorten the time needed to setup a new job (via the function illustrated in equation 5); thus improving the performance of all sub-elements in the flexible manufacturing system and obtaining feasible solutions for tough production scheduling.

8 CONCLUSION

This chapter proposes a methodology for optimizing sculptured surface CNC machining processes using CAM software and a novel intelligent algorithm based on the virus theory of evolution. The virus GA optimizes machining parameters in terms of quality characteristics according discrete machining phases such as machining time and remaining volume for roughing and final part surface deviation along with machining time for finishing. To overcome the trade-off existed between exploration and exploitation the algorithm was designed to self-adaptively change its searching capabilities regarding the problem's specialized features. To do so, genetic operators along with virus operators are integrated to the system so as to exploit both local regions of candidate solutions whilst searching for global optimum to the entire solution space. Automation routines undertake to retrieve information and execute repetitive tasks needed for the algorithm's simulation evaluations towards its convergence to optimum solutions for machining parameter values. The optimization approach takes into account single objective problem transformation through proper unbiased objective weighting, technological constraints and limitations to parameter values so as to formulate a practically viable and realistic optimization methodology. Evaluation cost of the optimization methodology is depended on the part's morphology and complexity, yet; is quite low if modern computers are available to run the application. Results obtained can be directly applied to industrial services and settings for the computer numerical control programming of sculptured surface machining.

REFERENCES

Anderson, N. (1970). Evolutionary Significant of Virus Infection. *Nature*, *227*(5265), 1346–1347. doi:10.1038/2271346a0 PMID:5455138

Bui, T. N., & Moon, B. (1994). A New Genetic Approach for the Travelling Salesman Problem, *Proceeding of the 1st IEEE Conference on Evolutionary Computing*, 1, (pp: 7-12).

C. A. T. I. A. V5 API Documentation (2005). *Dassault Systèmes*. Retrieved from http://catiadoc.free.fr/online/CAAScdDmiTechArticles/CAADmiTocActivity.htm

Choi, B. K., & Jerard, R. B. (1998). *Sculptured surface machining: theory and applications*. Dordrecht: Kluwer Academic Publishers. doi:10.1007/978-1-4615-5283-3

DeJong, K. A. (1975). *An Analysis of the Behavior of a Class of Genetic Adaptive Systems*. PhD Dissertation, Department of Computer and Communication Sciences, University of Michigan, Ann Arbor, MI.

Eiben, A. E., & Schippers, C. A. (1998). On evolutionary exploration and exploitation. *Fundamenta Informaticae*, *35*(1-4), 35–50.

Fogel, D. B. (1995). Phenotype, Genotype and Operators in Evolutionary Computation, *Proceedings of the IEEE International Conference on Evolutionary Computation*, ICEC '95, (pp:875-880), Perth, Australia, IEEE press, Piscataway, NJ. doi:10.1109/ICEC.1995.489143

Fountas, N. A., Kimpenis, A. A., & Vaxevanidis, N. M. (2012). Computational Techniques in Statistical Analysis and Exploitation of CNC Machining Experimental Data. In J.P. Davim (Ed.), Computational Methods for Optimizing Manufacturing Technology: Models and Techniques, (pp. 111-143). IGI-Global. doi:10.4018/978-1-4666-0128-4.ch005

Fountas, N. A., Kimpenis, A. A., Vaxevanidis, N. M., & Davim, J. P. (2012). Single and multi-objective optimization methodologies in CNC machining. In J. P. Davim (Ed.), *Statistical and Computational Methods in Manufacturing* (pp. 187–218). Berlin, Heidelberg: Springer-Verlag. doi:10.1007/978-3-642-25859-6_5

Fountas, N.A., Vaxevanidis, N.M., Stergiou, C.I., Benhadj-Djilali. R. (n. d.). Evaluation of 3- and 5-Axis Sculptured Surface Machining in CAM Environment through Design of Experiments (in press). *International Journal of Computer Integrated Manufacturing, 28*(3), 278-296.

Goldberg, D. E. (1989). *Genetic Algorithms in Search, Optimization, and Machine Learning*. Reading, Massachusetts: Addison-Wesley.

Holland, J. (1992). *Adaptation in Natural and Artificial Systems* (2nd Ed.). Massachusetts: MIT Press.

Holland, J. H. (1975). *Adaptation in Natural and Artificial Systems. University of Michigan Press*. Ann Arbor, Michigan: MIT Press.

Kalpakjian, S., & Schmid, S. R. (2008). Manufacturing Processes for Engineering Materials 5th Ed., Pearson; Prentice Hall.

Karagiannis, S., Iakovakis, V., Kechagias, J. D., Fountas, N. A., & Vaxevanidis, N. M. (2013, September 13-16). Prediction of Surface Texture Characteristics in Turning of FRPs Using ANN, *Proceedings of the 14th EANN Conference Engineering Applications of Neural Networks, Part I, CCIS 383*, EANN 2013, (pp: 144–153), Halkidiki, Greece. doi:10.1007/978-3-642-41013-0_15

Kersting, P., & Zabel, A. (2009). Optimizing NC-tool paths for simultaneous five-axis milling based on multi-population multi-objective evolutionary algorithms. *Advances in Engineering Software*, *40*(6), 452–463. doi:10.1016/j.advengsoft.2008.04.013

Krimpenis, A. A., Fountas, N. A., Noeas, G. D., Iordanidis, D. M., & Vaxevanidis, N. M. (2013, June 3-7). On the 3D Parametric Modeling of Manufacturing Systems. *7th International Working Conference "Total Quality Management – Advanced and Intelligent Approaches", 473-479, Belgrade, Serbia*

Kubota, N., Shimojima, K., & Fukuda, T. (1996). Virus-evolutionary genetic algorithm for a self-organizing manufacturing system. *Computers & Industrial Engineering*, *30*(4), 1015–1026. doi:10.1016/0360-8352(96)00049-6

Lopez de Lacalle, L. N., Lamikiz, A., Sanchez, J. A., & Salgado, M. A. (2007). Tool path selection based on the minimum deflection cutting forces in the programming of complex sculptured surfaces milling. *International Journal of Machine Tools & Manufacture*, *47*(2), 388–400. doi:10.1016/j.ijmachtools.2006.03.010

Nakahara, H., & Sagawa, T. (1989). *Virus Theory of Evolution*. Tokyo, Japan: Tairyusha.

Nash, J. F. (1951). Equilibrium Points in n-person games. *The Annals of Mathematics*, *54*, 289.

Ozturk, E., & Budak, E. (2007). Modeling of 5-axis milling processes. *Machining Science and Technology. International Journal (Toronto, Ont.)*, *11*(3), 287–311.

Quinsat, Y., & Sabourin, L. (2007). Optimal selection of machining direction for three-axis milling of sculptured parts. *International Journal of Advanced Manufacturing Technology*, *33*(7-8), 684–692. doi:10.1007/s00170-006-0515-5

Tamaki, H., Kita, H., Shimizu, N., Maekawa, K., & Nishikawa, Y. (1994). A Comparison Study of Genetic Codings for the Travelling Salesman Problem, *The 1st IEEE Conference on Evolutionary Computing*, 1, (pp: 1-6), Florida.

Vaxevanidis, N. M., Kechagias, J. D., Fountas, N. A., & Manolakos, D. E. (2013, July 3-5). Three Component Cutting Force System Modeling and Optimization in Turning of AISI D6 Tool Steel Using Design of Experiments and Neural Networks, *Proceedings of the World Congress on Engineering* WCE 2013, Vol I, (pp: 1-5), London, U.K.

Wang, J., Jensen, G., & Red, W. (2006). Global Finish Machining Using CM^2. *Computer-Aided Design and Applications*, *3*(1-4), 405–415. doi:10.1080/16864360.2006.10738486

Warkentin, A., Ismail, F., & Bedi, S. (2000). Multi-point tool positioning strategy for 5-axis machining of sculptured surfaces. *Computer Aided Geometric Design*, *17*(1), 83–100. doi:10.1016/S0167-8396(99)00040-0

Zeroudi, N., Fontaine, M., & Necib, K. (2012). Prediction of cutting forces in 3-axes milling of sculptured surfaces directly from CAM tool path. *Journal of Intelligent Manufacturing*, *23*(5), 1573–1587. doi:10.1007/s10845-010-0460-x

KEY TERMS AND DEFINITIONS

ANN: *Artificial Neural Network* (artificial neural networks -ANNs are models inspired by central nervous systems -in particular the brain- that are capable of machine learning and pattern recognition. They are usually presented as systems of interconnected "neurons" able to compute values from inputs by feeding information through the network).

API: Application Program Interface: Platform in which end users are able to develop their own customized programming routines to manipulate, automate or build software through existed automation objects, methods and properties.

CAD: *Computer Aided Design* (Computer aided design is the technology concerned with the development of new products in the form of 3D models with the use of computers).

CAM: *Computer Aided Manufacturing* (The use of computer software to control machine tools and related resources to plan, simulate and verify manufacturing operations for products).

CCT: Calculated Cycle Time: The processing time extracted as a statistical attribute after the end of a machining simulation performed in CAM environment.

CNC: *Computer Numerical Control* (The technology concerned with the control of machine tools through programming. CNC programs contain functions that specify machining parameters, determine cutting tool trajectories and handle commands such as spindle turning, coolant supply, etc).

EAs: *Evolutionary Algorithms* (Evolutionary algorithms -EAs, are heuristic search routines like Gas capable of providing optimum solutions to optimization problems using techniques inspired by natural evolution (genetic operators), such as selection, crossover and mutation).

GAs: *Genetic Algorithms* (Genetic algorithms -GAs are heuristic search routines which mimic the process of natural selection. GAs are applied in many scientific areas to generate useful solutions to optimization problems).

PPR: Products-Processes-Resources. The major components that constitute an action tree when modelling machining processes with the use of CAM systems.

QO: Quality Objective: An optimization target to a problem solving activity.

Chapter 5
Simulation of Manufacturing Processes via Virtual Reality

Mohamed-Amine Abidi
Lyon University, France

Pierre Chevaillier
Lyon University, France

Barbara Lyonnet
Nantes University, France

Rosario Toscano
Lyon University, France

Patrick Baert
Lyon University, France

ABSTRACT

In a world in continuous evolution, the different industrial actors need to be reactive to remain competitive and to conquer new market trends. To achieve this, they are constrained to improve their way of industrial management, both at the strategic level, to adapt to technological advances and follow market trends. In this chapter, we introduce a new simulation method that makes it easy to understand the results of a given simulation. This is of crucial importance because the design stage of a manufacturing system usually implies not specialist actors. The objective of the chapter is to present the main advantages of using the virtual reality (VR) to the manufacturing processes simulation. To this end, a state of the art will compose the first part of the chapter. In the second part, we address the issue of the contribution of the VR to the industrial simulation.

1. INTRODUCTION

To remain competitive in the increasingly unstable markets, the industrial actors often have to seek innovative solutions to improve their production, assist in the decision, update and adapt their means of production management, etc. This requires software tools to allow controlling and simulating various flows of production system. There are several tools to design and simulate various industrial processes and even the decision aid. We are particularly interested in the discrete event simulator of flow. These

DOI: 10.4018/978-1-4666-8693-9.ch005

tools are generally used by industrial engineers specialized in management and production engineering process simulation. This prevents the simple industrial actors to understand the simulation's results which present complex and specific data.

The objective of the reactivity in production is to regularly update production systems. Since the 1990s the increasing diversity and complexity of products has gradually led manufacturers to use computer software in various stages of production, so that the industry can control management systems, as well as to simulate different processes of production systems. With the extensive development of computer technology, these software packages are now available in all stages of an industrial system life cycle. This revolution opens the door to new technologies and to their integration into the creation process and in the simulation of industrial methods. Today, several researchers are working on projects for the integration of VR into design tools and simulation of industrial systems. VR is a technology that combines software tools and other equipment to set up an immersive interactive experience that simulates a real-world scenario in virtual mode, this technology is used in various fields such as: medical, nuclear, mechanical simulation, and computer industry etc.

The simulation field is a very large one, in fact, it is applicable to all industrial flows and even services, at all levels and in all phases of the life cycle of a production system (Berchet, 2000). Often, the software packages are used for modeling of industrial systems, improving workshops to control stocks capacity, simulating the different processes of production, validating the operational rules, etc. Flow simulation can also be used at the operating stage, in addition to planning tools to estimate or schedule delays for example. It also allows determining the optimal configuration of logistics system production values (Fontanili, 1999). In this chapter, we will focus mainly on the simulation of flow of production in manufacturing-type systems using a discrete event flow simulator in which the state variables change only at events such as the start or the end of a transaction, release of a workstation, the occurrence of a failure, the standby in a queue of parts, etc… (Bel & Kieffer, 2002). The simulation allows us to project ourselves into the future to quickly measure the consequences of an event coming from appearing. It also allows the decision support by simulating several correction scenarios in a rapid way, to limit the consequences of a critical event. The modeling and the flow simulation of industrial systems is a specialty of industrial engineering, it must therefore have sufficient knowledge to be able to establish a flow simulation for a production system. Therefore, modeling and simulation is not accessible to the public industrial users, this problem represents a disability for many industrial actors.

2. APPROACH COMPARISON

The typical problem with the representation of a simulation of a manufacturing system is to understand the simulation results. These results are generally understood only by experts. Although there are tools to represent them with 2D data visualization, to understand these results remains a lot of works of industry expertise, and the research works conducted by Dangelmaier *et al.* (2005) allow the implementation of a virtual reality system for 3D visualization of the complex results. The VR also allows simulating the configuration of machine installation, and the studies conducted by (Lindskog, Berglund, Vallhagen, & Johansson, 2013) show that industrial model VR can be used as a visual system when modeling and reconstruction of production systems and even testing the various setup configurations of the virtual model before the real installation of the system. Today, the 3D imaging methods and tools, i.e. the interactive virtual reality simulations appear in industrial activities, especially in project design or implementing

CAD models "Computer Aided Design" in industrial production (Sreng, Lécuyer, Mégard, & Andriot, 2006) (Jayaram, 2007), which is also the case for different industrial activities related to automotive and aerospace, etc…

The originality of this research is based on the design of a VR system that includes a flow simulator discrete event, which is generally a tool for the design and control of the production processes, and the optimization of production systems. This approach of VR allows us to develop an abstraction layer between the techniques of simulation flow and the actors concerned with the exploitation of complex data resulting from this simulation. The advantage of this layer is to have certain invisibility on different methods and techniques for the flows simulation that requires expertises in this industrial area. Indeed, this layer allows mastering of flow simulations through the VR tools, as well as to immerse the user in a scene integrating process upstream and downstream of workstation. It also allows us to have an experience feedback on the decision from the exploitation of the design. In conclusion, the simulation assisted by VR seems to us to be one of the most interesting approaches to simulate and model the production systems.

The objective of this chapter is to establish the general context of simulation of virtual manufacturing, to present a state of the art of the discrete event systems (DES), the simulation of DES, and the simulation of production flow. It also aims to introduce the different existing simulation tools, as well as a presentation of the VR, the existing projects of research in VR that treat the different industrial problems, the contribution of this new technology in the industrial context. And in the last part we address the issue of our contribution in this area through an architecture that allows implementing a solution for interactive and immersive simulation.

3. DISCRETE EVENT SYSTEMS

The discrete event systems (DES) are dynamic systems whose state evolves with the occurrence of events. Indeed, instead of focusing on continuous progress of events, we take into consideration their beginnings and their ends and their dynamic concatenations. Time in this type of system is encoded by the sequence of events, so it is discretized. Some examples of DES can be cited such as: communication networks, transportation networks and manufacturing systems. The study of discrete event systems can be conducted with tools or languages parallel programming or real-time. The finite-state machine, the dynamic algebraic models, the networks of queue and well on Petri networks, they are the basis of most of the tools of flow simulation of the discrete event systems. For example, Petri networks is a modeling tool particularly well suited to systems that incorporate parallelism and synchronization. Due to their mathematical modeling aspect, the Petri networks can be used to verify properties of the system such as blocking. Generally the modeling of an industrial system is based on mathematical and physical laws that govern its behavior. Thus, there are two main classes of models: knowledge models and representation models. Indeed, the processes described by equations involving the physical laws that govern their behaviors are called knowledge models. These models, also called white box models, have the advantage of having parameters directly related to the physical elements of the system. Thus, in the case where the model structure can be constructed with physical laws, but one or more parameters must be estimated from observations, model are called gray box model. Finally, the complex or poorly known systems are often equipped with a large number of sensors, which generate signals which can be used for the identification of a model from experimental data. The model obtained is called a representation model or black box model.

The discrete event systems are generally modeled based on events. In these systems the behavior is represented by transitions between different states of the system following the occurrence. The behaviors of these systems are usually modeled by networks of queues, Petri networks and the finite automata and their extensions and variants. The observability in such systems is treated by partitioning in advance in the model all the events into two subsets: a set of unobservable events and another with observable events include events precisely faults. Thus what are observed in a discrete event system are sequences of observable events. The diagnosability in discrete event systems is defined as follows: A system is diagnosable only if every occurrence of a fault is followed by a finite observable sequence that does not occur in the absence of fault. This definition is similar to the concept of signature defined for continuous systems. Also found in (Cordier, 2006) a study that shows the concept of signature can be formally defined for discrete event systems such as diagnosability which can be defined in the same way as in the continuous and discrete systems.

3.1. The Automats

Among the many tools that allow the study of discrete event systems, there are automata. An automaton is a state machine. It is represented by a set of states connected by transitions associated with events. An automaton is a finite state if the number of states is bounded. It can be seen as a machine having discrete inputs and outputs and which adapts its outputs in accordance with the changes of its inputs. If the next state machine can be determined without equivocally using its current state and symbol input, then the automaton is called deterministic, otherwise, the automaton is non-deterministic. An equivalence relationship exists between the two, in the sense that for any non-deterministic finite machine, an equivalent deterministic automaton can be found (Motwani & Ullman, 2007) (Lafortune, 2008).

3.2. The Networks of Queue

The networks of queues are parts of several formalisms for modeling discrete event systems. A queuing network is presented by the stations of several types of services (FIFO, sharing of server, etc.), and the sources of clients and the synchronization mechanisms between these stations. There are various simulation algorithms that use protocols "Chandy and Misra" optimized according to specific parameters of these parts. In such an algorithm, the process sends a null synchronization message when it enters a blocking period waiting for the messages. This has two advantages, the first is that the calculation procedures of predictions are performed during potential blocking periods, the second is that the number of messages circulating is limited.

3.3. The Petri Nets

Since the appearance of Petri nets in 1962 by Carl Adam Petri (Gesellschaft, 2006), this approach represents a powerful mathematical and graphical tool for designing discrete event systems. The model for this design allows analyzing properties and validation of their specifications. The Petri nets have the advantage of being a much more general model compared to automats, and enjoying much richer structures, these networks are perfectly adapted to the description of different types of discrete event system (Murata, 1989) (David & Alla, 1999) and (Giua & Seatzu, 2007). The Petri net is a directed graph comprising a finite set of places and a finite set of transitions, with the oriented arcs which ensure the passage of place and transition. The places (symbolized by circles) represent specific conditions for each status of system the information of type of resources number, status of a resource etc… The transitions (symbolized by lines) represent the actions that can cause the change of state of the system (discrete events). The directed arcs indicate causal relationships between states of system and events involved in its evolution. As each state change is caused by the arrival of an event, it logically follows that an arc can never link two nodes of the same type.

Figure 1. Example of a model of Petri nets

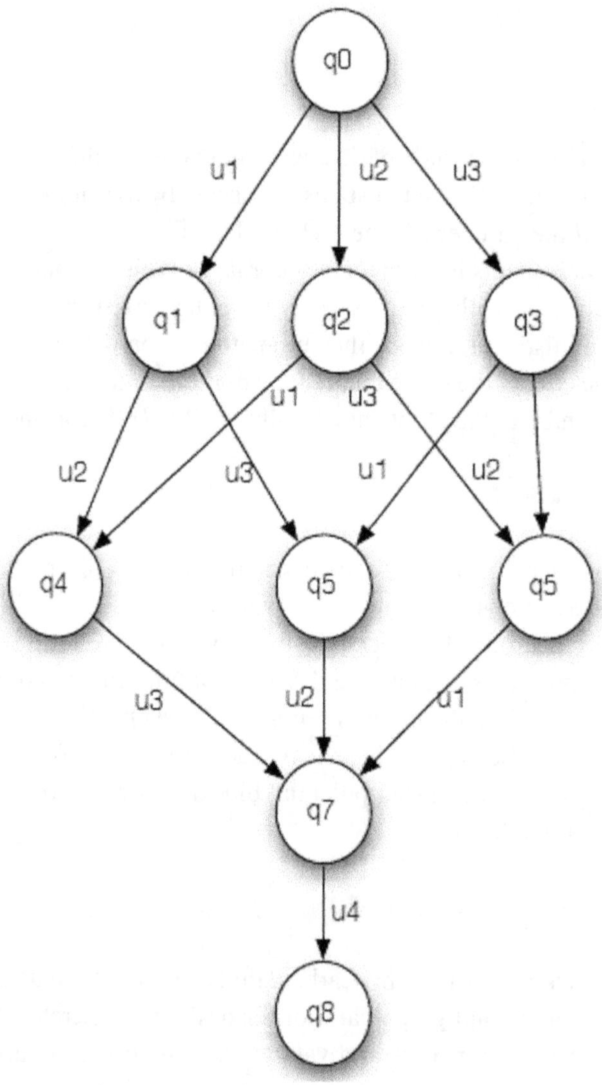

4. THE SIMULATION OF DISCRETE EVENT SYSTEMS

The simulation of discrete event systems precludes a continuous simulation. In this type of simulation time is cut into equal portions without which we cannot distinguish events. At each slice or time interval, we analyze the state of the system. According to (Claver, Pitt, & Gelinier, 1996), the simulation is an activation of the model studied in time, based on the rules of evolution or assumptions that we want to validate. The simulation is a technical description of the model over time, in order to know the dynamic behavior of the real system and even to predict the effect of different parameters or scenarios on the model. Gogg & Mott (1993) defines simulation as the art and science of creating a representation of a process or system in order to test and evaluate. Pegden *et al.* (1995) specifies that the modeling and simulation of flow presents a process of creating a model for a real system and conduct experiments on this model in order to evaluate different control strategies and to understand the operation. The flow simulation allows to reproduce the dynamic behavior of a real process of a software tool, without any risk of loss and saving time with the speed execution of the simulation, and answer type questions "What happens if ...? ".

4.1. The Production Systems

The first production systems were made by "Newell" and "Simon" in the 1950s. A set of generalities are obtained using all manner of industrial problems for the purpose of reinforcing the area of research, with changing to the specific production. The production systems have evolved rapidly due to the proliferation of expert systems. A production system is a computer program commonly used to provide a form of artificial intelligence, which consists primarily of a set of rules of behavior. These production rules are useful basic representations in automated planning. A production system provides the necessary mechanism for the implementation of production to achieve the objective of the system. A production system also contains a database, sometimes called working memory, which retains data on the current state of the system, and an interpreter of the rule.

4.2. The Evolution of Production Systems

A production system is an organization whose function is to provide engineered products or services. The economic context and market evolution have led the industrials to adapt their production systems to improve their industrial performance. The concept of industrial evolution refers to the ability of a factory which ensures the results of his organization and thus ensures the continuing evolution of the process. The establishment of the evolution of production systems was initiated by the automotive industry (Womack, Jones, & Ross, 1990). After the period of the traditional production, for several centuries, mass production developed from the late nineteenth century in major U.S. industries. After World War II, the Japanese launched a revolution in mass production by developing the principles of Toyota production system. The Toyota production system has meanwhile demonstrated its flexibility to respond to market fluctuations. These mutations have radically improved our fundamental ideas about how to produce some products or services. The term of « Lean production » was first named by Krafcik (1988) to describe the Toyota Production System (TPS). The Lean production is a management approach to enhance performance by eliminating waste. At the moment, the Lean approach seems to be an ideal solution. Its spectacular results in terms of increased productivity and reduced cost have created great interest in many companies

(Kilpatrick, 2003), (Shah & Ward, 2003), and (Demeter & Matyusz, 2010). Over the last ten years, Lean principles have been applied intensively in many sectors of activity, such as aerospace, electronics, mechanical subcontracting (Crute *et al.*, 2003), (Abdulmalek & Rajgopal, 2007), and (Baglin & Capraro, 1999). Furthermore, note that its implementation has been done in small and medium-sized companies (Achanga, 2006) as well as in large companies (Womack & Jones, 2005). In continuous improvement of the reactivity, the virtual reality and computer stimulations open interesting perspectives for optimization of current production systems. Optimization of material flow is a current issue of the organization of production systems. A simulation tool for visualizing flows, detecting inventory shortages and queues, should lead to a better system design and a better anticipation of technical and technological decisions. As suggested by the Lean principle of just in time, flexibility and reactivity of flows are crucial. Virtual reality would enable to take action upstream, if possible at the very start of the design phase. Thus the production systems would be more adapted to the different constraints of companies.

4.3. The Simulation of Production Flows

The flow simulation is a process that involves matching a system to a model that represents its characteristics. In this way, it studies the impact of change on the model and not on the system, by the ticket of programming of different rules and simulation parameters, and experiments provided by the progress of model and interpretation of results for the purpose of making decisions related to the system (Ait Hssain & Dunod, 2000). We can thus validate the changes before proceeding to their implementation. Various techniques are used for the simulation of the models, such as networks of queues, discrete event simulation, operational analysis or the Petri nets. Today, Petri nets (Gesellschaft, 2006) are supported by software tools which allow the system architect to proceed himself in the study of the behavior of the system as in the design phase and during its execution and during the life cycle of the system. Since the 1990s the increasing of diversity and complexity of products has gradually led companies to use the computer software in the production process. With the considerable development of computer science, these tools are now available in all stages of the life cycle of production systems. We often find that the computer software of design and operating are used separately and therefore there is no experience of feedback from the operating to the design. Our main objective is to establish a link between a tool of flow simulation and a virtual reality system while gradually refining the performance of the simulation model through dynamic exchange of data using the techniques of virtual reality.

The flow simulation allows:

- To observe the dynamic behavior of a system: simulate to understand.
- To experience the solutions of change: simulate to improve.
- Utilize the model to predict the behavior of system: simulate to anticipate.

In production and logistics, the simulation is mainly used for:

- Analyzing the physical and information flows.
- Analyzing resource states.

Synthetically, we can say that the simulation of discrete event flow:

Figure 2. Scenario of simulation of a production system.

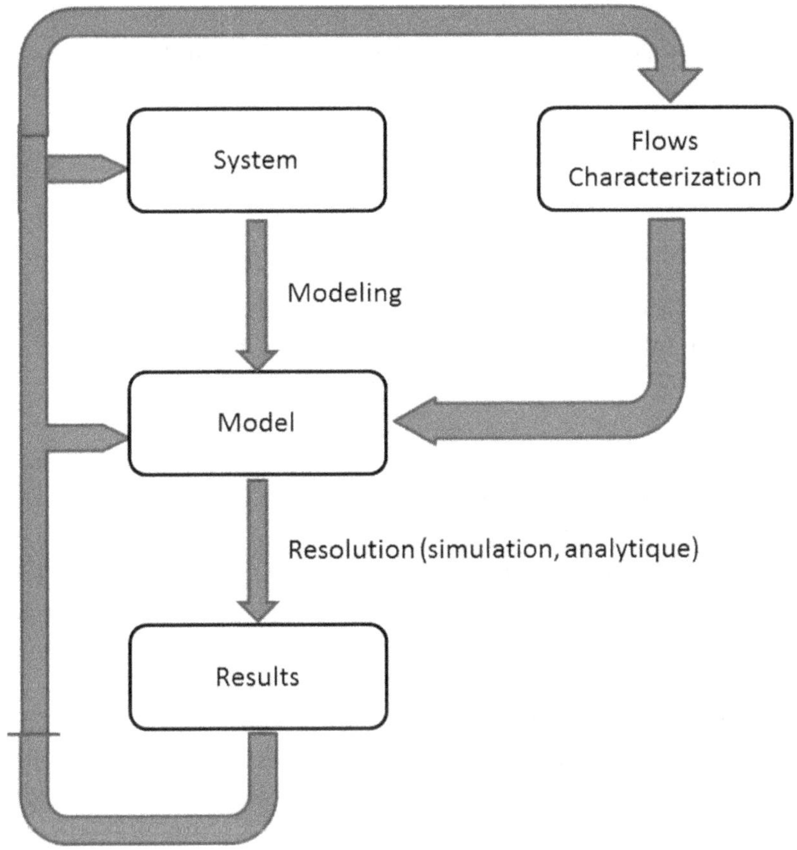

- Is a tool adapted for complex systems.
- Does not require mathematical modeling.
- Allows a level of detail that can be high.
- Presents a shape close to the real system.
- Provides arguments of communication and builds trust, through visualization.
- Offers the possibility to test different organizations.
- Helps to make better decisions.
- Generally production systems are based on the discrete event.

5. EXISTING TOOLS FOR THE SIMULATION OF INDUSTRIAL PROCESSES

To be reactive, must be competitive, is why manufacturers today have a vision to high production technologies. They compete simultaneously to achieve a common goal, which is the automation of manufacturing facilities and computerization of various functions of production, such as: design, workflow management of production and logistics. In this state of the art we focus on specialized software in workflow management and design of workshops. We start with the tools supporting the modeling functions of production systems and we detail the flow simulation software used most often in the design phase of workshops, the analysis tools, and the techniques used to exploit the results of the simulation. The simulation of a production system is the fact of playing with the various system parameters such as the size of the storage areas, priority rule etc., to arrive at a better configuration.

Our approach is to develop a virtual reality system that allows interfacing with a flow simulation via a tool of simulation. In the next chapter, we will show the architecture of our approach in a more detailed manner and an academic example that explains the interest of this approach. In this chapter, we show the architecture of our approach in a more detailed manner and an academic example that explains the interest of this approach.

Since the 1990s, the increasing demand for improving the efficiency and quality of production has gradually led industrials to seek solutions of computer software to master the various processes of production. With the considerable development of computer science, these tools are now available to control the different stages of the life cycle of a production system.

For the design stage, there are several tools of computer aided design (CAD) softwares, such as Catia, AutoCad, SolidWorks and ProEngineer etc. As well tools of CAPE (Computer Aided Production Engineering) are more specific to production systems such as Delmia, Tecnomatix etc, or the simulation tools flow, such as Witness, Arena, Promodel, Automod, eM-Plant, Quest etc. In industry, we used these simulation tools flows mainly in the design phase, for the design of a new production line or a new workshop. And by Klingstam, the virtual reality tools will do part of future of the production systems simulation (Klingstam & Gullander, 1999).

For the stage of data operating, there are several tools tailored to different levels of decision making such as APS and ERP used for medium and long term strategic decisions well as other operational tools of workshop for short-term scheduling. Before about ten years, tools of MES (Manufacturing Execution System) appeared, which have several functions and services more advanced compared to the simple supervision of model (the ISA 95 standard).

Figure 3. CAPE Tools: bridging the gap between product design and production (Klingstam & Gullander, 1999)

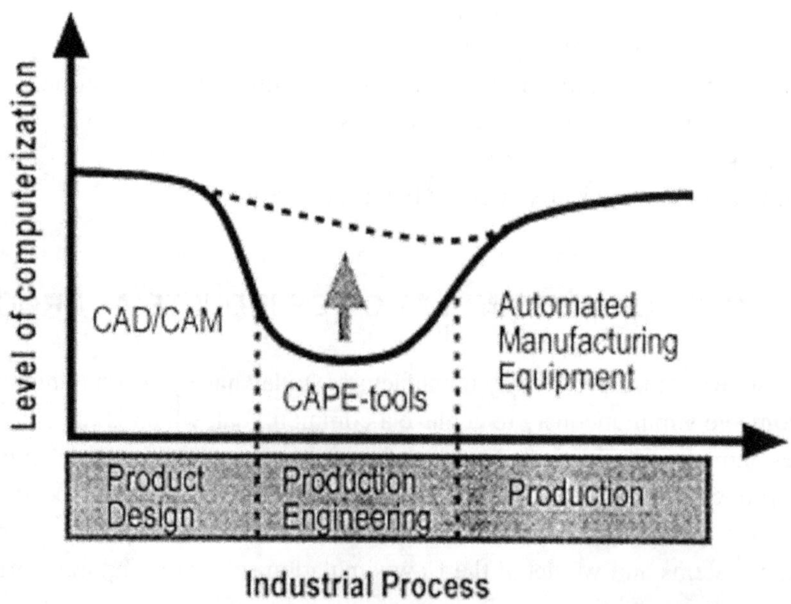

The main objective of our research to develop a system of interfacing that allows us to connect a flow simulation tool with a virtual reality system to be able to master the simulation models hiding all the technical aspects of the simulation through a data exchange protocol in real time and also the tools and techniques of virtual reality, and also for happen to implement a flow simulation tool interactive and immersive. In the remainder of this part we will describe the different tools and technologies of simulation, and in the second part of this chapter, we present various researches in virtual reality which treats industrial problems that will be discussed.

5.1. Tools of modeling and simulation « ARENA© »

Among the software tools most used in the modeling and simulation of industrial processes, we find that ARENA tool does not cease to evolve since the late 1990s. The ARENA modeling system is a flexible and powerful tool for industry analysts to create a simulation models for representing an industrial system. ARENA was published by (Takus *et al*, 1997); it is based on an object-oriented design to build a graphical model. The operation of this tool is simple; just we placed the graphic object called module in the ARENA scene to define the system components such as machines, operators and conveyors etc. ARENA uses the SIMAN simulation language to compile and run the models. After preparation of the simulation model, ARENA automatically generates the underlying model SIMAN to run in order to have the final result of the simulation. Graphical components used by analysts to create models are provided by the software. These components can also be tailored by the analyst to produce a modeling environment which is adapted to a specific field of application, the resulting collection of modules created by the user are contained within a model of application solution that can be shared by any users of ARENA.

5.1.1. Human machine interface of ARENA

ARENA contains a base collection of over 60 graphical components in the general context of the system. It was designed to provide modeling capabilities for all types of applications. ARENA offers modules specifically focused on aspects of manufacturing and industrial management, etc. For the manufacturing process, the tool provides specific components such as machine downtime, calendars of events, etc. ARENA interface consists of three panel modules: a common panel containing modules representing the fundamental process simulation such as departures, arrivals and service; support panel containing additional modules for actions and decisions; and the transfer panel, which is a panel of components that are used to model the transfers (or flows) of different entities throughout the system.

The animated 2D of simulation model is automatically included in the ARENA modules to allow a simplified representation of these models. To apply animations to a model, we can use the graphical symbols that are automatically provided when you manipulate an ARENA model and we can edit with the integrated module of the graphic tool. We can use also the icons in the graphics library of Arena or other external applications (e.g. clip art) to set up an animation of a simulation model.

5.1.2. Animation 2D with ARENA

The animation in ARENA can be launched simultaneously with the execution of the simulation or in post-processing mode. We create animations using the drawing tools of Arena, or from an AutoCAD

Figure 4. Human machine interface of ARENA

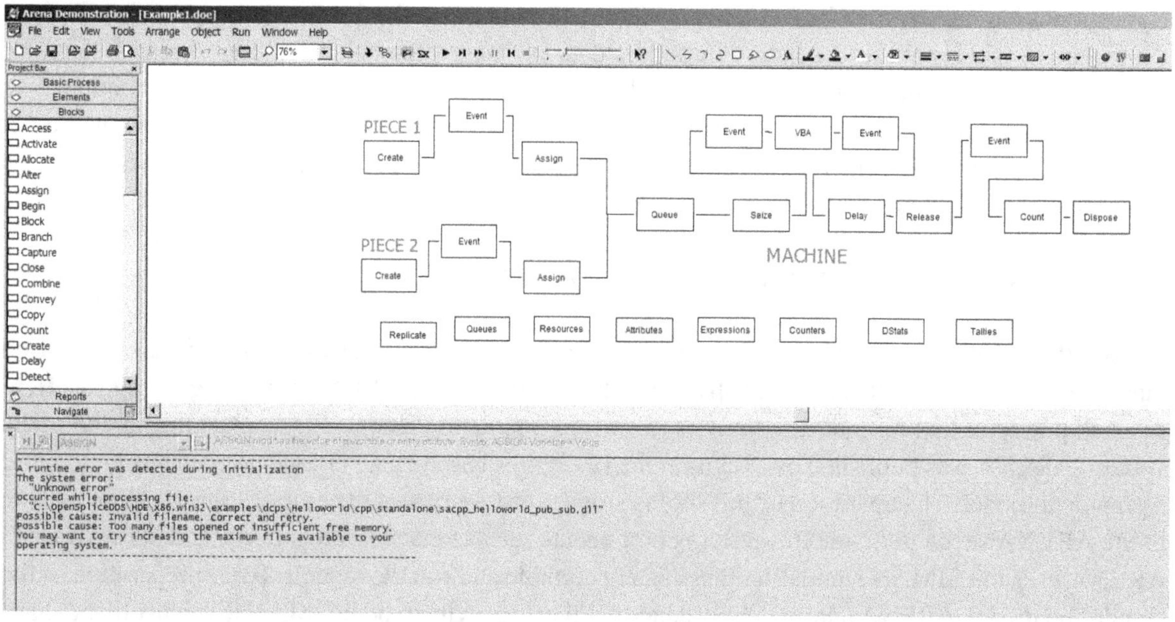

file or these animations can be created in other tools and imported into ARENA via Active X (Takus *et al*, 1997). Drawing tools of Arena include all standard objects such as rectangles, ellipse, arc, text, etc. and also provide a virtual unlimited palette of colors. The Arena interface was developed to allow direct import of CAD drawings and association of an animation and dynamic icons for the model. Once the drawings are imported into Arena, the user can enjoy the drawing tool integrated in ARENA for the implementation of the animation of model.

5.1.3. Custom development of AST'S with ARENA Professional Edition

The user can create his custom collection of graphical components using the features of the professional version of Arena. For example, an analyst in the automotive industry could develop an AST (Application Solution Template) personalized which allows simulating the welding lines.

5.1.4. The core of ARENA « SIMAN »

The technology which bases on ARENA is the SIMAN simulation language. The different modules of ARENA were created using blocks of modeling and components of the SIMAN technology. The modeling blocks of SIMAN are accessible to all users of ARENA. SIMAN offers to the user some flexibility and control of system logic. There are users who are used to program « with SIMAN » simulations directly in a text editor, they can still do that into ARENA. This tool offers a module allowing direct compile of the code which is an external file to the environment of the graphical modeling.

5.1.5. The Interactivity of ARENA with other tools

ARENA is a software complied with Microsoft Windows 95 and Windows NT. Arena was developed using the computer practices of Microsoft "MFC", ARENA is based on the object-oriented approach and is developed in C++. The open architecture of Microsoft allows us to integrate external data and even other applications with arena, via communication protocols such as TCP/IP, dynamic link libraries "DLL" or a data module Real-Time.

5.2. Tool of modeling and simulation « WITNESS »

WITNESS is a computer tool for modeling and simulation of production systems, among several other tools that are essential for industry. Following the flexibility and openness of simulation tools available today, we can now control all aspects of the organization of industrial engineering and manufacturing (Markt & Mayer, 1997). The WITNESS as well as several other software allow to simulate a factory from the initial design process to a fully functional factory. Witness is composed of several modules such as: MATFLOW (a system of flow planning) Witness Simulation Software, WITNESS OPTIMIZER. In the rest of this section, we will detail the different modules of Witness.

5.2.1. System of flow planning: MATFLOW

In the case where an industry is embarking on a conception of new installation of production system or redesigning an existing site, MATFLOW is a flow optimization system for production systems which allows us to manage the different logistic flows. This MATFLOW module (Markt & Mayer, 1997) is also integrated into AutoCAD, which allows us to save time and money on dealing with models of AutoCAD and then to export them directly to WITNESS. MATFLOW allows optimizing the configuration of the components by calculating the total debit of pieces. MATFLOW uses several techniques to manage the flow such as the calculation of pieces numbers with calculating their distance traveled, other characteristics of materials can be taken into consideration such as volume, weight or cost. So to summarize, MATFLOW allows to generate a simulation model, which can then be used to simulate the real production process and provided the necessary information of the operation of the system. MATFLOW uses "CAD layout" files, which are standard formats for the exchange of data between different tools in the field. This format provides a description of the different parts of the factory: calculation of production time, the size of batches for each machine and the calculation of the buffer size of the machines. AutoCAD allows us to prepare the factory model on a CAD layout plan, in the figure below the flows are represented graphically by arrows to indicate traffic of products. The optimization of CAD layout can be done manually or automatically. The manual method is to swap the positions of the machines or buffers; MATFLOW comprises an automatic optimization function that allows testing different scenarios.

In the case where an industry is embarking on a conception of new installation of production, the various parameters of the model are also transferable to WITNESS; they allow us to determine the capacity of the buffer lines and the machine capacity. Other features of MATFLOW allow the user to:

- Define groups of machines, or cells and units of productions or services.
- Represent the various inputs and outputs of machines.

5.2.2. Simulation study in WITNESS

Once the significant results of the optimization of the layout of MATFLOW are available, the simulation study process can be started using Witness, a real tool for modeling and simulation. The module MATFLOW supposes that there is a capacity for all flow-optimized. The next step is to evaluate the interaction between the different product lines that can share a resource. The optimization of manufacturing flow is provided by MATFLOW. WITNESS is used to simulate complete production systems. This allows industrial actors to design an installation to get a glimpse of how the production lines could work in reality. This is a good way to predict and solve any problems and inefficiencies that may be present on the production lines.

The Witness modeling module is able to model a variety of continuous and discrete events. The most basic elements of discrete modeling are: pieces, storage buffers, machines and conveyors. The pieces are simply objects that move from one machine to another on the production line. The buffers are areas of passive storage, where pieces must pass within a minimum delay and they cannot remain in the buffer longer. A piece can possibly be ejected from a buffer if it violates any of these conditions.

5.2.3. Graphical interface of WITNESS

The human machine interface of Witness is compatible with Windows. The main interface of the software consists of drop-down menus and toolbar. The operation of the simulation is controlled by a button bar below the screen (start the simulation; stop the simulation, etc…). The most activities of the simulation take place in the main window where we can find different elements of the simulation using the drag-and-drop mode, these widgets are predefined elements that can be used to formulate a simulation.

5.3. Tool of modeling and simulation « QUEST »

Today, all actors in the digital industry direct their research to the design and creation of tools that allows us to digitizing and simulate the industrial world. These montage technologies and virtual simulations allow to avoid or reduce the production of physical model thus to improve the quality and efficiency of montage and also to shorten the cycle of product development and to reduce the cost of production. There are several digitizing tools and industrial simulations such as « QUEST » by "Dassault Systems". In this part of chapter we present the particularity of this tool and some examples of work that uses it. QUEST is always proposed with DELMIA. The combination of tools DELMIA / QUEST is used to help the aerospace, automotive and other areas such as teaching in universities and to strengthen the culture of competitiveness in the aerospace industry (Bzymek *et al.*, 2008). QUEST offers a unique collaborative environment for industrial and manufacturing engineers, QUEST / DELMIA provides the best management practices workflow throughout the industrial process. QUEST / Delmia, we can create virtual factories and represented in 2D and 3D formats.

QUEST offers modeling tools (models predefined of machines, pieces models and other elements) that allows users to create a model of efficient simulation. Figure Below shows the human-machine interface of QUEST.

Using QUEST enables users to develop models for simulation and industrial installations, simulate process flows on these models and incorporate ergonomic constraint. The results of the simulation can be exploited in various graphics representation tools, so the data simulation can be extracted and exported

Figure 5. Human-Machine interface of WITNESS.

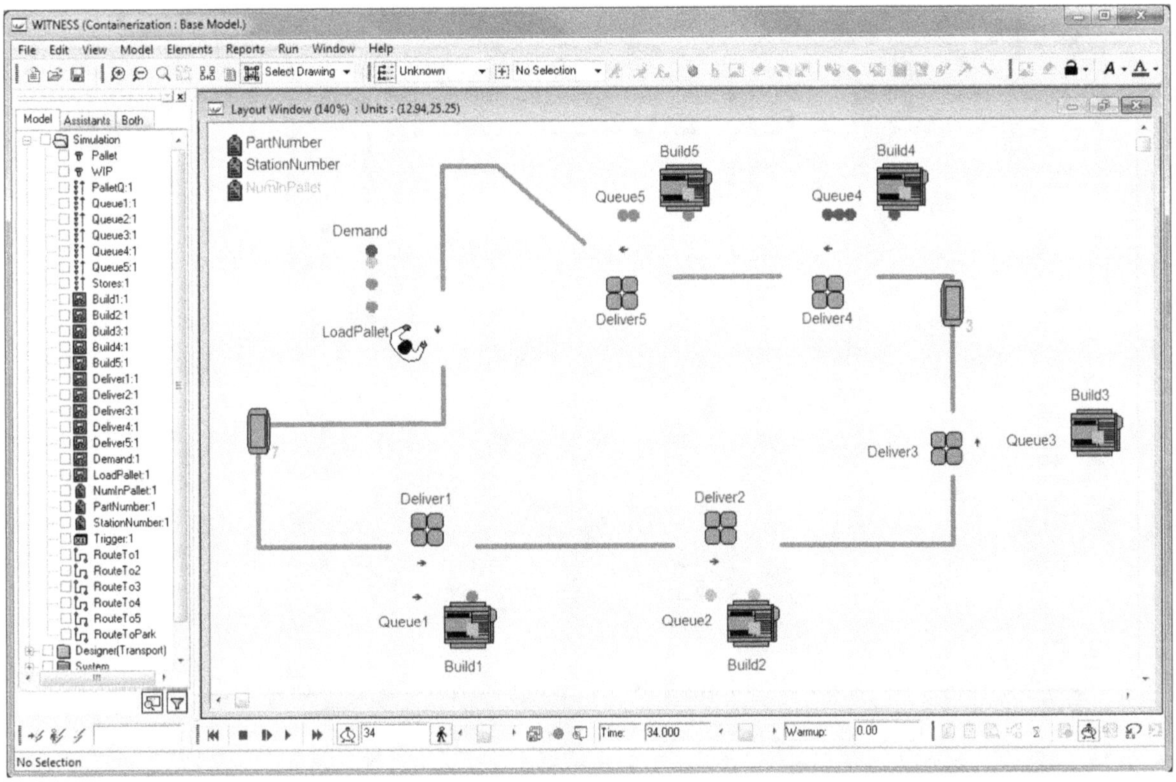

to various tools such as Microsoft Excel. QUEST provides possibilities to operate production variables in the virtual factory such as the duration of machining processes. These variables allow us to analyze the effects on different equipments of factory (Bzymek *et al.*, 2008).

The Simulation Control Language (SCL) and the Batch Control Language (BCL) can be incorporated into the simulation model to handle a complex production process. Full documentation allows us to use those languages that can be found in QUEST. Using this tool, users can view and prevent potential problems related to industrial flows, and improve existing processes. QUEST is powerful to design industrial models. In fact, it analyzes and visually represents data and complex manufacturing processes to facilitate the representation and the understanding of different manufacturing processes.

5.4. Reviewer and Our Proposed Approach

These tools of flow management are available today on the market and they presented many solutions for modeling, optimization and simulation of production processes. The users of these softwares have a particularly important role in this field: the experts can predict the evolution of charges of a production system throughout its operation, which allows to deduce several obstacles for simple users (René & Chevance, 2001). All these products tend to improve 3D rendering with each new version, confirming the market demand to disseminate the results to a wider audience. But they keep the following limitations:

Figure 6. Human-Machine interface of Delmia (Delmia, 2006).

- They require expertise in flow management, implementation and exploitation of results of flow simulation.
- Absence of interactions with the simulation: visual 2D represents the final results of the simulation in the form of an animated film. The only way to change the presentation is to re-configure the model of simulation.
- There is no consideration of human constraints (feasibility analysis and assessment of fatigue).
- There is no consideration of the dynamic behavior of the elements of the scene.

Our main objective is to develop a solution that combines a flow simulator with a virtual reality system while gradually refining the performance of the simulation model through dynamic data exchange using practical computings and virtual reality techniques. The challenge is to apply the concept of dimension I^3 "Interaction-Immersion-imagination" (Burdea & Coiffet, 1993) to develop an abstraction layer between the techniques of simulation and industrial actors on integrating virtual reality techniques. For this we propose to tackle this issue of flow simulation using virtual reality techniques to immerse the user in an industrial 3D scene that simulates different manufacturing processes.

6. VIRTUAL REALITY AND ITS CONTRIBUTION TO THE INDUSTRY

In this section, we will mainly introduce the virtual reality technology and its contributions to industrial design, numerical production, virtual training and the simulation of industrial processes through examples of previous research works that address issues related to the industry.

6.1. Virtual Reality

The Virtual reality as defined by Arnaldi, Fuchs & Tisseau (2003), is: « Scientific and technical domain exploiting computer science and behavioral interfaces to simulate in a virtual world the behavior of 3D entities that interact in real time with each other and with one or more users in pseudo-natural immersion by means of sensorimotor channels ». The immersion and interaction present the most important phenomena for applications VR, for this we should implement techniques which allow to immerse the user in a virtual world via sensory-motor interfaces whose purpose is to deal with virtual processes. The user receives a set of information through these sensory-motor interfaces such as: return immersive 3D on a big screen, force feedback through a haptic arm which reflecting the collision with virtual objects etc. The interaction with the virtual scene requires an interface of sensorimotor control as haptic arms for seizure of virtual objects, the motion capture system which can control a virtual avatar, data gloves etc. The VR technology is used in several domains such as video gaming, health, biology, military operations, education, learning and training etc. Today there are several types of virtual reality applications such as tele-presence applications for virtual conferences (Kantonen, Woodward, & Katz, 2010), the tele-operation (Tarault *et al.*, 2005) and industrial applications (e.g. validation ergonomic positions work (SIMCORE & CLARTE, 2013)), the immersive applications for visualization the specific results for the analysis of complex data and decision support (Eddy & Lewis, 2002).

Today, this technology became available for the industrial world, having been reserved for long time by major public research centers. Often considered as an extension of the simulation, the VR is distinguished by the fact that it offers a good level of immersion (display, touch etc.) to users. The VR allows the modern industry to simulate industrial processes in an efficient manner; it gives the opportunity to interact with the simulation model. A virtual simulation system includes a set of sensorimotor interfaces and controls the behavior of simulation in comparison with models programmed.

6.2. Virtual and Augmented Reality for Industrial Design

The aided design of virtual reality is a terminology of modern industry. This concept was introduced by Richir & Fuchs (2003). Today there are several research projects whose aim is to find solutions to meet the needs of virtual design. May be mentioned VR4D: Virtual Reality for Design of CLARTE (VR4D, 2013) which is a solution to help create virtual reality for design. The VR4D project presented a tool of virtual reality which allows a construction of small spaces because living areas integrated into land or air buildings represent an important industrial challenge. Several industrial organizations today promote the use of virtual reality systems in various stages of design and model validation. There are other projects which tackle the virtual design as VirtualiTeach (VirtualiTeach, 2013): This project integrates firstly a complete hardware platform based on the technologies of virtual and augmented reality, and secondly a set of software applications implementing an educational content inherent to the teachings concerned. The objective of this project is to follow up the design of mechanical parts and its car mounting. This is done by immersive situation in learning environments for visualizing the interaction with what is usually invisible such as the distribution of heat in a room with a coloring process.

Other researchers are using augmented reality technology to address issues of design and virtual simulation. Lee *et al.* proposes a design method of numerical production environments based Mixed Reality technology (Lee, Han, & Yang, 2011). This method allows integrating virtual components with others real, in a virtual production system with the aim of establishing a better configuration of a production system. Other researches targeted collaborative projects for the design in PLM environment (Product Life-cycle Management). This approach integrates knowledge based on engineering of industrial process; it uses multi-agent systems in platforms of virtual reality. Researchers at LSTB (Laboratory of Systems and Transport of Belfort) are working on the development of this approach. They propose an architecture for collaborative systems (Mahdjoub *et al.*, 2010) which allows aiding the design of industrial processes, and the analysis and simulation of these processes in a virtual environment are using techniques and tools of virtual reality. Lindskog et al. have used the mixed reality technology (virtual and augmented) and 3D imaging techniques to study the design of industrial systems (e.g. production lines and workstations) before their real implementation (Lindskog *et al.*, 2013). This allows evaluating the implementation of a new machine and its configuration before actually installing.

There are other types of systems that allow the prototype design. Choi et al. show in their paper a system for virtual prototyping (VP) for digital manufacturing of multi-material prototypes to facilitate rapid product development (Choi & Cheung, 2008). This system is based on an immersive environment "CAVE", whose objective is to allow a team of designers to analyze the quality of product design and to collaborate on their design work. Thanks to a simulation module, the system enables to evaluate and modify the models without any loss of materials prototyping. This type of system improves collaboration and communication of a team of designers working on product development.

Stark, Israel & Wöhler (2010), and Wiese *et al.* (2009) have developed a hybrid immersive environment for modeling virtual prototypes with VR technology. They noted that current modeling tools are only paper and CAD systems, even if they are complementary to some extent, there is still a lack of fluidity and intuitive interaction with the model. The figure below shows a designer that is about to model a prototype with the system of VR, and right figure shows the transformation of the model to a final prototype.

Today, there are several systems that support the model creation and modification in an environment of RA based on the use of interaction tools in 3D. Construct3D (Kaufmann & Schmalstieg, 2003), is a project developed for mathematical and geometric uses. The constructions of primitives can be created and displayed in a 3D space, as shown in the figure below.

Other environments of computer aided design and AR such as ARCADE, have been developed to facilitate the interactive design of model as shown in figure below (Ng, Ong, & Nee, 2010), where users can create new designs by modifying and combining virtual and real objects. The user can create and modify 3D models using methods of interaction based on markers. This design approach allows users to draw models; it's visualized and contextualizes then in a real design space. Ng, Oon, Ong & Nee (2011) have presented a design environment based on motion capture of fingers in the context of AR. Using gestures of fingers, the designer can visualize and manipulate a 3D model in an environment of AR, also evaluate the design and make changes that will be taken into account in the model as shown in the

Figure 7. Design of VR prototype (Stark, Israel & Wöhler, 2010).

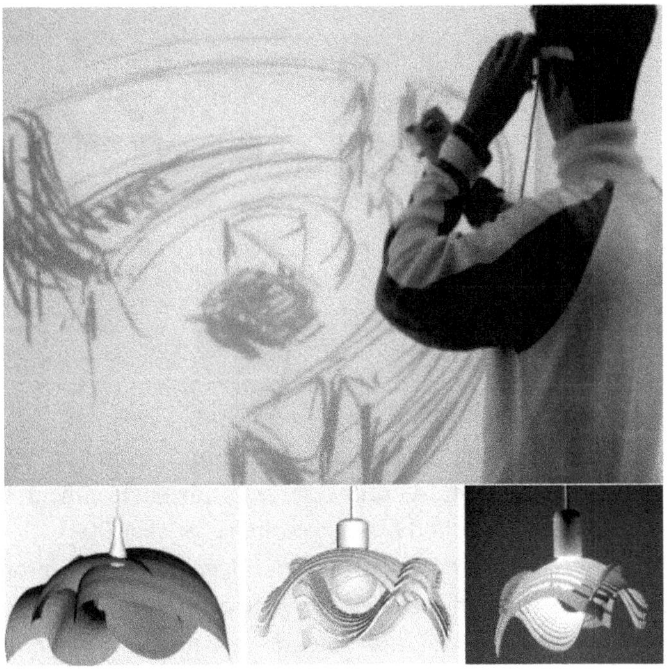

Figure 8. The system Construct3D (Kaufmann & Schmalstieg, 2003).

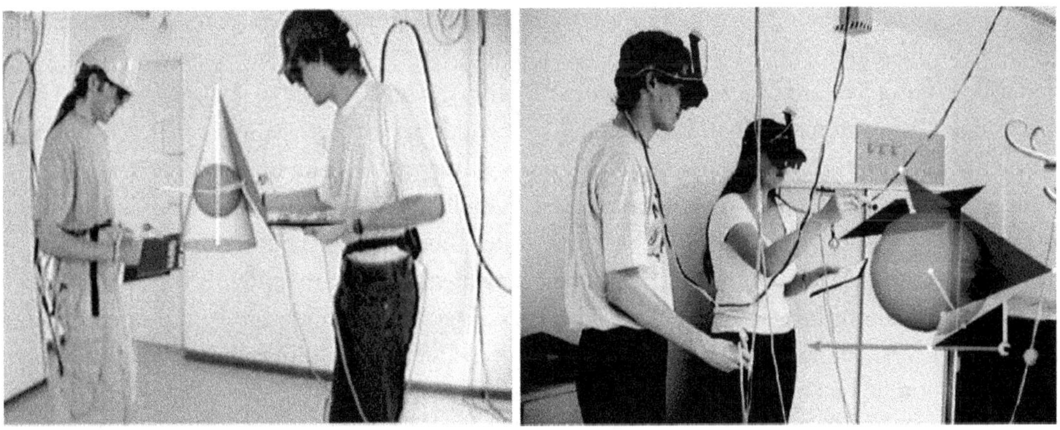

figure below: GARDE. This allows the user to contextualize the model in the work environment and to take into account the spatial information of the real stage for the implementation of the model. GARDE is a prototyping tool augmented and it benefits at the same time the benefits of a physical and virtual prototype, such as realism of the model, the ease of modification and contextualization.

In industrial environments textile type, we need an accurate representation of the virtual model, the virtual shadows can improve communication between users by providing locations and gestures of other users in a collaborative design system. The shadows do not allow us to locate the vertical position and

Figure 9. AR-3D modeling (Ng, Ong, & Nee, 2010) (Ng, Oon, Ong & Nee, 2011).

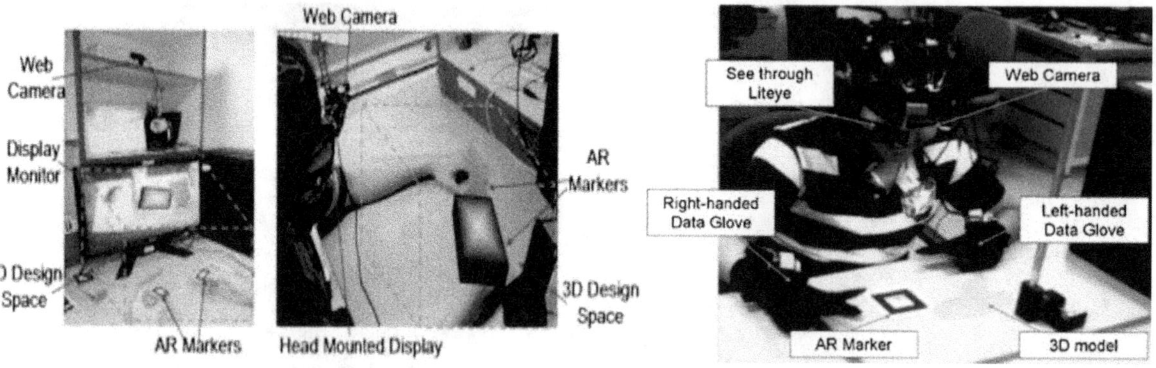

can only provide a direction approximate of pointing. Therefore, it is not easy for the user to choose a specific characteristic of a complex model. A collaborative design environment, including the visualization features VR and AR has been suggested by Chryssolouris *et al.* (2009).

To cope with the competitiveness in product design, Pouliquen *et al.* (2007) propose a solution based on virtual reality techniques. The researchers have been worked on the modeling of virtual hands to interact with models of a virtual environment; the authors integrate the requirements of health and safety. In his work, Pouliquen shows the design techniques of virtual hands which are coupled with gloves equipped with a motion capture system and a device of force feedback. With this system, the operator can interact with the models of a virtual environment in real-time.

Some specific searches to the design of industrial cables conducted by Raymond, James, & Graham (2009) led to the development of a VR system for modeling virtual cables. This system is called COSTAR (Cable Organization System Through Alternative Reality), which allows to design, assemble the plans of cables, to analyze automatically the design activities and to generate assembly plans in an immersive environment of VR. Bal & Hashemipour (2009) propose a methodology based on virtual reality to improve the process of design and implementation of holonic control systems in manufacturing practice. The main objective is to implement a holonic controller in small and medium-sized of manufacturers. As part of this research, Bal *et al.* have launched a case study in a manufacturing factory format of molding with a medium pressure. Thanks to the virtual factory, architecture of holonic command for a manufacturing system has been implemented for the dynamic scheduling of machines in a factory of average dimension, to ensure of the robustness necessary to treat the productive disorders.

6.3. Virtual and Augmented Reality for Numerical Production and Virtual Training

To manage the various processes of industrial production, industry actors are now using a new generation of technologies and innovative tools. Several researches in virtual and mixed reality address the issues related to industrial production in general, as well as visualization, manipulation, presentation and use of production models. Mention may be made to one work in this area, a project launched by the electronics giant Canon which is based on mixed reality technology (MR System, 2012). This work presents a system to aid the design and development of new industrial products, it allows designers to visualize

Figure 10. VR module for the control of transport operators (Bal & Hashemipour, 2009)

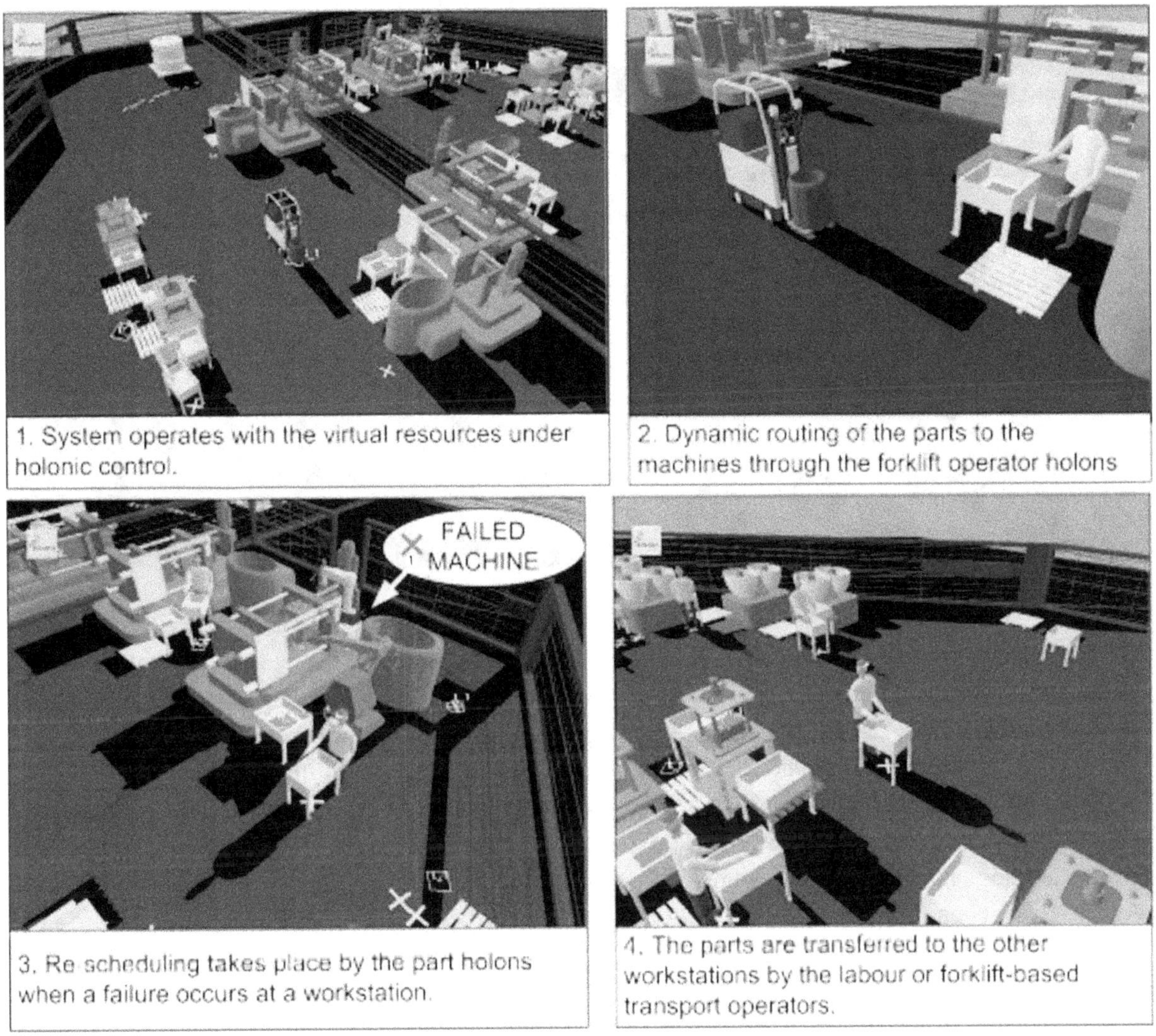

and interact with the product in collaboration with virtual three-dimensional objects placed in a real environment. The main objective of this system is to reduce the number of prototypes needed to develop a new product for mass production while compressing costs and minimizing environmental impact.

One of the key issues in using virtual reality for the simulation of complex industrial sites is to ensure the consistency of the models used at the different steps of the design process. Furthermore, reusing these models for the virtual training is needed to support various learning scenarios. Designing these environments independently to existing CAD models would result in unrealistic costs and extra workload. Recent works has been done to propose a unified approach of system modeling that ease the reuse of existing models for the production of informed virtual environments (Chevaillier *et al.*, 2011). This approach makes the virtual environment the single source of information, used (1) to support the real-time simulation of the environment, (2) as the shared knowledge-base of the virtual agents (e.g. Operators on a production line), and (3) as learning resources for educational agents (Marion, Querrec & Chevaillier, 2009). In this approach, not only geometrical files, but also the semantic modelings of

Figure 11. Canon MR System (MR System, 2012)

the industrial process (including procedures) are processed as input data. It is then possible to design non-specific VR simulation platforms and to develop generic educational agents as in (Buche, Bossard, Querrec & Chevaillier, 2010). This model-based approach is based on the standard modeling language SysML (Querrec, Vallejo & Buche, 2013). This approach has been successfully applied to the design of next generations of aircraft carriers (Marion, Septseault, Boudinot & Querrec, 2007). It was is possible to evaluate different configurations for the decks and to estimate system-level capacities and to validate takeoff, landing and refueling procedures before the building of the vessel has started. Very similar approaches have been followed by (Barot *et al.*, 2013), (Edward *et al.*, 2008) and (Gerbaud et al., 2008).

Other works deal with the problems of process planning and the production lines. The Boeing researchers have used the mixed reality technology for the implementation of a solution to assist in the design of cockpits (Caudell & Mizell, 1992). In general, such researches allow optimizing the methods of industrial production. This can be done by allowing for example operators to easily understand assembly procedures by displaying information on parts with images that indicate the order of assembly in a manufacturing line. As regards to the introduction of a finished product, the augmented reality provides the important techniques for prototyping process. For example, in the automotive industry, this technology has been used to evaluate the interior by superimposing different car interiors, which is generally available only with 3D models in the initial stages of development, on the model of the actual car (Fründ *et al.*, 2005).

In 2000, research conducted by Duffy & Salvendy (2000) has been presented as the first studies that address human and organizational aspects of the development of global products and virtual training in industries of manufacturing and services. Other works launched by Ladeveze *et al.* (2010) which present an interactive system of VR destined primarily for training. This system allows us to have a haptic real-time guidance for the assembly of CAD objects or tasks dismantling. The principle is to guide the operator through an artificial haptic orientation or choose not to follow the path and automatically search new routes for the purpose of training operators for assembling mechanical components of ALSTOM.

To address the needs of learning and training using simulation techniques in virtual reality, research launched by Manca, Brambilla & Colombo (2013) shows training approach based on virtual reality. This approach allows to improve the cognitive preparation of operators addressing the three components of the consciousness of situation (Experience: to master the factory and these units; comprehension: understanding their meaning and purpose, learning: how to project the current situation in the near future). This work discusses the benefits of integration and interconnection of a dynamic process simulator with a dynamic crash simulator to train operators (particularly field operations) to master the abnormal situations and to recognize the anomalies and malfunctions. The author has implemented this research in a deployed application on the VR platform as the figure below shows.

6.4. Virtual, Augmented Reality and the Simulation of Industrial Processes

The use of existing tools today for the simulation of various industrial processes, requires a good knowledge and expertise in the relevant field. Many industrial actors do not have this purely technical expertise. Therefore to allow them to participate in the analysis, optimization and validation of simulation models of different processes, many researchers sought to develop innovative solutions that allow for an abstraction between the methods and simulation techniques and non-technical stakeholders. Dangelmaier *et al.* (2005) propose a VR system to visualize a 3D simulation of a manufacturing line as shown in figure below. The objectives of this system are to assist users in the design phase of production models, validate these models and help optimize a production system.

There are other works that deal with specific issues such as production simulation in aerospace. Caggiano & Teti (2012) propose an innovative solution for digital simulation. The main feature of this solution is to aid the design and simulation of an existing cell of manufacturing components of aircraft engines to be improved by automated robotic deburring. Why the researchers put in place an application of 3D motion simulation is to illustrate the design of handling systems of materials. This system uses a discrete event simulation to analyze the different scenarios and improve the performance of the manufacturing cell.

Figure 12. The operator is immersed in the virtual environment (Manca, Brambilla & Colombo, 2013)

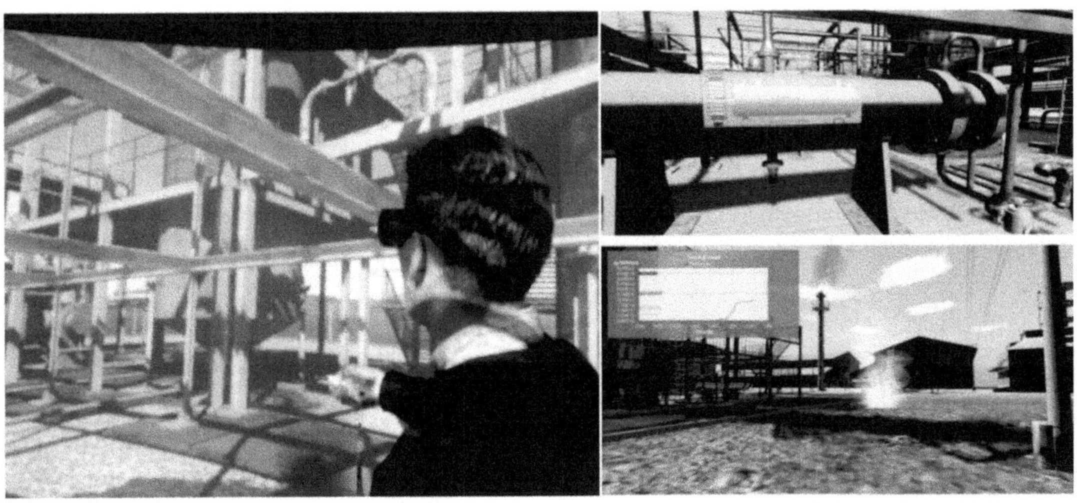

Figure 13. Visualization a simulation of a manufacturing system (Dangelmaier et al., 2005)

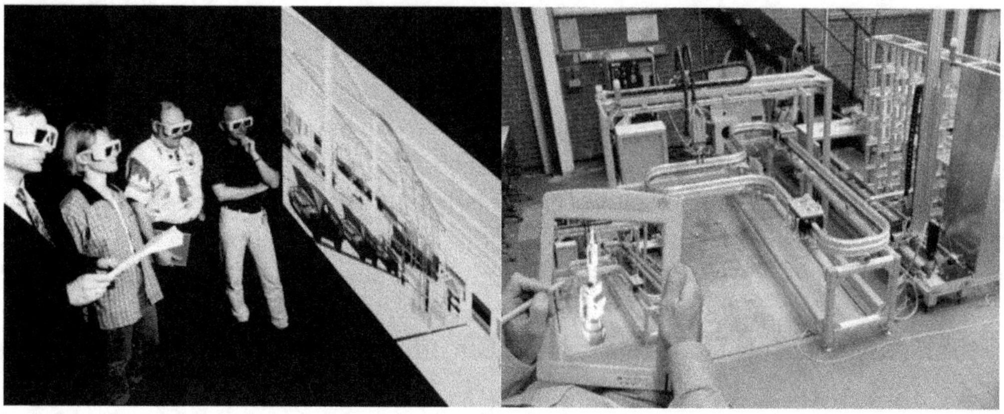

Multiple approaches and specific methods of design engineering and simulation of production processes are possible today thanks to various technologies. Tolio *et al.* (2013) seek to develop a new concept that offers the possibility of integrating various tools for design, simulation and analysis in a single solution. This concept allows supporting interoperability between these tools throughout the life cycle of production. The researchers in the field of architecture are generally concerned with solutions of simulation and learning of construction modes. Goulding *et al.* (2012) propose a VR solution for the simulation and learning of the most important techniques in construction of buildings. The main goal of this solution is to provide a work environment that allows different actors of industry (e.g. project managers, construction managers, architects, designers, suppliers and manufacturers) to have a multidisciplinary learning via a VR simulation of a construction site.

Lee, Han & Yang (2011) work on a simulation system based on mixed reality to simulate industrial operations in a virtual factory. This system enables manufacturers to launch virtual operations in order to analyze the feasibility of a solution in a real context and see their evolution over time. Leu *et al.* (2013) present in his article a state of the art which describes the various methods used to develop CAD models and the systems of simulation for the assembly of prototypes, planning and training. The author also present methods for the generation of CAD models from digital data acquired. This concerns the capture of movement, the assembly modeling, as well as human-machine interfaces. Leu et al. introduced a method for exchanging data between a CAD system and a VR system. This article described the implementation of these methods and provides examples of application of simulation based on a virtual model for the assembly of prototypes, planning and training. In a recent study addressing the treatment of simulation and control of different stages of PLM, Fillatreau *et al.* (2013) propose a solution of virtual reality that allows manufacturers to navigate in a 3D scene, and manipulate virtual components, whose purpose is to simulate and validate the various operations of a stage of PLM. The main objective of this system is to enable companies to improve the traceability of project reviews, decisions and documentation in all phases of the PLM.

Mujber, Szecsi & Hashmi (2004) propose in their article a classification of industrial applications VR, according to which there are three groups: design (modeling and prototyping), operations management (planning, simulation and training) and manufacturing processes (machining, assembly and inspection). This demonstrates a wide range of demand of VR in the industry. Today the VR is far from reaching all stages of PLM (e.g. validation of product requirements, validation of simulation models, simulation of flow, acceptance of products: are not really addressed by VR applications). The existing studies in the literature that address the integration of VR in different stages of PLM always focus on a particular stage.

Today, in the current market there are tools of simulation and planning of industrial platforms that are already marketed and use 3D technology. These systems are often used for the simulation of the fabrication and the implementation of a plant of factory in an industrial platform. We present in the figure below three commercial products, Tecnomatix Factory FLS (2013) by Siemens, Teamcenter Manufacturing Plant Simulation (2013) by UGS and MPDS4 Factory Layout (2013) by CAD Shroer. These products have a good market share in industry. All systems of VR quoted in this part of state of the art are quite similar because they all provide a visual platform with a 3D rendering. These tools are designed to meet the following industrial needs: design of prototypes, validation of maintenance process and machining, the validation of the ergonomics of various industrial operations, simulation of plans of the furnishings, the simulation of the manufacturing process, the evaluation of the establishment of industrial components in a virtual platform, the simulation of machining and assembly operations etc. These systems 3D simulation are used to visualize and refine the theoretical results before setting up the actual process.

Several researchers are beginning to implement the results of their research in the design and evaluation of manufacturing systems through augmented reality applications. Gausemier & Frund (2002) have established a system of AR which allows designers to define and evaluate their models of manufacturing systems utilizing a webcam and markers in a real environment, as shown in the figure below. Other similar systems are presented by (Doil & Schreiber, 2003). These AR systems follow the same concept of VR systems.

Marc, Belkacem & Marsot (2007) propose a method for validation of simulations, presented as a tool to three layers (Simulator, Simulated, Simulation situation), the tool consists of a projection system and a force feedback interface. The validation method of the proposed simulations in this paper is developed in a complementary module of the system; the internal validation is based on the capability of the simulator to reproduce the features of the real system. The validation includes an apparent validation (realism control) and functional validation (controlling the behavior of real technical system).

In the world Aerospace, Harris & Morgenthaler (2004) present with their team of experts in visualization 3D their development of simulation tools for virtual prototyping. For realizing the tests on flight operations and visualize in real time the design and planning of the missions of aerospace as the two figures below shows. There are other innovative solutions using immersive 3D visualization to simulate planning, optimization of phases of technical design of future aerospace missions.

Takeda *et al.* (2008) have developed a simulator for the maintenance system remotely using software of robotic 3D simulation. This simulator is connected to a manipulator control system that was developed

Figure 14. Simulation tools of manufacturing processes

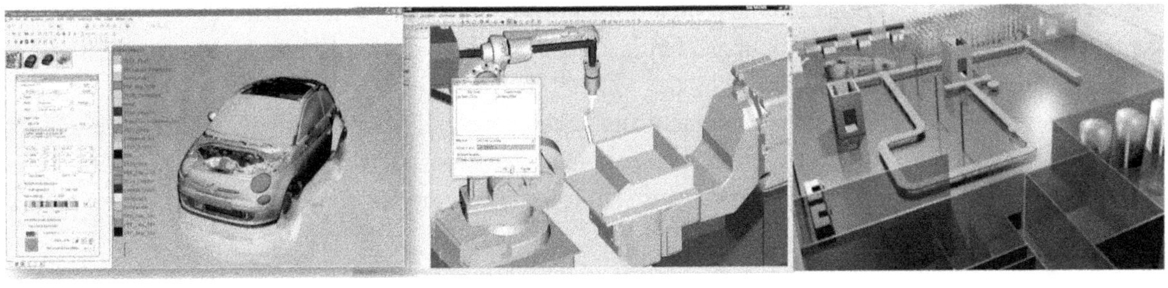

Figure 15. Design of manufacturing systems in AR (Gausemier & Frund, 2002)

Figure 16. The functioning of deployment of Titan IV / Milstar (Harris & Morgenthaler, 2004)

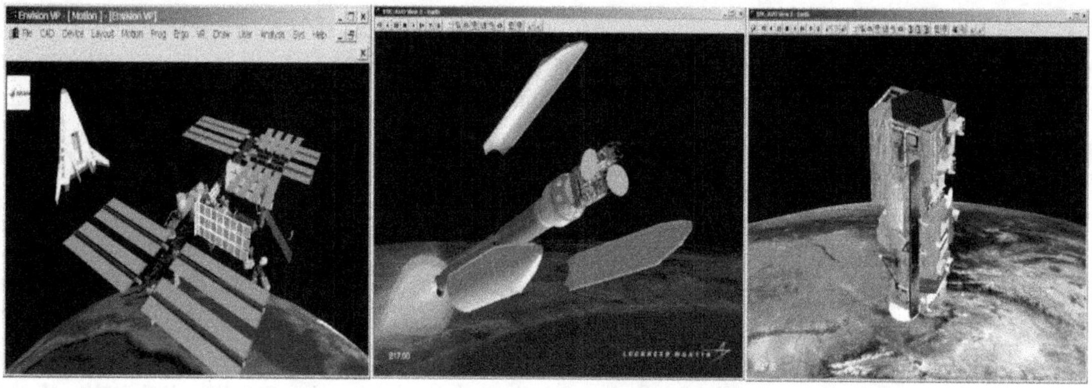

as part of the maintenance system. It allows reconstructing positions of the manipulator using position data transmitted by the engine through a local network. It can also be used for operator training. In the context of research in the field of design and numerical simulation, Xinhua, Youhui & Qi (2011) propose a system of virtual montage that is based on 3D simulation tool « Delmia ». The technique is to create a virtual prototype and simulate the whole assembly process in a virtual environment « Delmia ». This system allows avoiding or reducing the production of physical models, therefore it can not only improve the quality and efficiency of the montage, but also to shorten the cycle of product development and to reduce production costs.

The design process of automated manufacturing systems typically involves physical prototypes, which are expensive, to validate the interactions between hardware and software components. Weber-Jahnke & Stier (2009) propose a VR system which aims to improve the prototyping process by making virtual prototypes. The approach proposed in this paper presents a combination of a VR engine capable of simulating in 3D laws of physics rigid bodies with industrial modeling software to control the prototype simulated using virtual sensors.

Today, virtual reality tools for the simulation of manufacturing processes present a lack of realism and interaction. Despite the remarkable progress in the development of algorithms for collision detection for convex objects, only the collision detection algorithms for non-convex objects remained limited. Tesic & Banerjee (1999) propose a methodology to encapsulate the characteristics of virtual objects for collision detection and automatic procedures for creating virtual objects and their separation into convex pieces. This methodology is suitable for interactive simulation, it can be used in many contexts such as: the virtual assembling, the simulation of industrial robot, also in processes where the modeling precise has a high level of collision detection such as robotic paint, robotic welding and the operation of machining to numerical control.

There are other sectors of industry that requires the use of techniques of simulation and evaluation such as construction. According to Li, Chan & Skitmore (2012), the construction is probably the most dangerous sector in Hong Kong, accounting for 76% of fatal accidents in the region, twenty times more than any other industry. Li *et al.* proposed a safety assessment method presented by a 4D interactive evaluation system that provides an improved security technique. This system allows interacting with risky scenarios and a range of possible operations for treaties, the method provides a grid of evaluations of operation chosen by users.

The strategic constraints of health type and legality in the industrial world are pushing researchers to study and simulate the noise of machinery in a factory environment. Aurich *et al.* (2012) proposed a new concept to study the level noise in an industrial context, the establishment of a method of virtual reality using the acoustic measurements of a real factory to have a realistic simulation. This approach allows to reduce the downtime during the improvement process and to test the different scenario of improvement. To represent the simulation results, the researcher uses a CAVE (cave automatic virtual environment). This system provides modules to check the noise level taking into consideration the various regulations.

7. CONTRIBUTION TO THE SIMULATION OF VIRTUAL MANUFACTURING

The main contribution of this chapter is to present an innovative approach, also an architectural design of our VR system which allows simulating the flows of production, to estimate the states of a manufacturing process, to validate the operations of assembly and to interact in real time with a simulation of the considered manufacturing process.

This system is implemented in our VR platform, where the user will be totally immersed in a virtual scene in 3D. This system allows the user to move in the virtual stage and to change the point of view of the virtual camera through the capture of movement. To manipulate and interact with the objects in the virtual scene we use the VR joystick. The interactive VR system is connected to a flows simulator as shown in figure below (Figure 22), which allows to visualize a simulation of flows of production in the course of execution and to interact with it by changing its configuration and its parameters on the VR platform. This system democratizes the use of simulation tools by rendering accessible to any industrial actors, the techniques and methods of the simulation of production systems. Such an approach makes it possible for the industrials to have a global vision of their systems of productions to estimate the weak points so as to be capable of bringing corrective actions.

7.1. Architectural Design

The virtual reality environment is a mixture of sensory-motor systems, a 3D technology and the development of computer programs. The development of an innovative system based on virtual reality requires the design of an architecture that describes the different parts of the system, the overall integration of modules and devices of the virtual reality platform and ensures interoperability in the system.

We present in figure below an architectural diagram that summarizes the overall structure of the system to achieve. This architecture allows the system to be broken down into four main parts: the simulator of production flow, the data exchange protocol, a 3D modules and a platform VR.

- *Production flow simulator*: The industrial engineers use specific tools to design and simulate the production process. In our approach, the Arena tool will be integrated into the virtual reality system proposed, in order to have immersive and interactive simulations. The integration of this tool requires the development of a communication layer between Arena and our VR system.
- *Protocol for data exchange*: For a good analysis of the results of a simulation of flow through the VR system, it is necessary to develop a communication protocol to ensure data exchange in real time between the Arena and the VR system tool.
- *Modules 3D*: The implementation of an immersive and interactive simulation in a VR environment requires the development of processing modules of virtual objects (Moving avatar, Following the axis of camera view, handling 3D objects via Flystick, Menu-virtual, animate 3D objects and interaction-scenario). These modules run around a 3D engine (e.g. Unity3D, Virtools).
- *VRPN controllers*: it's the software part of the platform VR. This section integrates the various controllers that allow connection between devices VR and 3D engine. The development of these controllers is based on the framework VRPN (Virtual Reality Peripheral Network: This frame-

Figure 17. Architecture of our VR system for the simulation of production flow

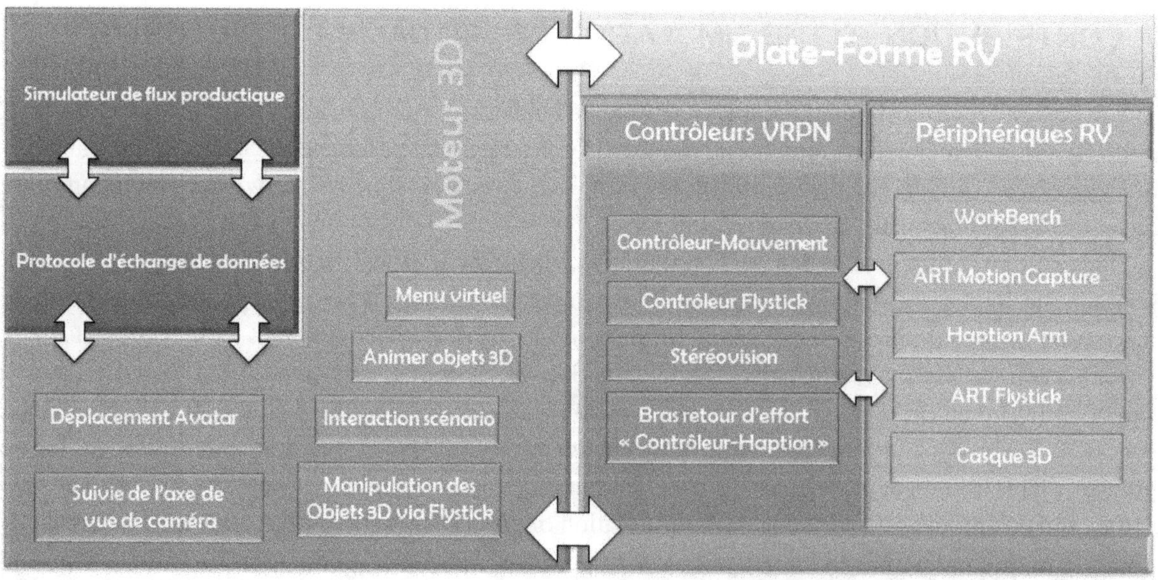

work is used to connect devices of VR in a cluster). For example the "Controller-Movement" has a connector between the "Motion Capture ART" and the 3D engine, this connector is used to transmit and represent the motion capture data in the 3D engine via the network.

- *Peripherals VR*: This block integrates the sensorimotor part of the platform VR. This module understands the different devices and their connections.

7.2. Functioning of System

The architecture presented above provided a comprehensive overview of the system that we want to develop. This architecture also presents the different possible connections between the various system modules. This system will allow to immerse the user in a virtual 3D simulation and to interact with it via the peripherals of VR platform. The application allows the user to move in a virtual environment using a motion capture system (sensors and IR cameras). For the capture and manipulation of industrial objects in a virtual environment, you can use a Flystick (of ART) or a Wiimote. The main feature of our system is to run a simulation of production flow in a virtual reality scene. To this end, we developed 3D processing modules that allow to represent the results of the simulation tool (e.g. Arena or Wintess) in a 3D environment through a transmission module and data synchronization. This system allows for more interaction with the 3D environment to edit and re-setup the simulation run by the engine of the simulator tool. The interaction with the 3D scene can be performed using a haptic force feedback (e.g. Haption). The use of this type of interface provides indeed a good fluidity of interaction with virtual objects and adds sensation to the experience of virtual simulation.

7.3. Technical Environment

The development of this approach requires the use of software and hardware tools specific for virtual reality. We present in this section the main tools that are used throughout this project.

7.3.1. Hardware Tools of VR Platform

The VR platform of our laboratory includes the following elements:

- *WorkBench*: it is a virtual reality device that allows visualization in relief on a dual-screen device as shown in figure below. This equipment allows immersing the user in a virtual reality experience of high level. This device facilitates the work on virtual prototypes for the functions of formation, assembly of technical parts or the assistance in the maintenance. The Workbench consisting of two perpendicular screens, including a horizontal, is located approximately 1 m from the ground.
- *ARTrack "motion capture system"*: designed by the ART Company. This system consists of four infrared cameras, an outfit "mocap" with a set of retro-reflective passive markers (see figure below) and a controller computer. This concept allows recovering 3D position and orientation of all markers in the scene of the VR platform. To control virtual objects, the system offers a Flystick equipped with buttons, two analog axes for orientation and an active marker for its location in the real environment.

- *Virtuoso 6D*: "Haptic arm" the system is designed by the company "Haption", it consists of a computer controller and an arm of force feedback. This device is a bidirectional sensorimotor, it allows to user to exert a force on the arm and also to have a force feedback following an action.

7.3.2. Software Tools of VR Platform

For the development of our system, the following software will be used: Arena®, Unity3D® and Microsoft Visual® C ++, as well as specific libraries for virtual reality as VRPN for the development of different modules of the system.

- *Arena* ®: software of design and simulation of industrial production process, this tool is based on the SIMAN engine to simulate processes with discrete events.
- *Unity3D* ®: This is a 3D engine, it can be called as: a development package of 3D environments. This tool will allow us to develop industrial 3D scenes, the modules of interacting with objects in the scene and rendering simulation scenarios.
- *Microsoft Visual C ++*: software for development projects in computer language C ++. It will be used to develop interfaces between devices VR and 3D engine. These interfaces will allow us to generate a layer "DLL: Dynamic Link Library", which establishes a connection between Unity3D and VR platform.

Figure 18. VR Platform at ENISE.

8. PERSPECTIVES

In recent years, several research works have been done to virtualize certain industrial processes such as simulation of assembly of pieces, simulation of production flow etc. To deal with such problems, we encountered some technical and scientific obstacles such as: the development of a data exchange protocol between a simulation tool and a VR system, the choice of a part of the simulation to virtualize, the behavioral programming operation of a virtual factory and the coupling between the UML diagrams and the others based on discrete events. Our work was part of a project called Optiflux in collaboration with Renault Trucks. This company is interested in the results of this work and begins to run the simulation to virtualize their production lines.

For our future work, we aim to develop a virtual reality system for online simulation. This system will connect a platform VR with an industrial information system (e.g. ERP) and a central system of control/command of machines. The objective of this system is to provide real-time the current operation of the factory, to analyze and simulate this over time and to take control over the machines of different production lines for reconfigure and improve the production processes.

9. CONCLUSION

Today, virtual reality is used in many industrial settings such as training, decision support, helping with maintenance; simulation of industrial operations, etc. In this chapter, a new approach is developed for popularizing the manipulation of simulation of industrial processes. This approach allows for the user to visualize and interact with an industrial simulation, and it allows for the industrial actors to visualize the functioning of their factories with a 3D visualization. This approach is based on the VR technology and the behavioral programming; this has enabled us to develop an immersive and interactive VR system to bring the user into a 4th dimensions "sensation" via specific peripheral devices of VR. The main goal of this work is to enable different industry actors to participate in the analysis and improvement of industrial processes.

REFERENCES

Abdulmalek, F. A., & Rajgopal, J. (2007). Analyzing the benefits of lean manufacturing and value stream mapping via simulation: A process sector case study. *International Journal of Production Economics*, *107*(1), 223–236. doi:10.1016/j.ijpe.2006.09.009

Achanga, P., Shehab, E., Roy, R., & Nelder, G. (2006). Critical success factors for lean implementation within SMEs. *Journal of Manufacturing Technology Management*, *17*(4), 460–471. doi:10.1108/17410380610662889

Ait Hssain, A., & Dunod, E. D. (2000). *Optimisation des flux de production: méthodes et simulation.* Dunod.

Arnaldi, B., Fuchs, P., & Tisseau, J. (2003). *Traité de la réalité virtuelle*. Les Presses de l'Ecole des Mines de Paris.

Aurich, J. C., Yang, X., Schröder, S., Hering-Bertram, M., Biedert, T., Hagen, H., & Hamann, B. (2012). Noise investigation in manufacturing systems: An acoustic simulation and virtual reality enhanced method. *Journal of Manufacturing Science and Technology*, *5*(4), 337–347. doi:10.1016/j.cirpj.2012.09.010

Baglin, G., & Capraro, M. (1999). *L'Entreprise Lean Production ou la PME compétitive par l'action collective*. Presses Universitaires de Lyon.

Bal, M., & Hashemipour, M. (2009). Virtual factory approach for implementation of holonic control in industrial applications: A case study in die-casting industry. *Robotics and Computer-integrated Manufacturing*, *25*(3), 570–581. doi:10.1016/j.rcim.2008.03.020

Barot, C., Lourdeaux, D., Burkhardt, J.-M., Amokrane, K., & Lenne, D.BAROT. (2013). V3S: A Virtual Environment for Risk-Management Training Based on Human-Activity Models. *Presence (Cambridge, Mass.)*, *22*(1), 1–19. doi:10.1162/PRES_a_00134

Bel, G., & Kieffer, J. P. (2002). *Pilotage assisté par la simulation discrète, Méthodes du pilotage des systèmes de production, 99 - 127*. France: Hermès Science Europe Ltd.

Berchet, C. (2000). *Modélisation pour la simulation d'un système d'aide au pilotage industriel*. France: National Polytechnic Institute of Grenoble.

Buche, C., Bossard, C., Querrec, R., & Chevaillier, P. (2010). Pegase: A generic and adaptable intelligent system for virtual reality learning environments. *International Journal of Virtual Reality*, *9*(4), 1–13.

Burdea, G., & Coiffet, P. (1993). *La Réalité Virtuelle*. Paris: Hermès Sciences Publications.

Bzymek, Z. M., Nunez, M., Li, M., & Powers, S. (2008). Simulation of a machining sequence using delmia/quest software. *Computer-Aided Design and Applications*, *5*(4), 401–411. doi:10.3722/cadaps.2008.401-411

Cad Shroer MPDS4 Factory Layout. (2013, August). Retrieved from: http://www.cadschroer.com/index.php?screen=1&ziel=ProductsMPDS&thema=mask1.php&id=351&land=com

Caggiano, A., & Teti, R. (2012). Digital Manufacturing Cell Design for Performance Increase. *Procedia CIRP*, *2*, 64–69. doi:10.1016/j.procir.2012.05.041

Caudell, T., & Mizell, D. (1992). Augmented reality: an application of heads-up display technology to manual manufacturing processes, *Proceedings of the Hawaii International Conference on Systems Science, HI, USA*, 659-669. doi:10.1109/HICSS.1992.183317

Chevaillier, P., Trinh, T. H., Barange, M., Devillers, F., Soler, J., De Loor, P., & Querrec, R. (2011). Semantic modelling of virtual environments using MASCARET. *Proceedings of the Fourth Workshop on Software Engineering and Architectures for Realtime Interactive Systems, IEEE VR*, Singapore.

Choi, S., & Cheung, H. (2008). A versatile virtual prototyping system for rapid product development. *Computers in Industry*, *59*(5), 477–488. doi:10.1016/j.compind.2007.12.003

Chryssolouris, G., Mavrikios, D., Pappas, M., Xanthakis, V., & Smparounis, K. (2009). In L. Wang & A. Y. C. Nee (Eds.), *A Web and Virtual Reality-based Platform for Collaborative Product Review and Customisation* (pp. 137–152). Collaborative Design and Planning for Digital Manufacturing. doi:10.1007/978-1-84882-287-0_6

Claver, J.-F., Pitt, D., & Gelinier, J. (1996). *Gestion de flux en entreprise: modélisation et simulation.* France: Hermès Science Publications.

Cordier, M. O. (2006). Comparing diagnosability in continuous and discrete-events systems. *In Proceedings of the 17 the International workshop on principles of Diagnosis.* 55-60.

Crute, V., Ward, Y., Brown, S., & Graves, A. (2003). Implementing Lean in aerospace-challenging the assumptions and understanding the challenges. *Technovation, 23*(12), 917–928. doi:10.1016/S0166-4972(03)00081-6

Dangelmaier, W., Fischer, M., Gausemeier, J., Grafe, M., Matysczok, C., & Mueck, B. (2005). Virtual and augmented reality support for discrete manufacturing system simulation. *Computers in Industry, 56*(4), 371–383. doi:10.1016/j.compind.2005.01.007

Dangelmaier, W., Fischer, M., Gausemeier, J., Grafe, M., Matysczok, C., & Mueck, B. (2005). Virtual and augmented reality support for discrete manufacturing system simulation. *Computers in Industry, 56*(4), 371–383. doi:10.1016/j.compind.2005.01.007

David, R., & Alla, H. (1999). *Du Grafcet aux réseaux de Petri.* France: Hermès Science Publications.

Delmia/Quest User Manual. (2006). Delmia/Quest.

Demeter, K., & Matyusz, Z. (2010). (in press). The impact of lean practices on inventory turnover. [Corrected Proof.]. *International Journal of Production Economics.*

Doil, F., & Schreiber, W., A. T. P. C. (2003). Augmented Reality for Manufacturing Planning. *Proceeding of the Workshop on Virtual Environment,* 71-76.

Duffy, V. G., & Salvendy, G. (2000). Concurrent engineering and virtual reality for human resource planning. *Computers in Industry, 42*(2-3), 109–125. doi:10.1016/S0166-3615(99)00065-2

Eddy, J., & Lewis, K. (2002). Visualization of multidimensional design and optimization data using cloud visualization, *ASME Conference Proceedings,* 899-908. doi:10.1115/DETC2002/DAC-34130

Edward, L., Lourdeaux, D., Lenne, D., Barthes, J. P., & Burkhardt, J. M. (2008). Modelling autonomous virtual agent behaviours in a virtual environment for risk. *International Journal of Virtual Reality, 7*(3), 13–22.

Fillatreau, P., Fourquet, J.-Y., Bolloc'h, R. L., Cailhol, S., Datas, A., & Puel, B. (2013). Using virtual reality and 3D industrial numerical models for immersive interactive checklists. *Computers in Industry, 64*(9), 1253–1262. doi:10.1016/j.compind.2013.03.018

Fontanili, F. (1999). Intégration d'outils de simulation et d'optimisation pour le pilotage d'une ligne d'assemblage multi-produit à transfert asynchrone. University of Paris XIII, France.

Fründ, J., Gausemeier, J., Matysczok, C., & Radkowski, R. (2005). Using Augmented Reality Technology to Support Automobile Development. *Lecture Notes in Computer Science, 3168*, 289–298.

Gausemier, J., & Frund, J. M. C. (2002). AR-Planning Tool Design Flexible Manufacturing Systems with Augmented Reality. *Proceedings of 8th Euro graphics Workshop on Virtual Environment*, 19-25.

Gerbaud, S., Mollet, N., Ganier, F., Arnaldi, B., & Tisseau, J. (2008). GVT: a platform to create virtual environments for procedural training. *Proceedings of IEEE VR Conference*, 225-232. doi:10.1109/VR.2008.4480778

Gesellschaft, F. R. I. (2006). Carl Adam Petri und die Petrinetze. *Informatik-Spektrum, 29*(5), 369–381. doi:10.1007/s00287-006-0107-7

Giua, A., & Seatzu, C. (2007). A system theory view of Petri nets. *Advances in Control Theory and Applications, 353*, 99–127.

Gogg, T. J., & Mott, J. R. A. (1993). Introduction to simulation. *In Proceedings of the 25th conference on winter simulation*, 9-17. doi:10.1145/256563.256571

Goulding, J., Nadim, W., Petridis, P., & Alshawi, M. (2012). Construction industry offsite production: A virtual reality interactive training environment prototype. *Advanced Engineering Informatics, 26*(1), 103–116. doi:10.1016/j.aei.2011.09.004

Harris, E. N., & Morgenthaler, G. W. (2004). Planning, implementation and optimization of future space missions using an immersive visualization environment (IVE) machine. *Acta Astronautica, 55*(1), 69–78. doi:10.1016/j.actaastro.2003.11.002 PMID:15786593

Jayaram, S., Jayaram, U., Kim, Y., DeChenne, C., Lyons, K., Palmer, C., & Mitsui, T. (2007). Industry case studies in the use of immersive virtual assembly. *Journal of Virtual Reality, 11*(4), 217–228. doi:10.1007/s10055-007-0070-x

Kantonen, T., Woodward, C., & Katz, N. (2010). Mixed reality in virtual world teleconferencing, *Proceedings of the IEEE Conference on Virtual Reality*, 179-182.

Kaufmann, H., & Schmalstieg, D. (2003). Mathematics and Geometry Education with Collaborative Augmented Reality. *Computers & Graphics, 27*(3), 339–345. doi:10.1016/S0097-8493(03)00028-1

Kilpatrick J. (2003). Lean principles. *Utah manufacturing Extension Partnership.*

Klingstam, P., & Gullander, P. (1999). Overview of simulation tools for computer aided production engineering. *Computers in Industry, 38*(2), 173–186. doi:10.1016/S0166-3615(98)00117-1

Krafcik, J. F. (1988). Triumph of the lean production system. *Sloan Management Review, 30*(1), 41–52.

Ladeveze, N., Fourquet, J.-Y., & Puel, B. (2010). Interactive path planning for haptic assistance in assembly tasks. *Computers & Graphics, 34*(1), 17–25. doi:10.1016/j.cag.2009.10.007

Lafortune, C. S. (2008). *Introduction to Discrete Event Systems* (2nd Ed.). Boston: Springer - Verlag.

Lee, J., Han, S., & Yang, J. (2011). Construction of a computer-simulated mixed reality environment for virtual factory layout planning. *Computers in Industry, 62*(1), 86–98. doi:10.1016/j.compind.2010.07.001

Lee, J., Han, S., & Yang, J. (2011). Construction of a computer-simulated mixed reality environment for virtual factory layout planning. *Computers in Industry*, *62*(1), 86–98. doi:10.1016/j.compind.2010.07.001

Leu, M. C., ElMaraghy, H. A., Nee, A. Y., Ong, S. K., Lanzetta, M., & Putz, M. et al. (2013). CAD model based virtual assembly simulation, planning and training. *Manufacturing Technology*, *62*(2), 799–822.

Li, H., Chan, G., & Skitmore, M. (2012). Visualizing safety assessment by integrating the use of game technology. *Automation in Construction*, *22*(0), 498–505. doi:10.1016/j.autcon.2011.11.009

Lindskog, E., Berglund, J., Vallhagen, J., & Johansson, B. (2013). Visualization Support for Virtual Redesign of Manufacturing Systems. *Journal of Procedia*, *7*(0), 419–424.

Lindskog, E., Berglund, J., Vallhagen, J., & Johansson, B. Visualization Support for Virtual Redesign of Manufacturing Systems. In: Procedia 7, Nr. 0, S. 419 – 424, 2013. doi:10.1016/j.procir.2013.06.009

Mahdjoub, M., Monticolo, D., Gomes, S., & Sagot, J.-C. (2010). A Collaborative Design for Usability approach supported by Virtual Reality and a Multi-Agent System embedded in a PLM environment. *Computer Aided Design*, *42*(5), 402–413. doi:10.1016/j.cad.2009.02.009

Manca, D., Brambilla, S., & Colombo, S. (2013). Bridging between Virtual Reality and accident simulation for training of process-industry operators. *Advances in Engineering Software*, *55*, 1–9. doi:10.1016/j.advengsoft.2012.09.002

Marc, J., Belkacem, N., & Marsot, J. (2007). Virtual reality: A design tool for enhanced consideration of usability "validation elements". *Safety Science*, *45*(5), 589–601. doi:10.1016/j.ssci.2007.01.004

Marion, N., Querrec, R., & Chevaillier, P. (2009). Integrating knowledge from virtual reality environments to learning scenario models. A meta-modeling approach. *Proceedings of the International conference of Computer Supported Education*, 254-259, Lisbon, Portugal.

Marion, N., Septseault, C., Boudinot, A. & Querrec, R. (2007). GASPAR: Aviation management on aircraft carrier using virtual reality, *Proceedings of Cyberworlds*, Hanover, Germany.

Markt, P. L. & Mayer, M. H., (1997). WITNESS simulation software: a flexible suite of simulation tools, 7(10), 711-717.

Motwani, J. R., & Ullman, J. (2007). *Introduction to Automata Theory, Languages, and Computation.* USA: Prentice Hall.

Mujber, T., Szecsi, T., & Hashmi, M. (2004). Virtual reality applications in manufacturing process simulation. *Journal of Materials Processing Technology*, *155*, 1834–1838. doi:10.1016/j.jmatprotec.2004.04.401

Murata, T. (1989). Petri nets: Properties, analysis and applications. *Proceedings of the IEEE*, *77*(4), 541–580. doi:10.1109/5.24143

Ng, L. X., Ong, S. K., & Nee, A. Y. C. (2010). ARCADE: A Simple and Fast Augmented Reality Computer-Aided Design Environment Using Everyday Objects. *Proceedings of IADIS Interfaces and Human Computer Interaction 2010 Conference*, 227–234.

Ng, L. X., Oon, S. W., Ong, S. K., & Nee, A. Y. C. (2011). GARDE: A Gesture-based Augmented Reality Design Evaluation System. *International Journal of Interactive Design and Manufacturing*, *5*(2), 85–94. doi:10.1007/s12008-011-0117-9

Pegden, C. D., Sadowski, R. P., & Shannon, R. E. (1995). Introduction to Simulation Using SIMAN. New York, NY, USA: McGraw-Hill, Inc.

Pouliquen, M., Bernard, A., Marsot, J., & Chodorge, L. (2007). Virtual hands and virtual reality multimodal platform to design safer industrial systems. *Computers in Industry*, *58*(1), 46–56. doi:10.1016/j.compind.2006.04.001

Querrec, R., Vallejo, P., & Buche, C. (2013). MASCARET: create virtual learning environments from system modelling. Engineering Reality of Virtual Reality, *SPIE Electronic Imaging Conference*, San Francisco, CA, USA.

Raymond, C. W., James, M., & Graham, R. (2009). Automated design process modelling and analysis using immersive virtual reality. *Computer Aided Design*, *42*(12), 1082–1094.

René. Chevance., J. (2001). Méthodologie en matière de performance des systèmes. Techniques de l'ingénieur. Techniques de l'ingénieur, Paris.

Richir, S. & Fuchs, P. (2003). Réalité virtuelle et conception Méthodes. *Techniques de l'ingénieur.*

Shah, R., & Ward, P. T. (2003). Lean manufacturing: Context, practice bundles, and performance. *Journal of Operations Management*, *21*(2), 129–149. doi:10.1016/S0272-6963(02)00108-0

SIMCORE & CLARTE. (2013, August). Ergo-wide: Analyse ergonomique en réalité virtuelle. Retrieved from: http://www.simcore.fr/Ergonomie.asp

Sreng, J., Lécuyer, A., Mégard, C., & Andriot, C. (2006). Using visual cues of contact to improve interactive manipulation of virtual objects in industrial assembly/maintenance simulations. *Journal of Visualization and Computer Graphics*, *12*(5), 1013–1020. doi:10.1109/TVCG.2006.189 PMID:17080829

Stark, R., Israel, J. H., & Wöhler, T. (2010). Towards Hybrid Modeling Environments-Merging Desktop-CAD and Virtual Reality-technologies. *Annals of CIRP*, *59*(1), 179–182. doi:10.1016/j.cirp.2010.03.102

System, M. R. a product design support system of CANON, (2012, May). Retrieved from: http://www.actinnovation.com/innovation-technologie/canon-mr-system-realite-augmentee-4831.html

Takeda, N., Kakudate, S., Nakahira, M., Shibanuma, K., & Tesini, A. (2008). Development of a virtual reality simulator for the ITER blanket remote handling system. *Fusion Engineering and Design*, *83*(10-12), 1837–1840. doi:10.1016/j.fusengdes.2008.05.042

Takus, D. A., Profozich, & David, M. (1997). ARENA SOFTWARE TUTORIAL, *Winter Simulation Conference*.

Tarault, A.; Bourdot, P.; Vézien, J. & Sacari. (2005). An immersive remote driving interface for autonomous vehicles, *Computational Science – ICCS*, 3-15.

Teamcenter Manufacturing Plant Simulation. (2013, August). Retrieved from: http://www.robertirobotics.com/ugs/efactory/emplant/_files/fs_tecnomatix_em_plant.pdf

Tecnomatix Factory Layout Simulation. (2013, August). Retrieved from: http://www.plm.automation.-siemens.com/zh_cn/Images/7656_tcm78-64291.pdf

Tesic, R., & Banerjee, P. (1999). Exact collision detection using virtual objects in virtual reality modeling of a manufacturing process. *Journal of Manufacturing Systems*, *18*(5), 367–376. doi:10.1016/S0278-6125(00)87639-6

Tolio, T., Sacco, M., Terkaj, W., & Urgo, M. (2013). Virtual Factory: An Integrated Framework for Manufacturing Systems Design and Analysis. *Procedia*, *7*, 25–30.

VR4D: Virtual Reality for Design, (2013, August). Retrieved from: http://www.lecolededesign.com/fr/actualites/bdd/actualite/1247

VirtualiTeach. Assistance de Conception, CLEMI, CLARTE & CEA-List, (2013, August). Retrieved from: http://www.ac-creteil.fr/retrouvezlactualite-avril2013-pedagogieconcrete.html

Weber-Jahnke, J. H., & Stier, J. (2009). Virtual prototyping of automated manufacturing systems with Geometry-driven Petri nets. *Computer Aided Design*, *41*(12), 942–951. doi:10.1016/j.cad.2009.06.012

Wiese, E., Israel, J. H., Zöllner, C., Pohlmeyer, A. E., & Stark, R. (2009). The Potential of Immersive 3D-sketching Environments for Design Problem-solving. *Proceedings of 13th International Conference on Human–Computer Interaction*, 485–489.

Womack, J., & Jones, D. (2005). System Lean: Penser l'entreprise au plus juste. Village mondial, 2ème édition, Paris.

Womack, J. P., Jones, D. T., & Ross, D. (1990). The Machine that Changed the World. *The International Journal of Human Factors in Manufacturing*, *4*(3), 1522–7111.

Xinhua, L., Youhui, L., & Qi, L. (2011). Virtual Assembly and Simulation of Vibration Sieve Based on a Human-Interface Environment. *Procedia Engineering*, *15*(0), 2988–2992. doi:10.1016/j.proeng.2011.08.562

KEY TERMS AND DEFINITIONS

3D Interactions: A form of human-machine interaction where users are able to move and perform interaction in 3D space. Both human and machine process information where the physical position of elements in the 3D space is relevant.

3D Visualization: A variety of technologies that make images and movies appear more lifelike in print, on the computer, in the cinema or on TV. Known as "stereoscopic imaging" and "3D stereo", people sense a greater depth than they do with 2D and feel they could reach out and touch the objects.

Augmented Reality: A live direct view of a physical, real-world environment whose elements are augmented by computer-generated sensory input such as sound, video, 3D objects.

Industrial Design: A process of design applied to products that are to be manufactured through techniques of mass production.

Simulation of Industrial Processes: Process simulation is a model-based representation of production, evaluation, chemical, physical, and other technical processes and unit operations in software.

Virtual Manufacturing: The modeling of manufacturing systems using audiovisual or other sensory features to simulate or design alternatives for an actual manufacturing environment, or the prototyping and manufacture of a proposed product using computers.

Virtual Reality: An immersive multimedia or computer-simulated life, replicates an environment that simulates physical presence in places in the real world or imagined worlds. Virtual reality can recreate sensory experiences, which include virtual taste, sight, smell, sound, and touch.

Section 3

Industrial Applications of Robotics, Automation and Control

Chapter 6
Conceptual Process for Designing High-Technology Products:
Case Study of a Litter-Collecting Robot

Arsalan Safari
Qatar University, Qatar

ABSTRACT

In this chapter, a systematic and practical design process and methodology is presented and applied to design a new high- technology product: a litter-collecting robot. Although considerable theoretical and practical models have been developed in product design and development, there are still limited effective models on the practical design process on a detailed level. This chapter elaborates on recent relevant research in the design methodology field and try to improve the details of product design process and apply it to a litter-collecting robot design. The detailed and practical approach demonstrated on the design of a high- tech product in this paper, can be applied effectively to the design process of industrial products.

INTRODUCTION

The main goal of this paper is to improve and utilize an advanced design process and methodology in order to design a high technology product to do some specific functions in commercial, residential, and light industrial environment. Although there is substantial academic research on product design and development process, there is limited documented comprehensive research on the practical and wide application of optimizing the product design process. The case study of this research is a new litter-collecting robot. Although this conceptual design process solution may not be the optimal solution for this product, because of time constraints, the main target here is to show an effective practical process in actual product design which is applicable in system design process as well.

DOI: 10.4018/978-1-4666-8693-9.ch006

Many studies argue that a critical success factor in product design and development process is following a structured design approach (Brown, Schmied, & Tarondeau, 2003; and Ernst, 2002), and firms that execute a formal approach might have higher rate of survival if they have a formal approach to product design and development (Baker & Hart, 1999).

Bailetti and Litva (1995) focused on the importance of customers requirement information in design process and tried to test the design managers' prospective on the source of customer requirement information. They highlighted the importance of creating a model to ensure the customers' information from various sources are consistent for design. Pahl and Beitz (1993) categorized the design process into the four different stages: first, tasks' clarification and design specifications and development; second, conceptual design; third, embodiment design; and fourth, detail design. Clarification includes the requirements gathering (the demands and desires) and defining limitations, which comes to a specification list. Function logic is a way of completing the first two stages. Suh (1988) explained the axiomatic approach design in his study in three phases: first, problem definition that comes in the functional requirements and constraints of design; second, conceptualizing and devising a solution creatively; and last, analyzing and determining if the proposed design solution is rational and consistent with the predefined problem. As aforementioned, function logic is a method to complete the first two stages by formalizing the second stage of the axiomatic process using the first stage feedback. The analytical stage is assisted by supported functions with the information of the allocation list. In addition; Sturges et al. (1997) believe detailed synthesis methods which usually follow the conceptual design process need specifications of input. The function logic in design is suggested as a tool for systematically promoting innovation, managing the information of conceptual design, and producing such specifications.

Tomiyama et al. (2009) categorized all type of design theories and methodologies as "math-based methods", "methodologies to achieve concrete design goals", "process methodologies" in practical use, and traditional design methodologies in academic world and then elaborated the gap between practical and educational ones.

This study considers most of the theories in the present design process and tried to provide an improved and detailed method of product design on a litter-collecting robot, a new high-tech industrial product. The present approach involves more rigorous attention to and more comprehensive coverage of details in the design. In this design process, apparent inconsistencies and contradictions in each of the existing available commercial/ prototype designs are identified, and several guidelines for removing inconsistencies and adapting to changes are provided. The problem is defined and analyzed by using generic brainstorming techniques through a survey team that reviews user responses. The problem identified by users are categorized under several headings, including function/need, change/drivers, people, environment, market window, technology, materials and manufacturing, usage locations, form/style/aesthetics, quantity/cost, parallel products, standards/guidelines/ protocols/laws.

Second, in the requirement engineering section, the production characteristics are analyzed from different perspectives of product design specifications and an engineering framework for proposed design is provided. Several ideas through a team of experts generated, evaluated, and improved to have an acceptable concept for the litter-collecting robot. At the end of this part, a system identification matrix (SIM), main and subsystems, inputs, outputs, and product architecture schematic (PAS) of the litter-collecting robot are provided. Traditionally, a robot is designed as a tool for achieving a specific objective. However, a common platform among robots with different purposes is desired to minimize the cost and time required for robot system development.

PROBLEM DEFINTION AND ANALYSIS

The objective is to design an unconventional product from a human prospective so that the robot is to be designed to be an artificial (simulated) living creature with all five senses: hearing, speech, smell, taste, and touch. So that, as with human beings, the "creature" can actually interact with its environment. We found existing commercial robots and identified some inconsistencies. The samples were examined and existing commercial designs were photographed; apparent inconsistencies and contradictions among existing designs were identified and recorded; guidelines for removing inconsistencies and adapting to changes were then developed.

Customer feedback is crucial in the problem definition of the product design process. We created a questionnaire for potential customer input. The questionnaire is distributed among randomly selected customers who are asked to complete the questionnaire thoughtfully and to provide ideas about product attributes and uses desirable for potential future buyers of the new product. The questionnaire is designed so that users can provide feedback on each question according to a priority range, 1-10= lower to highest priority to the customer. Further, users are also able to include specific detailed concerns if necessary in spaces provided for each question. For interested readers the questionnaire is in Appendix A.

Our method of design relies on customer feedback. So we conducted a convenience survey.[1] Also, due to cost and time constraints we only collected ten responses. For demonstration purposes, our sample suffices to show the details of our improved product design process. Responses from potential users were ranked in Table 1.

From Table 1, it is clear that the main priority is safety. Therefore, the litter-collecting robot must not pose any potential threat to a person or any potential damage to electrical devices, furniture and other equipment in the workplace and at home. As well, it should not emit radiation, which could be a present or future health risk. Function/need is the second highest priority among users. This is expected, because the robot should fulfill its intended purpose. Unexpectedly, the environment and the level of technology are lowest priority. Next we developed the problem by using generic brainstorming techniques with a team of four experts. The problems are categorized under the following headings:

Table 1. Survey results

Priority	Factors
#1	Standards/Safety
#2	Functions/need
#3	Cost
#4	Usage location
#5	Forms/style/aesthetics
#6	User friendly
#6	Market window (durability)
#8	Environmental effects
#9	Technology type (i.e. latest technology)

- **Function/Need**
    - The robot should be simple in design
    - The robot should be able to accurately identify litter, pick it up, and dispose of it in the garbage
    - The robot should be able to effectively define the shape/size/weight/nature of litter
    - The shape/size of the robot should be identified
    - The robot should be light in weight
    - The robot should contain modules for its functions
    - The robot should be safe for children and pets
    - The robot should be easy to operate
    - The robot should contain bags that are easy to change (or if possible contain no bags)
    - The robot should be durable
    - The robot's height should be minimal for safe access beneath furniture
    - A folding robot may need minimum space to store it
    - The robot should contain extendable/detachable hands for the easy collection of litter
    - The robot should operate easily on both smooth surfaces and carpeted areas
    - The robot should be able to operate effectively in dusty/wet/smoky/polluted/corrosive areas
    - The robot should be able to easily recharge itself (battery) for at least 12 hours
    - The robot should be easy to operate and easy GUI programming
    - The robot should be controlled fully by a powerful remote
    - The robot's litter collecting box should be removable and easy to remove
    - The robot should work in cold, hot, and warm conditions
    - The robot should be able to replace its own bag/litter collection box
- **Changes/Drivers**
    - The main objective is to make a robust product with a minimal ("zero") failure rate for its designed life period
    - Another objective is to create an unconventional product (robot) that operates effectively with minimal cost; also, the robot should have an award-winning appearance.
    - Help people for litter collecting in unstable places and increase of comfort level
- **Users**
    - Any buyer should be able to operate the robot easily and get full value in the product for the product's life cycle
    - In terms of functional issues, users should find the result of functional issues by guidelines and instructions
    - Component cost should be minimized and vendors should be able to keep the cost as low as possible
    - The product should be easy to produce with a design that can accommodate changes in user needs during the life cycle of the product
    - The design of the packaging used with the product should result in minimal cost and damage during shipping
    - The product should be easy to sell (and therefore attractive to potential customers) because of its innovative design, low cost, high quality, and maintainability
    - The product should be easily disposed
    - The product should be easy to use for disabled people

- The product should be easy to use for elderly people
- The product should be easy to use where children play
- Should be helpful for janitors

- **Environment (Context, Infrastructure)**
    - The product should not contain or emit any potentially dangerous level of radiation, noise, heat, or gas
    - The product should provide some added features like arranging papers/arranging books/segregating the biodegradable litter from non-bio degradable litter (if possible)
    - The main hardware components should be reusable and other components should be recyclable
- **Market Window**
    - The life of the product should be at least three years and it must remain fully functional during this period
- **Technology**
    - The software should be robust
    - The hardware should be robust
    - The robot should be able to communicate with other similarly-designed robots to enhance productivity
    - Calibration of the system should be made automatically in adjustable mode
    - The robot should be able to illuminate its working area so that it does not need indoor power and it conserves energy
    - The robot should contain a variable controlled speed
    - The robot should be able to be equipped with an optional solar panel as an additional power system
    - The robot should contain Bluetooth technology to recognize and possibly talk with other robots to share work in one operational area
    - The robot should be able to sense litter and distinguish it from other materials/objects
- **Materials and Manufacturing**
    - The robot should be easy to manufacture and assemble
    - The robot should contain as few parts as possible
    - The robot should contain as many interchangeable parts as possible to minimize maintenance costs
    - The robot should be sturdy and durable
- **Usage Locations**
    - The robot should be used for hotel, hospitals, shopping centers, airport (indoor) cleaning, school (indoor area) cleaning, and the public community center
    - The robot should be used for dangerous areas where it's difficult for people to reach (corrosive/polluted) areas
    - The robot should be used for top security areas
- **Form/Style/Aesthetics**
    - The robot should be user friendly
    - The height of the robot should be adjustable so that the most people can access to the components
    - The robot's appearance should be pleasing to the eye
    - The robot should be adjustable (height/width/thickness)

- **Quantity/Cost**
  - The purchase price should be competitive
  - The operating cost should be as low as possible
  - The return on investment should be optimal
  - Parallel conventional products
  - Similar products must be examined. The designed features of the robot should match or preferably exceed those of conventional litter collecting devices (e.g. vacuum cleaner) on the market in terms of function, cost, and appearance.
- **Standards/Guidelines/Protocols/Laws**
  - The robot should meet the US Occupational, Safety and Health Administration (OSHA) requirements
  - The robot should be able to detect potentially dangerous situations (e.g. contacting electrical cables) and respond accordingly as per guidelines specified.

Theses aspects of robot design are discussed in the section below in terms of an engineering and economic perspective.

REQUIREMENTS ENGINEERING

Requirements engineering is the next step of the litter collector robot design in which we formalize, analyze and then find engineering characteristics of our product from different aspects while considering customers' needs. One standard method used in requirements engineering is Quality Functional Deployment (QFD). QFD is designed to help planners focus on characteristics of a new or existing product or service from the viewpoints of market segments, company, or technology-development needs. QFD helps to transform customer needs (the voice of the customer) into engineering characteristics. Herzwurm, Schockert, and Mellis (2000) elaborated the use of QFD in systems design and development. In this process, unlike the QFD and previous requirement analysis methods, there is a team of design expert who are analyzed the customers' needs into different aspects of design and re-evaluated the main factors of the product design before moving directly to product characteristics section.

1. Product Characteristics

According to Phal and Beitz (1993), task clarification and providing list of requirements are important in design. They believe abstracting and creating function structures often causes difficulties due abstract representation, however, we believe that for each aspect of design, at least an abstract is required to help designers to understand the general specification of the subject (product).

a. Functionality of Product (Robot)

i. Primary Function (Degrees Of Freedom, Control System, and Software)

Linear, extension, rotation, and twisting motions in the robot, robotic controllers, and degrees of freedom for the robot are the primary function to be performed (Arthur, 1985).

ii. Secondary Function (Robot Sensors)

The robot sensors should be able to simulate effectively the five senses of human beings (touch, smell, sound, taste and sight). These sensory systems process inputs of environmental parameters, including distance, light, sound, strain, rotation, magnetism, scent, temperature, pressure, and altitude. Therefore, the robot should be equipped with computer sensors to facilitate responses to the environment. Sensors allow the robot to detect objects and variations in the environment. The robot thus should be able to adjust its behavior based on the environmental information it receives.

iii. Third Function of Robot

Actuators are tools that facilitate movement and change system commands into actions. Actuators can be categorized into three main types: electric, hydraulic and pneumatic. Electric actuators are "simply electro-mechanical devices", such as stepper motors, solenoids and electric motors which facilitate movement through the use of electronically controlled gear systems. Electric motors are the most common actuator type. Hydraulic actuators facilitate robot movement through pressurized fluid flow involving a series of valves and pumps. The hydraulic fluids usually consist of reasonably non-compressible oils. Hydraulic actuators are applied where a lot of power is required for moving heavy objects. Finally, Pneumatic actuators force the pistons' movement using compressed gas through pumps and valves. Both pneumatic and hydraulic actuators use a series of valves, pumps and pistons to generate movement in the robot Grippers. Compressed gas is needed in area where electric devices are too dangerous. As well, hydraulics can be messy when leaks occur.

b. Durability

Since a robot requires little energy, it can be operated efficiently (as fuel efficient machine).

c. Lifespan

The life cycle of a robot should be long. However, it should be maintained and serviced at regular intervals. The robot should be subjected to an extensive evaluation procedure. It should be fully capable automated machine. The robot should be ideal for companies and retailers that require additional help without the need to hire more personnel.

d. Stability

The base of robot also has to be sturdy so as to maintain a law centre of gravity. Since it is litter-collecting robot, the loads to be carried are generally light.

e. Reliability

Reliability is an important factor in product design. According to Dhillon (2007), there are four types of failure in robot reliability and its safe operation; random component failure, hardware or software failure, and human errors. Therefore, it is needed to take into consideration of program failure, slips, and fall trip, wheels jammed in the robot design. Looking into the fault trees helps to make the robot reliable.

The functions of the robot are either remotely controlled by the computer systems or preprogrammed in the robot computer system. The computer system therefore must reliably program the robot not only to effectively pick up litter but also to efficiently differentiate between litter and other objects in surroundings. Dhillon (2007) indicated some methods of reliability analysis which can be effectively used in robot reliability study. These methods are failure modes and effective analysis (FMEA), parts count method, Markov method, and fault tree analysis.

f. Packaging

The litter-collecting robot is fragile, so during transportation it should be packed in foam packing to minimize moisture and shock damage. In addition, aluminum metal parts are anodized.

g. Quality

- Culture and context issues, customer requirements
- An easy-to-read instruction manual is required to benefit customer satisfaction with the product and to ensure that customers use and maintain this product correctly.
- Aesthetics
- As with any product, robots should be aesthetically pleasing. Because robot is a product of which people like to have in good shape.
- Environmental issues
- The robot should be user friendly to ensure good housekeeping and enable the cleanup of bio-hazardous wastes, including cans, paper, diapers, and etc.
- Standardization
- Most pioneering activities in robotics have been conducted independently of each other at various locations throughout the world. It seems that the most important factor in robotics now is the standardization which is important to consider in design process.
- Documentation should have the details of litter-collecting robot including the classification, details of how to use, all instructions, details on parts, preventive maintenance, and customer service centers locations, the robot's advantages in relation to other litter-collecting devices ((e.g. vacuum cleaner), the list of robot components, the list of internal systems/subsystems, and the future condition of the litter-collecting robot.

h. Variation/Robustness

Litter-collecting robot uses a repetitive series of actions. Thus, the robot construction should be robust, yet simple with welded components. In addition, the joints of robots should be able to pivot.

i. Affordability

The overall product cost, including that for design, manufacturing, distribution, operation, and disposal (environmental cost) of the robot, should be comparable to or lower than that of similar cleaning devices to ensure affordability. Two similar products are the Lego Mindstorm which is a general use robot costing $250; the second, is the Litter-Robot, by Automated Pet Care Products Inc., which automatically cleans cat litter. The cost is \$315. The manufacturing process of the proposed litter-collecting robot can be similar to the first one, but the power is stronger with more complicated grippers compared to the Lego Mindstorrn product. Finding the final price need more cost analysis; however, this cost estimate is affordable and desirable for consumers considering that they have been paying more for the other established products.

i. **Fabrication Cost**: The cost of fabrication is related mostly to the number of robots to produce in a period of time. The total costs of robots can be reduced by increasing the number of product. Usually for the mall batch size, it is not feasible to operate a manufacturing plant and it is better to order and provide these parts through other manufacturing plants. Determining the optimal number of products to produce during a given period (e.g. a year) required a feasibility study and marketing research.
ii. **Storage/Distribution Cost:** There are two main costs associated with the robot's inventory cost: 1. Storage cost and 2. Investment cost (monetary value of raw materials or final product in storage during the given period of time). Robot is as expensive product but it does not require a large area or special conditions for storage. Therefore the investment cost is greater than the storage cost. The optimal number of raw materials or robots to keep in storage can be calculated quickly using inventory planning and control techniques. However, the cost of storage and distribution is minimal compared with that for product manufacturing.
iii. **Operating Cost:** The robot operating cost is minimal, and it must be assembled just for its initial use. In addition, a special program is included in the processor so that the robot can conduct its primary duty of litter collection effectively. However, program options are available to enable flexibility for the customer to change program.
iv. **Environmental Cost:** The product does not entail any environmental impacts or any environmental cost largely because the robot is constructed of small plastic and metal parts that can be disposed of or recycled easily. Most of the robot's parts are similar to toy parts made from special polymers which are recyclable.

j. Fabricability

The ease of fabrication of the product is an important factor as well. We should consider manufacturability of the robot components as well as assimilability of these components to make the final product.

k. Manufacturability

The main parts of the robot should be polymer parts, which are preferred because of their high modulus of elasticity and light weight (low density). According to Ashby (2005) for material selection in the mechanical design process, for different function of robot, we have different metrics (performance index) which we should maximize it:

- For stiffness of robot parts (as a tie, tensile strut) during loading, the metric=E/ρ
- For stiffness of robot parts (as a beam, loaded in bending), the metric= $E^{1/2}/\rho$
- For supporting the buckling by robot parts, the metric= $E^{1/2}/\rho$

where E is elasticity and ρ is density of the material. According to Ashby (2005), metal alloys, polymers (composites), and woods (parallel to grain) materials with the maximum of E/ρ and $E^{1/2}/\rho$ values are best to select for robot parts. For the purposed robot, we find are best (based on Ashby 2003 and its related graph in Appendix C). For the proposed robot, metals are too heavy and wood is inflexible and degrade quickly. Thus, polymers such as Acrylonitrile Butafiene Styrene (ABS), polyethylene (PE) or polystyrene (PS) are the preferred materials for the purpose litter-collecting robot. Injection and mold compression are feasible method for manufacturing polymer parts for the robot. For a large batch size (mass production), an automatic process (line layout) is more economical effective.

l. Assembleability

An effective assembly process for the proposed robot is crucial because of the many parts that connect. Therefore, an assembly map should be provided for assembly line.

m. Installability

No installation process is required, and in fact the robot is assembled in the plant. Once purchase, the robot is ready to pick up litter at home, the office, or the job site.

n. Usability

The robot is a product designed to be use friendly. The buyer just needs to read first instructions in order to use the machine correctly, charge the battery, and know when to conduct maintenance and capacity of loading. The warranty and supporting services are also important to consider in this part.

o. Maintainability

Due to the potential need for repair or replace parts, maintenance is an important in design. A specialized technician is not needed to replace parts; however, instructions to assist users with parts replacement should be included in the product package by the manufacturer. Nevertheless, some sensitive and electronic parts may have to be changed by a qualified technician. Further, since the robot is a complicated product with many electrical and mechanical parts, it is not possible for all parts to be readily accessible to customers. That is, store/outlets with certain parts may have a limited geographic distribution in some regions.

p. Safety

The litter-collector robot is protected against electronic shock, fire, poison, explosion, and other such dangers. However, the instructions are emphasized: 1. the robot should be operated/ stored out of the reach of children; 2. the battery should contain a warning sign/message on it and be accompanied with instructions; 3. the design process should eliminate or at least minimise common hazards in product development. For example, edges, sharp corners, and all potentially hazardous sources, such as chemicals, radiation, very cold/hot temperatures ($-100^{\circ}C < T < +100^{\circ}C$) should be considered in the design.

Such situations/conditions where robot use is potentially dangerous or discouraged should be mentioned on the package and in the instructions. Indeed, it is also important with robot usage to be aware of the Asimov (1994) proposed robotics' laws:

- Law 1: "A robot may not injure humans or through inaction, allow humanity to come to harm".
- Law 2: "A robot may not injure a human being, or, through inaction, allow a human being to come to harm, unless this would violate a higher order law".
- Law 3: "A robot must obey orders given it by human beings, except where such orders would conflict with a higher order law".
- Law 4: "A robot must protect its own existence as long as such protection does not conflict with higher-order law".

q. Marketability

Generally, the life cycle of a high-tech product such as a robot is limited because modern technology advances rapidly, so the inherent changing life cycle should be factored into the design process. A further complication is the difficulty of marketing such a specialized product. Nonetheless, the successful marketing of similar products (indicated above), indicate a promising outlook for a market for these kinds of high-tech products. This optimism is largely a result of automation, information technology and industrial high-technology development. Therefore, seven years is an acceptable period for initial design.

2. Production Design Specifications

According to Ulrich and Eppinger (2004), the purpose of the product design specification (PDS) is to ensure that the design addresses the customer needs. Based on Ulrich and Eppinger (2004) we decided that the customer needs should be clearly stated so that they are independent of the way that the final design is actually implemented. Internal market research (in the previous section) indicates that the litter-collecting robot should be designed as a general purpose robot for use in houses, hotels, airports, hospitals, schools, libraries, community centres, and other indoor places. The robot must be able to readily search for litter, correctly identify it, easily pick it up, and accurately place it inside garbage containers/bags. When the bag is full, the robot should replace the bag.[2]

CONCEPT DESIGN AND EVALUATION

1. Ideation process

A team of four experts in product design were consulted for the conceptual design of the product, based on the requirement engineering results. The result of the ranking of all main design factors are in Table 2.

To systematically generate ideas for product design, the ten attributes in Table 2 were combined in pair wise combinations. There are 9!(9*8*7*6*5*4*3*2*1= 362,880) possible combinations. The "Riff by Attribute" method is used because robot characteristics are selected on the basis of just one factor. The "Riff by Attribute" method is the identification of key characteristics of a conventional solution in

Table 2. Ranking of all factors by 4 different design experts

Factors	Exp1	Exp2	Exp3	Exp4	Total
Functionality(F)	10	9	7	9	35
Durability (D)	4	1	4	6	15
Quality (Q)	7	8	5	7	27
Affordability(A)	8	7	6	10	31
Manufacturability(M)	3	4	8	4	19
Installability(I)	1	3	3	1	8
Usability(U)	6	5	2	5	18
Maintainability(N)	2	2	1	2	7
Safety(S)	9	10	10	8	37
Marketability(K)	5	6	9	3	23

design. To narrow down the options and generate ideas, the most important factors are considered first. The three main important factors which are based on the ranking in Table 2 are safety, functionality, and affordability. Therefore, there are 9*8*7=504 pair wise combinations.

The combinations listed below feature only two of the three factors. Further, safety has the most weight and affordability has the least weight of the three factors. For brevity, the highest 20 pair combinations have been selected amongst the 504 pair wise combinations. Below are the twenty most important combinations for idea generation.

1. SF: It should have three to four wheels, heavy base and strong arms to pick up objects
2. SA: It should be cost effective but strong materials for robot's parts (polymers were found to be preferable)
3. SQ: It should have long life cycle, support uneven loading (considering safety factor)
4. FA: It should be cost effective but strong materials for robot's parts (polymers were found to be preferable)
5. FQ: It should have reliable programming, having sensitive sensors to accurately recognize objects and their shape
6. AQ: It should have mass production, having quality control system for manufacturing
7. SK: It should have American standard (OSHA), Canadian standard for robot (Z434)
8. SM: It should prepare emergency means a head of time to compensate for the relatively low reliability of an object
9. SU: It should put clear instructions and warning message for the robot, and sensitive sensors to recognize objects from human
10. SD: It should use materials (metal alloys, some composites) which have resistance in diverse temperatures (-50°C<T<+50°C)
11. SN: It should be easy to repair, easy access to the joints, providing enough maintenance guide in the robot package
12. FK: The robot is designed to lost for seven years, it should be re-designed

13. FM: It should fulfill required strength, stiffness of robot parts to have yielding, buckling, and fracture resistance, and sensors should be reliable for manufacturing process and heat treatment of polymers
14. FU: It should have clear and easy access of buttons, easy programming, and manual guide for different languages
15. FD: Wear and tear should be minimized during operation
16. FN: It should provide proper maintenance documentation in the package, preventive maintenance checklist, and easy to access for maintenance
17. AK: It should use composite (polymer) as a low cost materials with high elasticity and light weight, 7 years marketable, colorful package
18. AM: It should be mass produced (through automation)
19. AU: It should have easy and low cost training (CDs) and instruction package,
20. AD: It should have quality control system during manufacturing for durability assurance of the robot

2. Concept Generation

After providing product design specification (PDS), the expert design team was asked individually to brainstorm and find the best real design solution for the robot. The team members considered the main factors, pair wise combinations, details of requirements and constraints. They mainly used analog reasoning while designing the exterior shape of the robot and the interior structure. The sketches from the four team members follow separately. Based on a review and analysis of the four concepts conducted in a group meeting, a decision matrix for the concepts was developed. Then a sketch of the design concepts was created as can be seen below.

a. Concept A

- Tall robot with a polymer base
- Four roll wheels
- Dimensions W200 × L500 × H600 mm
- 2 arms and hands on left and right sides
- Rechargeable battery & its plug
- Polymer body (structure)
- Bumper on the base
- Litter storage in the top
- Head turning 360° with 2 sensors
- Different sensors in different places
- Power on/off switch

b. Concept B

- Robot with rectangular box shape with a polymer body
- Four roll wheels
- Dimensions W200 × L500 × H500 mm

Figure 1. Concept A diagram

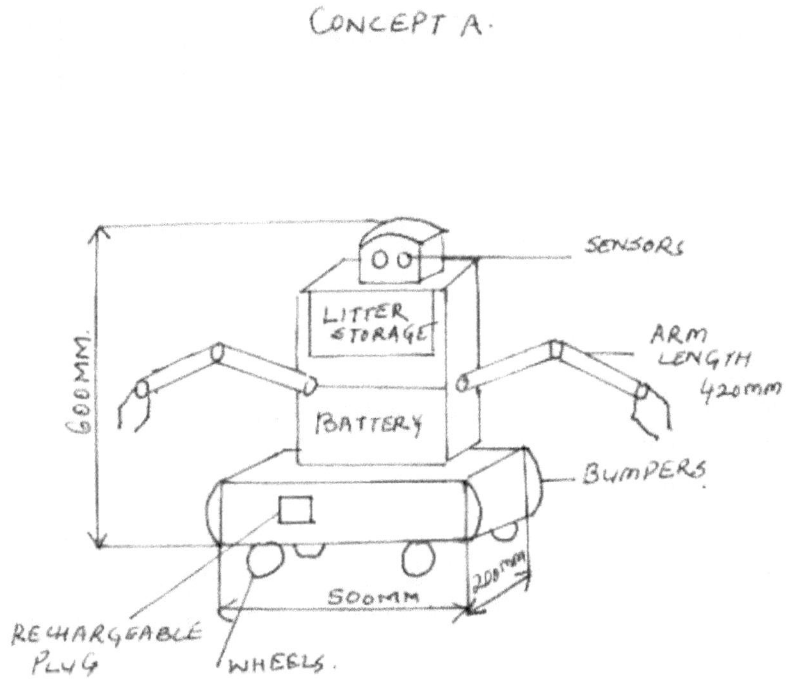

Figure 2. Concept B diagram

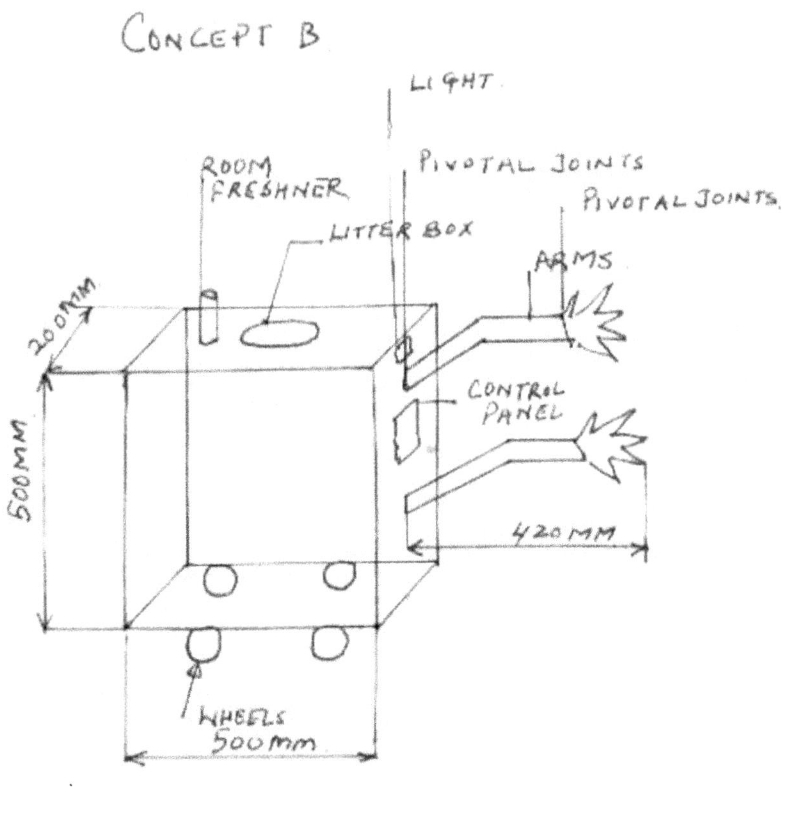

- 2 arms and hands in front of the robot
- Rechargeable battery & its plug
- Polymer body (structure)
- Litter storage in the box shape
- Different sensors in different places
- Power on/off switch
- Room freshener

c. Concept C

- Tall robot with two legs, a box shape, and a pyramid shape on top
- Four roll wheels
- Dimensions W200 × L500 × H600 mm
- 2 arms and hands left and right sides
- Rechargeable Battery
- Polymer body (structure)
- 2 litter storage areas in the two legs (front)

Figure 3. Concept C diagram

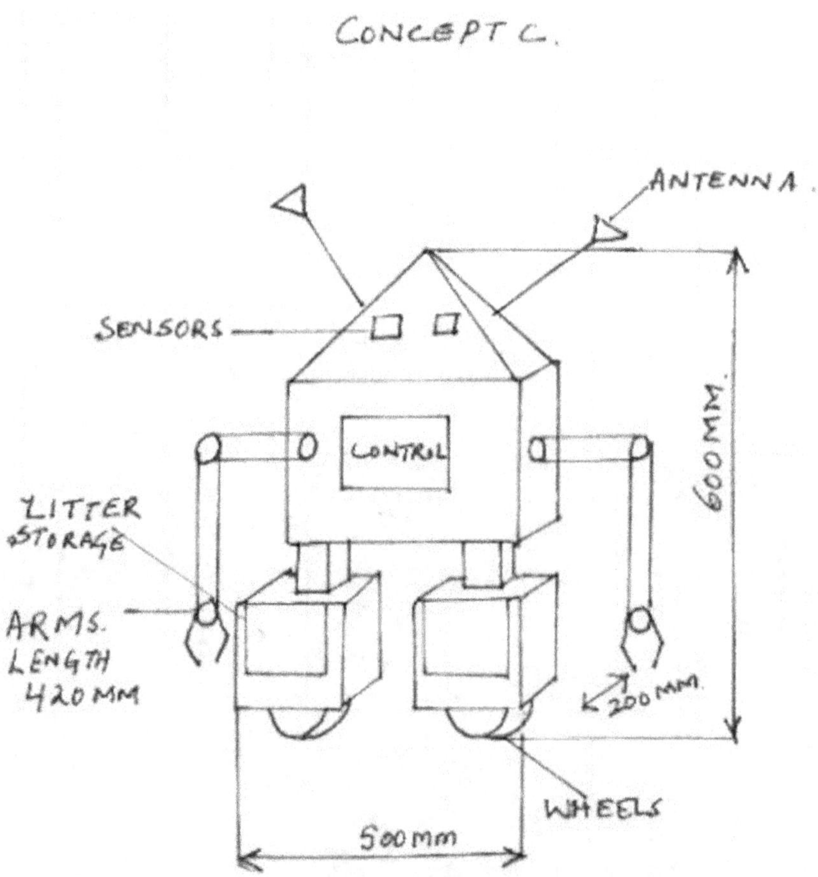

- Different sensors in different places
- Power on/off switch
- Two long antennas

d. Concept D

- Short, tortoise-shaped robot
- Three roll wheels
- Dimensions W500 × L500 × H200 mm
- 2 arms and hands on the top
- Rechargeable battery
- Polymer body (structure)
- Litter storage in the top
- Head turning 360° with 2 sensors
- Different sensors in different places
- Power on/off switch

Figure 4. Concept D diagram

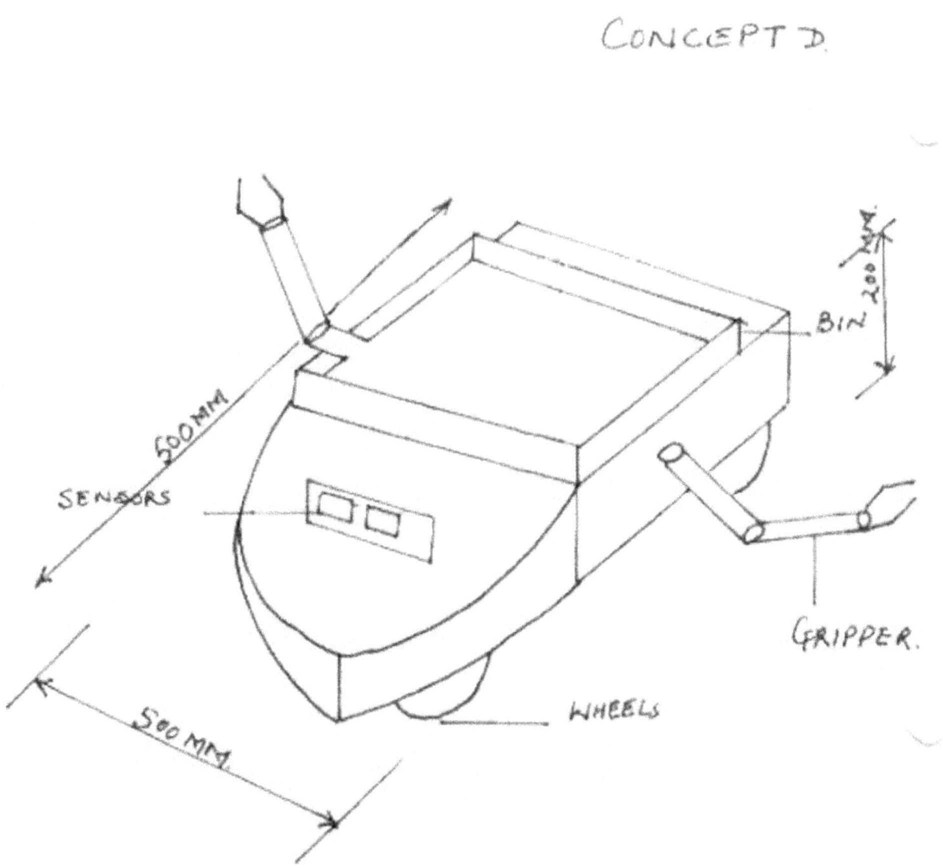

3. Concept Evaluation

According to the comparison matrix in Table 3 and the decision matrix for concepts in Tables 4 and 5, concept D is ranked highest. However, it was not selected as the final and best concept. Instead, the advantages of concepts A, B, and C were analysed and applied to modify and improve concept D1 and to minimize its disadvantages as much as possible. The final concept, concept D1, attained the highest score as shown in Table 5. A sketch of D1 follows Table 5.

4. Concept Modification

Concept D1 is the final conceptual design of litter collecting robot. The litter collecting robot uses energy and information controlled by a computer program and an initial command to start up and collect litter. The robot is composed of a heavy base body, three wheels, and two sturdy arms and hands. The arms/hands are attached to the head of robot and able to swivel 360° horizontally in order to move garbage around the robot without movement of wheels. Sensors for ultrasonic, light, sound, and mechanical inputs are located at various parts of the robot. The sensors permit the robot to find and transport litter, to detect and avoid obstacles and to recognize special sounds and unexpected events.

a. Concept D1

- Short, tortoise shaped robot
- Three roll wheels (2 in the front and 1 in the back)
- Dimensions W500 × L500 × H200 mm
- 2 arms and hands on the top

Table 3. Concept design descriptions

Concept	Description	Pros	Cons
A	Tall robot with height 600mm, length 500mm, and width 200 mm. 4 wheels & 2 hands which it can rotate 360° A litter bin in the top	Flexibility and high speed Head rotating 360° Easy to move	Difficult to turn sides fast by wheels Large size Needs high production cost
B	Robot with rectangular box shape with height 500mm, length 500mm, and width 200 mm 4wheels & 2 hands A litter storage in the box shape,	Easy to manufacturing Fast	Arms are just in one direction (one side), Couldn't turn back & sides Easily Not good looking Limited view of sensors
C	Robot with two legs but two wide wheel Tall, with 600 mm height, 500mm length, and 200 mm wide Two small place for litters in the legs Pyramid head and two hands	Wide view of the sensors Good shape Fast	High cost of production Unsuitable litter storage Unsteady robot (just 2 wheels)
D	Tortoise shape with three wheels Short robot with two hands on the top A rectangular place in the back of head for litters 500 mm length, 500 width, 200 height	Small and light Less material, easy to make Fast Head rotating 360°C	Low capacity of litter storage

Table 4. Decision Matrix for 4 Concepts

Factor (Weight)	Concept A		Concept B		Concept C		Concept D	
	Rating	Score	Rating	Score	Rating	Score	Rating	Score
Functionality (18%)	2	0.36	0	0	2	0.36	2	0.36
Durability (4%)	1	0.04	-1	-0.04	-1	-0.04	1	0.04
Quality (13%)	1	0.13	1	0.13	-1	-0.13	1	0.13
Affordability (16%)	1	0.16	-2	-0.32	-2	-0.32	2	0.32
Manufacturability (9%)	1	0.09	1	0.09	2	0.18	1	0.09
Installability (2%)	1	0.02	0	0	1	0.02	2	0.04
Usability (7%)	2	0.14	0	0	-1	-0.07	2	0.14
Maintainability (1%)	1	0.01	-1	-0.01	0	0	1	0.01
Safety (20%)	2	0.4	0	0	2	0.4	2	0.4
Marketability (11%)	1	0.11	0	0	-1	-0.11	1	0.11
Total	1.46		-0.15		0.29		1.64	
Rank	**2nd**		**4th**		**3rd**		**1st**	

Table 5. Decision Matrix for 5 Concepts

Factor (Weight)	Concept A		Concept B		Concept C		Concept D		Concept D1	
	Rating	Score	Rating	Score	Rating	Score	Rating	Score	Rating	Score
Functionality (18%)	2	0.36	1	0.18	2	0.36	2	0.36	2	0.36
Durability (4%)	1	0.04	-1	-0.04	-1	-0.04	1	0.04	1	0.04
Quality (13%)	1	0.13	1	0.13	-1	-0.13	1	0.13	1	0.13
Affordability (16%)	0	0	-1	-0.16	-2	-0.32	2	0.32	2	0.32
Manufacturability (9%)	1	0.09	2	0.18	1	0.09	1	0.09	2	0.18
Installability (2%)	1	0.02	2	0.04	-1	-0.02	2	0.04	2	0.04
Usability (7%)	2	0.14	0	0	-1	-0.07	2	0.14	2	0.14
Maintainability (1%)	-1	-0.01	1	0.01	-1	-0.01	1	0.01	1	0.01
Safety (20%)	2	0.4	1	0.20	1	0.20	2	0.4	2	0.40
Marketability (11%)	1	0.11	0	0	-1	-0.11	1	0.11	1	0.11
Total	1.28		0.54		-0.05		1.64		1.73	
Rank	3rd		4th		5th		2nd		1st	

- Rechargeable battery
- Polymer body (structure)
- Bumper on the base
- Litter storage in the top
- Head at a higher height turning 360°
- Different sensors in different places
- Power on/off switch

Figure 5. Concept D1 diagram

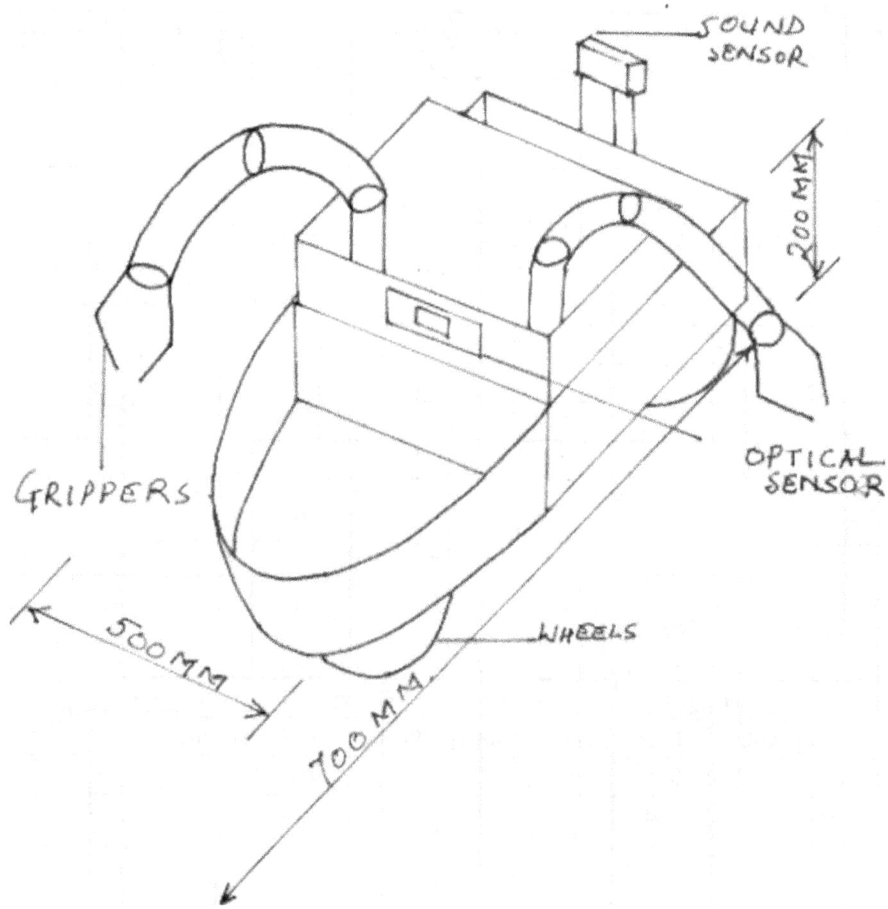

SYSTEM DESIGN AND EVALUATION

The process of system design begins after finalizing and completing design concept. In the system design section, the concept's principal systems are identified and the interfaces between the systems are established. These interfaces will be the main requirements of the subsystem. The system identification matrix (SIM) for the litter collecting robot is first created to determine the relationships between this robot's subsystems and each PC and FR. Then, the top level systems and sub-systems are combined together into a primary product architecture schematic (PAS).

1. Inputs/Outputs of the System

Table 6 details input and output of the robot's system.

2. Identify Subsystems

The robot components are subdivided into the hierarchy of robot's sub-systems.

Table 6. Input/output of our system

No.	Inputs	Outputs
1	A disorganized place having lots of litter	An organized place having no litter
2	Control and program	Walking and looking around the specified area for litter, collecting & putting them at predefined area
3	Dissatisfied persons working this area (Robot workspace)	Satisfied person in this working area (Robot workspace)
4	Ultrasonic sensor	Energized to look around for litter and differentiate between litters, recognize people working around robot and other useful stuff.
5	Actuator	Useful rotation of robot to pick up litter as requirement
6	Sensor motor	To fully functionalize the robot
7	Battery	To energize the robot to successfully carry out the function (kinetic energy movements)
8	Sound sensor – hardware	
9	Light sensor – hardware	
10	End of arm tooling(changeable)	To pick up litter of different size as per requirements
11	Wheels	To move around the working area smoothly
12	Reusable software and software preprogrammed to suit as per different situations	Better cleaned space and more satisfied customer
13	Remote control	To control the system without supervision
14	Person having a Remote/command controller for robot	Command execution by robot and stop automatically when work finish
15	Bottom of microcomputer	Execution of preloaded software program
16	Empty garbage bin at the back of robot before collecting litter	Full garbage bin at the back of robot after collection of litter
17	Electrical cables having no power for data carry out	Carryout data for useful means
18	High cost to clean area (if manpower would have used)	Negligible cost to clean areas
19	No satisfaction for visitors (in a nasty environment)	Highly satisfied visitors (in a cleaned environment)
20	Damage to environment	Environment perfection
21	"No hurt" to persons and usable indoor contents	"Hurt" to person and usable indoor contents
22	"DIRTY" night	"CLEAN" morning
23	"NO TECH" environment	"HIGH TECH" environment
24	Gainful environment	No gainful employment
25	No initial cost to system	High initial cost to systems
26	High safety concern due to exposure of people to hazardous environment	No safety concern for people
27	No damage to environment at the end of tenure	Environmental damage at the end of its life cycle (when disposal of robot came into picture)
28	Dependency on people by cleaning company	"No more dependency' on people by cleaning company
29	Possible information leak at highly secretive area by untrusted people	No chance for leak of information from a closely guarded work site
30	No risk to un attentive children	Personal risk for un attentive children while robot is in operation
31	No efficient services to customers	Efficient services to customers

continued on following page

Table 6. Continued

No.	Inputs	Outputs
32	No energy consumption	Energy consumption is high while in operation
33	'No proper display of image' to customers	'Display of image' to customers.
34	Neighbors' friendly	Neighbors' envy
35	Never help the organization when they are ideal	Can help the organization in putting the proper thing at proper place at proper time when robot has no other litter collecting work to do.

a. Robot Structural Sub-System

- Main body frame
- Support structures for the frame
- Fasteners to join body parts

b. Robot Sensor Sub-System

- Sound sensor activated by sound and a voice command.
- Ultrasonic vision sensor to detect litter
- Touch sensor for tactile sense of litter
- Light sensor to distinguish litter and other objects
- Interfacing hardware/electronic components to synchronise sensors

c. Robot Motor Drive Sub-System

- Two servo motors
- Timer belts

d. Power Subsystem

- On-board power supply
- A rechargeable power supply
- Interface subsystem, robot system computer for remote command and execution of commands
- Blue tooth
- Programming data from main computer by USB
- Programming on robot computer
- Controller

e. End of Arm Tooling

- Cylinders operated by servomotor
- Gripper (size and shape depends upon nature of the work environment and the litter)

f. Driven Unit of Robot

- Front wheel (rubber)
- Back wheel (nylon)
- Drive shaft
- Brake (connected to servomotor)

3. System Identification Matrix

The system identification matrix (SIM) is created based on the first level of analysis of the product design (see function requirements in section 2-1). All FRs and PCs are allocated to one or more subsystems, and every subsystem is designed to support certain functions of the main product.

4. Product Architecture Schematic

Product architecture schematic (PAS) and Workflow diagram for the product should be developed to illustrate inputs, outputs, internal transactions, flows, and a general overview of the system. The PAS and workflow diagrams of the litter-collecting robot are provided in Figures 6 and 7 in Appendices B and C, respectively.

Table 7. System Identification Matrix (SIM)

Functional requirement /Subsystems	Robot Structure	Robot Sensors	Servomotor	Power Unit	Hardware	Software & Interfacing	End of Arm Tooling	Driven Unit
Safety			√	√			√	
Litter collection								
Durability	√		√		√			
Quality		√		√		√		√
Affordability								
Fabric ability								
Install ability								
Usability			√			√		
Maintainability		√	√		√		√	
With stand Frequent usage						√		
Resist rust	√				√			
Resist electric shock		√		√				√
Prevent current from leakage		√		√				
Less breakdowns		√				√	√	

CONCLUSION

Every design is intended to make our life better and easier. This study discussed the systematic and practical process for product design which is improved and developed based on current research in this field. The litter-collecting robot case study discussed in this paper constitutes an advanced tool in terms of mechanical and electrical specifications. The main target was to demonstrate this practical design approach to a case study involving the design of a litter collecting robot. It is applicable to other industrial products.

The first part of this study is research and a comprehensive survey on user's ideas about this product. According to this survey, product users are concerned primarily with safety. Other user priorities in order are functionality, quality and usage location. Six other factors, including form/style, people, market window, environment and technology are important to users. It means that the kind of technology used for robotic operation and environmental effect are of little concerned.

In the second part, based on users' needs, the characteristics of the robot are explained which includes engineering, marketing, manufacturing, and environmental aspects of the product design. The results of this section are in the product design specification (PDS) which is used in the conceptual design. A team of four experts in product design ranked main factors and generated concepts through 20 pairs of the main important factors (out of 9! =362,880 pair combinations) with considering the details of requirements and constraints. They mainly used analogical reasoning while designing the exterior shape of the robot and the interior structure to do functions. Based on a review and analysis of these four individuals' concepts conducted in a group meeting, a decision matrix for these concepts was developed. The advantages of all concepts were analyzed and applied to modify and improve a new concept to minimize disadvantages as much as possible.

In the last section, in the system design, the system identification matrix (SIM), inputs, outputs, main and subsystems are identified and they are combined together into a primary product architecture schematic (PAS). This design process and case study illustrate an improved systematic and practical approach in product design that is valuable for researchers and professionals in industrial environment and can be applied readily to the design of other industrial products.

REFERENCES

Ashby, M. F. (2005). *Materials selection in mechanical design, MRS BULLETIN.* Cambridge, UK: Cambridge University Press.

Asimov, I. (1994). The Robots of Dawn, Spectra publication, Washington D.C., USA

Bailetti, A. J., & Litva, P. F. (1995). *Integrating Customer Requirements into Product Design, the Journal of Product Innovation Management, 12 (1), Baker, M. & Hart, S. (1999). Product Strategy and Management.* Harlow, UK: Prentice Hall.

Brown, K., Schmied, H., & Tarondeau, J. C. (2003). Success Factors in R&D: A Meta-Analysis of the Empirical Literature and Derived Implications for Design Management. *Design Management Journal, 2*, 72–87.

Critchlow, J. (1985). *Introduction to Robotics*. Michigan, USA: Collier Macmillan.

Davila, T. (2000). An Empirical Study on the Drivers of Management Control Systems, Design in New Product Development. *Accounting, Organizations and Society*, *25*(4–5), 383–409. doi:10.1016/S0361-3682(99)00034-3

Dhillon, B. S. (2007). *Applied Reliability and Quality*. London: Springer-Verlag.

Ernst, H. (2002). Success Factors of New Product Development: A Review of the Empirical Literature. *International Journal of Management Reviews*, *4*(1), 1–40. doi:10.1111/1468-2370.00075

Hundal, M. S. (1991). Use of Functional Variants in Product Development, *ASME Design Theory and Methodology Conference*, Miami, Florida, USA.

Matthiassen, B. (1997). *Design for robustness and reliability – improving the quality consciousness in engineering design*. Lyngby: Department of Control and Engineering Design, Technical University of Denmark.

Pahl, G., & Beitz, W. (1993). In K. Wallace, (Ed.), *Engineering Design: A Systematic Approach*. New York: Springer-Verlag.

Shore, H., & Arad, R. (2003/2004). Product robust design and process robust design: Are they the same? *Quality Engineering*, *16*(2), 193–207. doi:10.1081/QEN-120024007

Sturges, R. H., O'Shaughnessy, K., & Kilani, M. I. (1990). Representation of Aircraft Design Data for Supportability, Operability, and Producibility Evaluations, *Carnegie Mellon University Engineering Design Research Center Report,* 01-30-90.

Sturges, R. H., O'Shaughnessy, K., & Reed, R. G. (1993). Systematic Approach to Conceptual Design. *Concurrent Engineering*, *1*(2), 93–105. doi:10.1177/1063293X9300100202

Suh, N. P. (1988). The Principles of Design. New York: Oxford University Press.

Taguchi, G. (1986). *Introduction to quality engineering – designing quality into products and processes*. Tokyo, Japan: Asian Productivity Organization.

Tennant, G. (2002). *Design for Six Sigma: launching new products and services without failure*. Hampshire, UK: Gower Publishing Limited.

Tomiyama, T. (2005). Meijer BR, Directions of Next Generation Product Development. In H. A. ElMaraghy & W. H. ElMaraghy (Eds.), *Advances in Design* (pp. 27–35). London: Springer.

Tomiyama, T. (2006). A Classification of Design Theories and Methodologies. *Proceedings of the 2006 ASME IDETC*, Paper No. DETC2006-99444, ASME. (CD-ROM) doi:10.1115/DETC2006-99444

Tomiyama, T., Gu, P., Jin, Y., Lutters, D., Kind, C., & Kimura, F. (2009). Design methodologies: Industrial and educational applications. *CIRP Annals-Manufacturing Technology*, *58*(2), 543–565. doi:10.1016/j.cirp.2009.09.003

Ulrich, K. T., & Eppinger, S. D. (2004). *Product Design and Development* (3rd Ed.). McGraw Hill, Irwin.

Whitehead, T. N. (1954). *The design and use of instruments and accurate mechanism*. New York, USA: Dover Publications.

KEY TERMS AND DEFINITIONS

Conceptual Design: Conceptual design is a preliminary draft to showing ideas and to convert product specifications and requirements to new solid models.

Process Design: Process design is creating and developing a sequence of activities that transform inputs to outputs considering customer expectations, output specifications, costs and other constraints.

Product Architecture: Product architecture is a scheme that shows the functional elements of a product and the interfaces between these elements.

Product Design: Product design is an effective way of designing and developing of ideas through a process that leads to new products.

System Design: System design is the process of identifying the components, modules, structure, inputs, outputs, interfaces, and data of a system based on the specified requirements.

System Identification Matrix (SIM): System Identification Matrix is a matrix that shows the relation between subsystems and functional requirements (the functional requirements are met by each subsystem).

Workflow: Workflow is a scheme that illustrates a series of activities that converts inputs to outputs.

ENDNOTES

1. Survey sampling is about the sample selection of the target population for conducting a survey. The process includes several steps: defining the population of concern, identifying a set of measurable events or items (sampling frame), determining a sampling method for choosing events or items from the target population, defining the sample size, implementing the sampling plan, and sampling and data collection.
2. We considered four constraint categories: functional, environmental, life cycle, and structural. Once we reviewed product specifications, the technical constraints of the robot were determined. Since they are too technical, they are not included here. Within the four main categories of constraints, there are many specifications in the category of functionality/performance, environment, product life span, life in service, shelf life, cost, quantity, maintenance, marketing, packaging, size and weight restrictions, shipping, manufacturing processes, aesthetics, ergonomics, customer requirements, quality and reliability, company constraints, processes, safety, testing, legal, installation, documentation, and disposal fields.

APPENDIX A

Potential Customer Survey

(All questions are rated on a scale of zero to ten; with zero meaning "of no importance" to ten, "of vital importance".

a. Function/Need

i. What priority (main characteristic) is most important in a litter collecting robot? (check one only)
   - Do you want to save time?
   - Do you want to enhance comfort level?
   - Are you physically able to pick litters?
   - Do you want to reduce manpower and automate litter picking?

ii. Does the shape/size/weight of the robot matter to you?

iii. How does the other functionality of the product matter to you (in addition to collect litters)?

b. People

i. How important is user friendliness to you as compared to other factors?

c. Environment (Context, Infrastructure)

i. Do you want to clean your work place on a continual basis?

ii. How important is product disposal after its life to protect your environment?

d. Market Window

i. How important is product durability to you?

e. Technology

i. What type of command/control do you need?

ii. How important is energy saving to you?

f. Usage Locations

i. What is your preferred location to use this product (Hotel/hospital/shopping center/school/house/ public common place/any other location-please specify)?

g. Form/Style/Aesthetics

i. How important are form/style/face to you?

h. Quantity/Cost

i. How important are post purchase expenditures to you?
ii. What type of value addition is important for you by using this product (please specify)?

i. Standards/Guidelines/Protocols/Laws

i. How important is safety to you?
ii. How important is the ergonomic design of products to you?

APPENDIX B

Workflow Diagram

Figure 6. Workflow Diagram for the Litter Collecting Robot

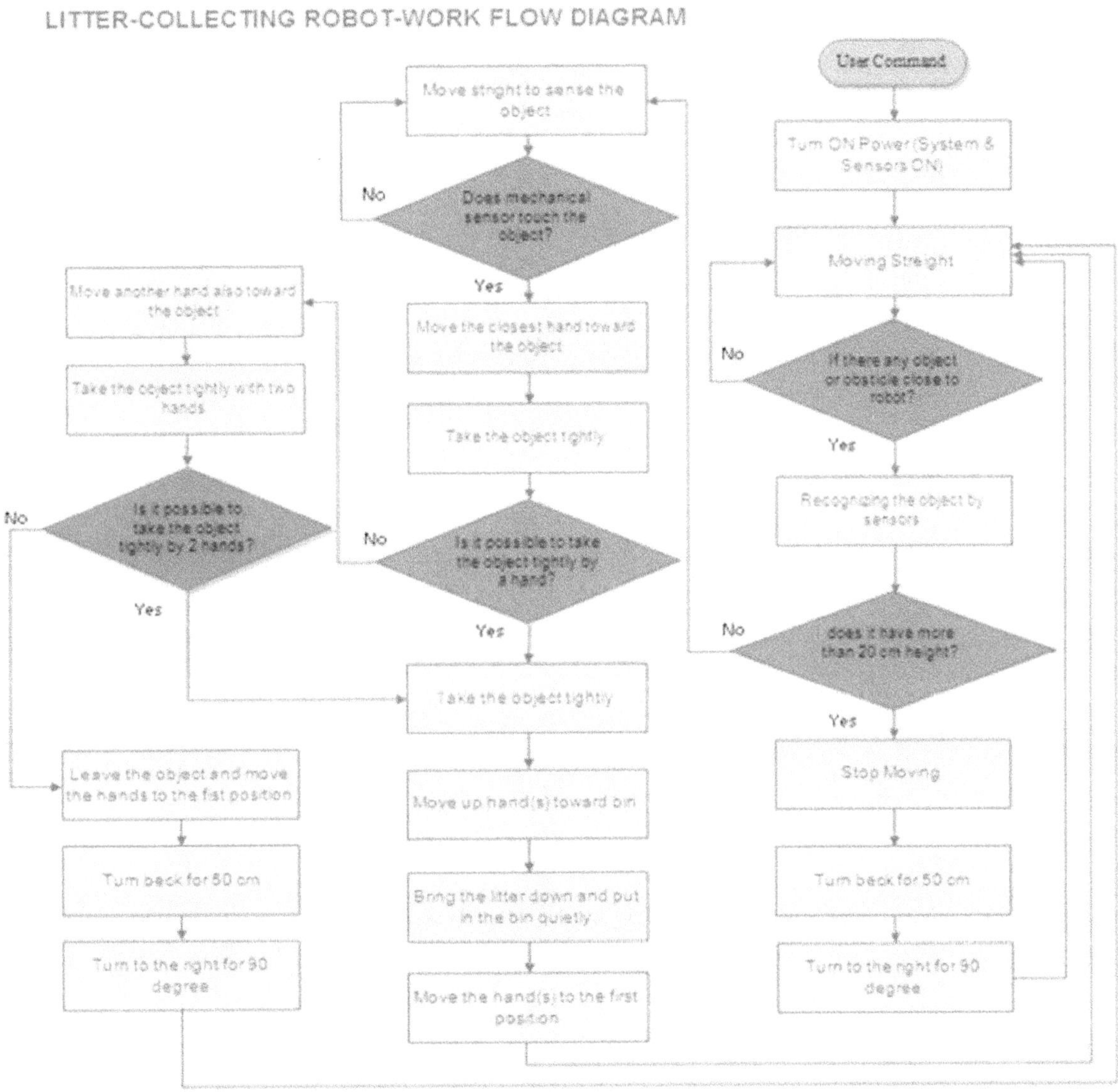

APPENDIX C

Product Architecture

Figure 7. Product Architecture Schematic for the Litter Collecting Robot

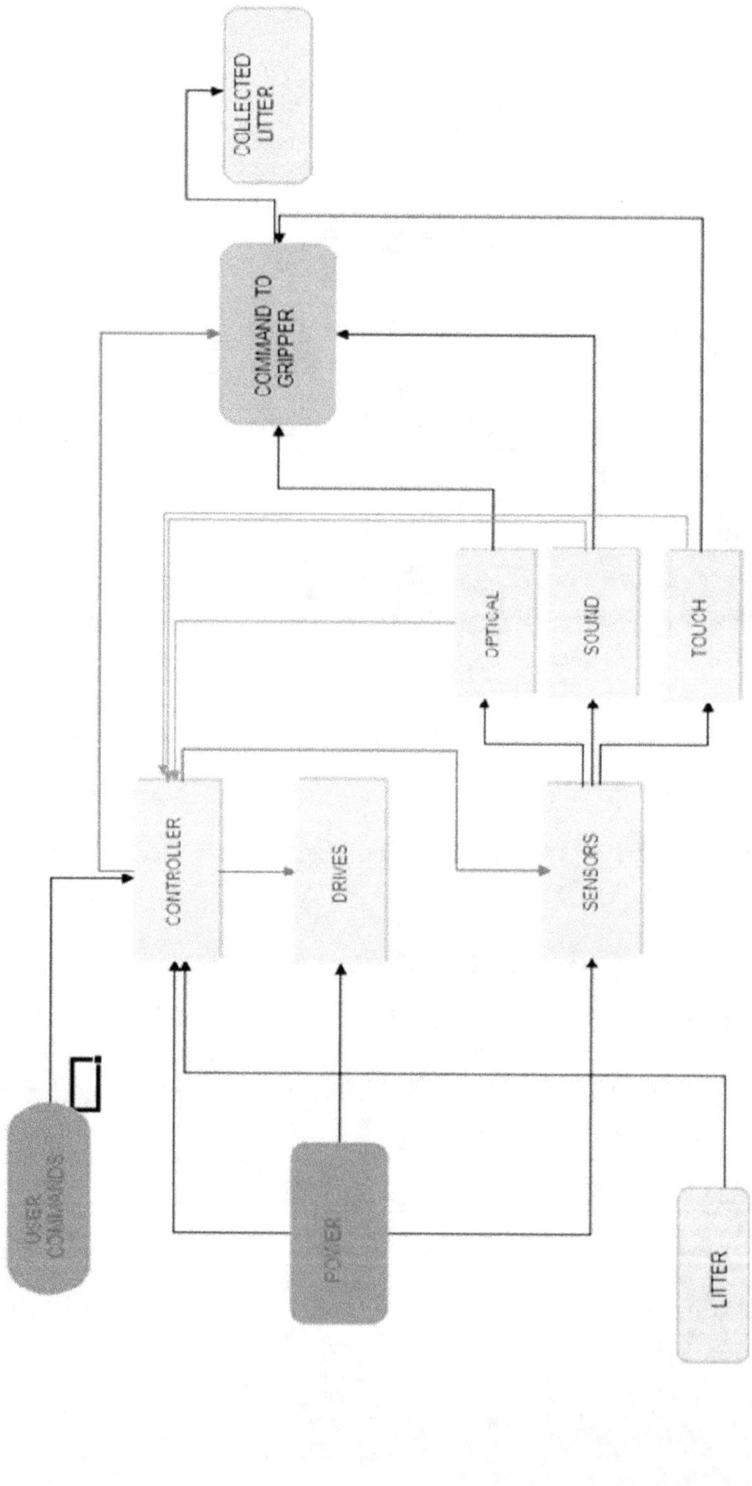

Chapter 7
Investigation of Optimum Conformations and Structure Analysis of RL and LR Nests using Ramachandran Plot

Sumukh Deshpande
University of Hail, Saudi Arabia

Saikat Kumar Basu
University of Lethbridge, Canada

Pooja Purohit
Almas International, Saudi Arabia

ABSTRACT

We have surveyed polypeptides with the optimal conformations of nests which are the common anion-binding motifs comprising 8% of the amino acids which are characterized by a structural depression or a hole. Using automated bioinformatics algorithm, novel ring structure of the nest has been found. Using automated algorithm, models of polypeptides were made in-silico (computationally) and oxygen atoms are inserted along the extension of the NH groups. These sophisticated algorithms allow insertion of atoms along the NH group at the correct distance which causes extension of the group thus forming hydrogen bond. Optimal conformations of these structures are found from these customized models. This study chapter provides a demonstration of an important discovery of optimum conformations of RL and LR nests by the use of sophisticated bioinformatics automation pipeline and a unique application of automation and control in bioinformatics.

INTRODUCTION

Robotics and Automation has revolutionized current technologies. Automation is widely used in the area of scientific research. One of the major applications of automation lies in the area of bioinformatics (Ochoa et al., 2011). One of the bioinformatics techniques involves the usage of automated algorithms and control in the development and discovery for optimum structures of protein structure motif called

DOI: 10.4018/978-1-4666-8693-9.ch007

as nests. These structural motifs form polypeptide main chain structure consisting mainly of α-helix and β-sheet which forms secondary, tertiary and quaternary structures in proteins. But apart from these two major conformations, other conformations also exist. Some of them are 3_{10}Helix, α-sheets, nests, niches, asx turns, β bulges, β turns (type I and I', type II and II'), ST staples, ST turns, Schellman loops (Watson & Milner-White, 2002a; Duddy, 2004). Nests are formed when the main chain NH groups of three successive residues bind a common anion-binding motif. In nests, when main chain NH groups bind an anionic atom it forms a depression and or a hole in which the anion fits thus forming a nest structure (Torrance, 2009; Watson & Milner-White, 2002a; Watson & Milner-White, 2002b; Milner-White, 2004; Milner-White & Russell, 2005; Milner-White, 2006). Nests are more common structures among all the other conformations and 8% of the amino acids take part in forming nests. Like nests, a structure known as niche also exist. A niche is formed main chain CO (Carbonyl) groups of three successive residues bind a cationic atom or group often HOH or NH_2 (Torrance, 2009). Nests are defined by two alternative enantiomeric main chain polypeptide conformations with four characteristic main chain dihedral angles of two successive amino acid residues: φ_i,ψ_i and φ_{i+1},ψ_{i+1}. Negative φ values are considered to be right-handed (R) and positive φ values left-handed (L). The two types of conformations RL and LR are found in native proteins (Watson, 2002; Watson & Milner-White, 2002; Milner-White, 2004). Nests are also found in compound or overlapping formation where a combination of right handed and left handed residues make compound nests such as RLRL nests, LRLR, RLLR. Some of the examples of compound nests include P-loops of nucleotide tri-phosphate binding proteins with five consecutive NH groups facing into a concavity forming an LRLR structure (Watson, 2002; Watson & Milner-White, 2002; Milner-White, 2004; Milner-White & Russell, 2008).

From previous work done, it was found that all enantiomeric peptides form ring structures (Watson & Milner-White, 2002). Enantiomers are those molecules which are optical isomers of each other. These enantiomers are found at $-90^0, 0^0; 90^0, 0^0$ in Ramachandran plot. When LR conformations are considered, most of residues form mirror peptides. Mirror peptides are those molecules are almost linear-shaped repeating dipeptides. They are defined by $\varphi_i = -\psi_{i+1}$ and $\varphi_{i+1} = -\psi_i$.

Previous work on nests included finding average parameters in nest subclasses which include α-turns, Asx-nest, ST-nest, paperclip, type I β-turn, type I β-bulge loop, type 2 β-bulge loop (Watson & Milner-White, 2002a). From the parameters calculated average values were taken. For RL, φ_i,ψ_i was -94,-3 and $\varphi_{i+1},\psi_{i+1}$was 76,22. For LR, φ_i,ψ_i was 74,16 and $\varphi_{i+1},\psi_{i+1}$was -77,-20 (Watson & Milner-White, 2002). These values were used to construct the first two models of rings.

A description of the rings in the Ramachandran plot provides an illustration of various conformations found in the plot when an analysis will be performed (Fig. 1). When we take oligopeptides having certain φ, ψ angles, various conformations such as rings, alpha helices, beta sheets and linear structures are found in the plot.

Using certain bioinformatics algorithms and tools, an automated pipeline can be constructed which will help us to model polypeptides using specific combinations of φ, ψ angles. Certain parameters were used for modeling helices which were helical radius, helical climb and helical angle. Helical radius is the radius of the helix from the helical axis. Helical angle is the angle subtended by the dipeptide on the helical axis. Helical rise or helical climb is the distance along the helical axis per dipeptide. These three features were used to determine the behavior of peptides in Ramachandran plot. Determination of the behavior of peptides was achieved using MATLAB software (MathWorks, 2012). The pipeline also included a program which allowed the user to choose the φ, ψ angles of the peptide. This allows enhanced control, flexibility as well as customization over the automated pipeline which utilizes certain

Figure 1. Various conformations in Ramachandran plot

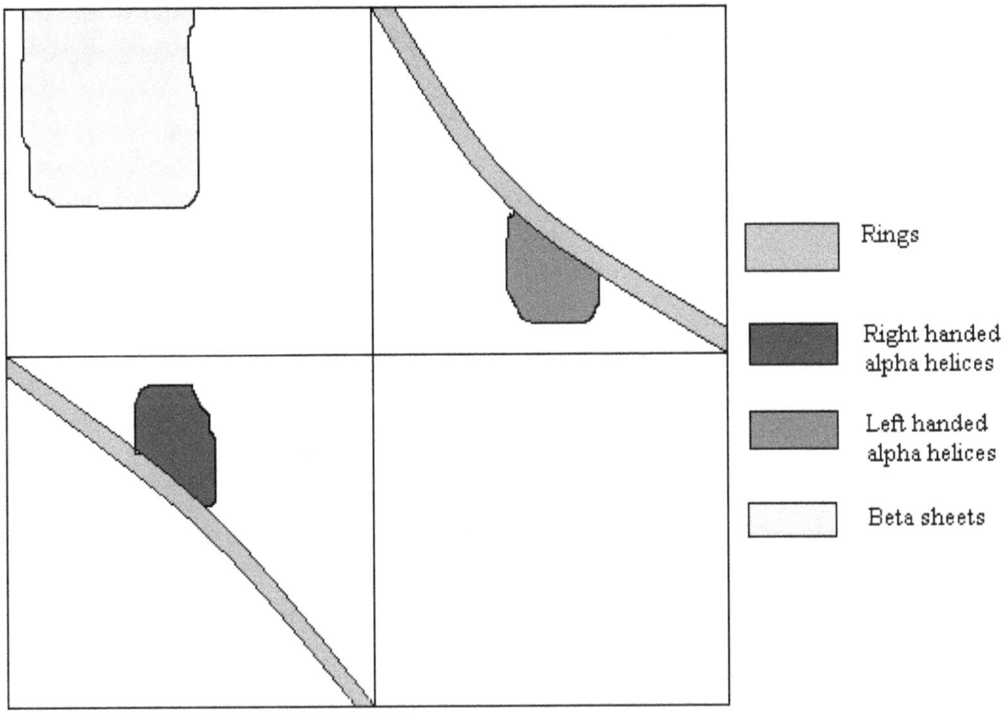

parameters such as φ, ψ angles as input and displays a contour plot describing features in the respective plot. This contour plot was plotted on Ramachandran plot illustrating the areas in the plot where the various angles, radius and climb can be found.

MATERIALS AND METHODS

Models of polypeptides were made with Ramachandran Plot Explorer version 1.0. For this analysis 8 amino acid alanine polypeptide was constructed for this research. The backbone torsion angle of each residue has to be changed in order to make that specific model. Since a polypeptide model is made up two sets of dihedral angles namely φ_A,ψ_A and φ_B,ψ_B in which φ_A,ψ_A is α_R (helix), and φ_B,ψ_B is α_L helix. The combination of φ_A,ψ_A and φ_B,ψ_B gives rise to RL nest or LR nest depending on which is required. Here we study RL nests, so we require that combination of angles. The values φ_A=-90, ψ_A=0 were used for the first residue and the values in the region of φ_B=70, ψ_B=20 for the second residue were varied. For φ_A,ψ_A the residues are 1, 3, 5 and 7. For φ_B,ψ_B angles the residues are 2, 4, 6 and 8.

For insertion of an oxygen atom along the NH group, a python program was used which was runnable on PyMol software (PyMOL Schrödinger, 2010). Two atoms were inserted at position NH(i) and NH(i+2) namely, O(i) and O(i+2). For creating a complete polypeptide model for the analysis, two oxygen atoms were inserted at two positions along NH group on the main chain. They were added along the extension of the main-chain NH group at the correct hydrogen bonding distance for H···O of 1.75 A. New oxygen atom was along 5th NH group position at a distance of 1.75 A.

Many polypeptide models were made by fixing φ_i, ψ_i at -90⁰,0⁰ and varying φ_{i+1}, ψ_{i+1}. The φ_{i+1} value was varied from 45⁰ to 100⁰ and the ψ_{i+1} value varied from -20⁰ to 45⁰. The distance and bond angles between oxygen atoms O(i) and O(i+2), NH(i) and O(i+2) were measured using PyMol (PyMOL Schrödinger, 2010).

The deviation of NHO bond was measured using trigonometric functions and formulas. The diagrammatic illustration is provided in Fig. 4 which describes the angular deviation of the NHO bond. To determine angular deviation, three figures are required viz. NH(i) to NH(i+4) length, length from NH(i) to O(i) and length from O(i) to NH(i+4). S is calculated by addition of three figures and dividing the sum by 2. Next the area of triangle was calculated by herons formula:

$$Area = \left[s(s-a)(s-b)(s-c)\right]^{\frac{1}{2}} \quad (15)$$

When the area is calculated, the height of the triangle i.e. length of O(i) to Z is calculated by the formula, h = (2 * area)/base. When the height is known, length of NH(i)Z is calculated by Pythagoras Theorem knowing distance from NH(i) to O(i) and height O(i)Z since O(i)ZNH(i) is a right angled triangle. This is given by the formula: NH(i)Z=%(NH(i)O(i))^2-(O(i)Z)^2)]$^{1/2}$, where NH(i)O(i) is distance between NH(i) and O(i) and O(i)Z is distance between O(i) and Z. When NH(i) to Z distance is known, angle Y is calculated by the formula: sin Y = NH(i)Z/NH(i)O(i) . Since angle Y and Z is known, the actual deviation angle X is calculated by X = 180 – angle Y – angle Z.

Schellman loops are the commonest motif to incorporate a nest, which is an RL nest (Watson & Milner-White, 2002a; Duddy, 2004). They were examined, with particular reference to the nest. The i+1 residue of the Schellman loop hydrogen bonds to the i+4 residue whereas the ith residue bridges the i+3 and i+5 residues. However, many of them occur at the C-termini of the α-helices and have a different hydrogen bond arrangement.

RESULTS

Each of the modeled octapeptides were examined using PyMol and its conformation was studied. We already know that, in a Ramachandran plot, rings are found across a diagonal as shown in Figure 7. To the upper right hand side of the rings diagonal, α_R helices are found and to the lower left hand side of the rings, α_L helices are found. In the repeating dipeptides, the φ_i,ψ_i angles are fixed at -90⁰,0⁰.

1. Structural Analysis of Nests

For the nest in proteins the average values for RL nests are α_R with φ=-94, ψ=-3 and α_L with φ=76, ψ=22 while those for the LR nest are α_L with φ=73, ψ=20 and α_R with φ=-83, ψ=-22. This was the basis for the values chosen for modeling (Fig. 2). Since these values are the average values, we take the average around these values and start our analysis.

Nests used for binding an oxygen atom have a depression in which the oxygen atom fits and binds both the NH groups to form a nest structure. This is the ideal structure of the nest as previously discussed (Watson & Milner-White, 2002)]. The ideal nest conformation comprises of two successive NH groups bonded to an anion (Fig. 3). Alternate NH groups bind the anion atom at an N-H···O angle of 180⁰ thus forming a perfect nest structure.

Figure 2. (A) Position of RL nests, and (B) LR nests in Ramachandran plot.

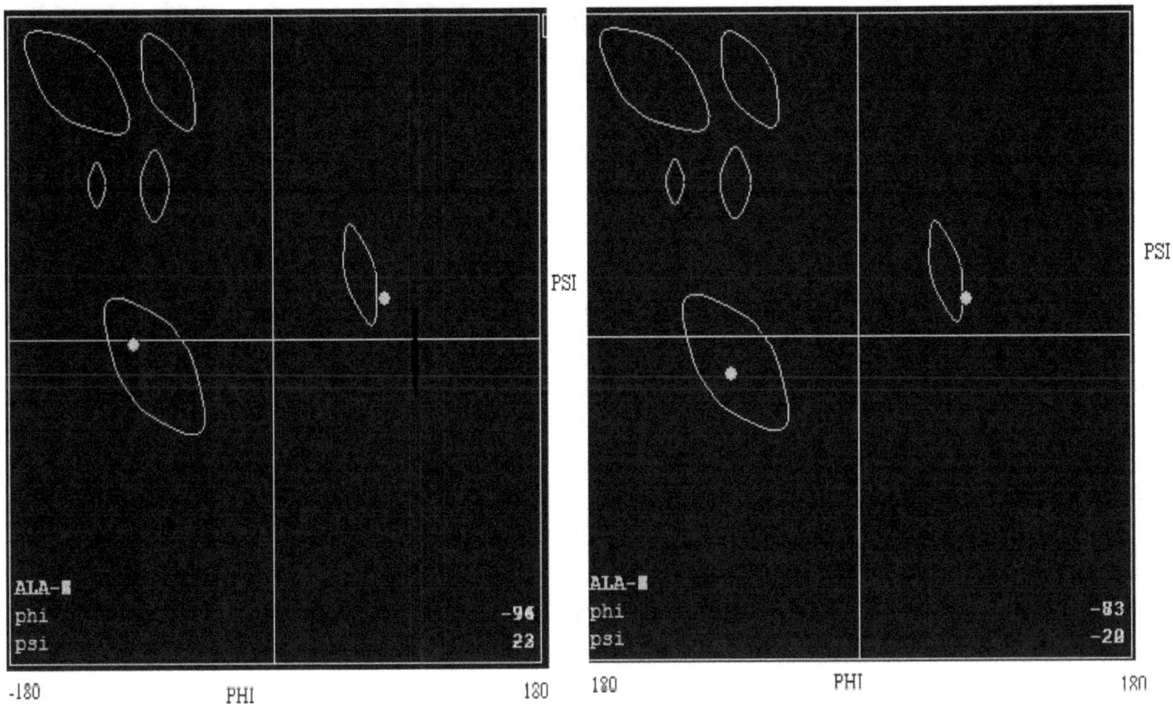

A **B**

Figure 3. (A) Structure of an ideal nest in molecular form where blue shaded molecules represent NH bonded to yellow coloured OH atoms, and (B) Simplified representation of nest where two NH bonds bind an oxygen ion.

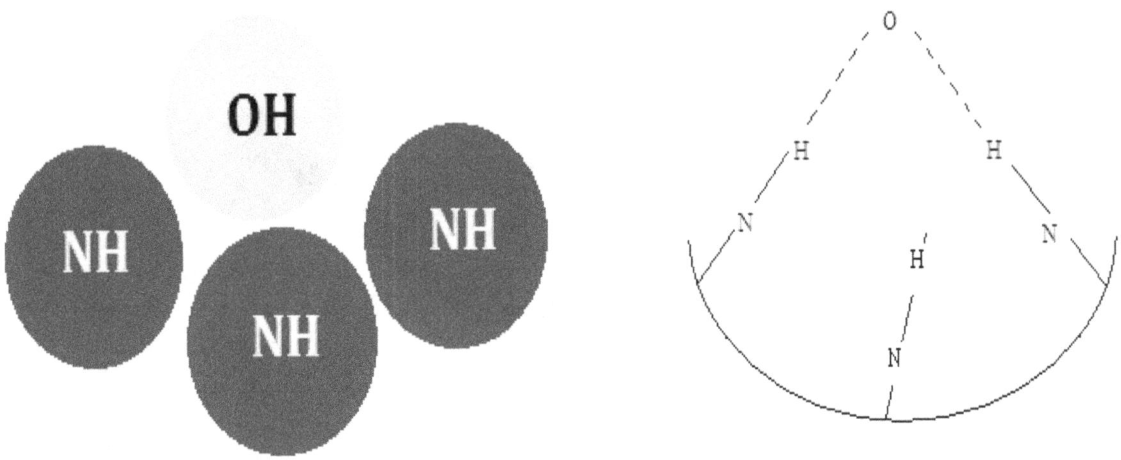

A **B**

Our first major finding was the shape of the nest. Since the shape of the nest as discussed earlier was thought to be a ring structure having one reference plane for all the main chain atoms (Torrance, 2009; Watson, 2002; Milner-White, 2004; Milner-White & Russell, 2005; Milner-White, 2006). But upon analysis the shape of ring structure was different (Fig. 4). RL or LR nests have a "Crown" or "zig zag" conformation. This ring polypeptide structure consisted of 8 residues in which the first residue was the α_R and the second is α_L. It repeats to form an RLRLRLRL structure where the α_R is on the top side of the crown and the α_L is on the down side of the crown, the RLRLRLRL being (up, down, up, down, up, down, up, down).

The oxygen atom was expected to bridge two NH groups with N-H···O angle of 180^0. But the analysis done on rings gave striking result. The results showed that in a ring, the main-chain NH groups face to the left direction with an angular deviation of 30^0 (Fig. 5). This refers to the projection of the NH groups onto the plane of the ring and measurements as within that plane. The oxygen atoms (O3 and O5) are inserted at NH groups number 3 and 5 as shown in Figure 16A. NH3 binds to O3 and NH5 binds to O5. The O3 and NH5 distance is 4.29 A. But when O5 to NH3 distance is calculated, it comes out to be 1.71 A which is approximately same as O5 to NH5 distance. This means that O5 bridges NH3 and NH5.

The distance between the oxygen atoms bound to NH groups is important for finding optimal nest conformations in the rings. Ideally, they coincide, so the distance should be zero. From Fig. 5, O3 binds to NH3 and O5 binds to NH5. If the distance between the NH group and the oxygen atoms (O3 and O5) is optimal, then O3 bridges NH1 and NH3; O5 bridges NH3 and NH5. It means that an oxygen atom bridges two NH groups and one NH group binds to two oxygen atoms.

Figure 4. The zig zag conformation of nest

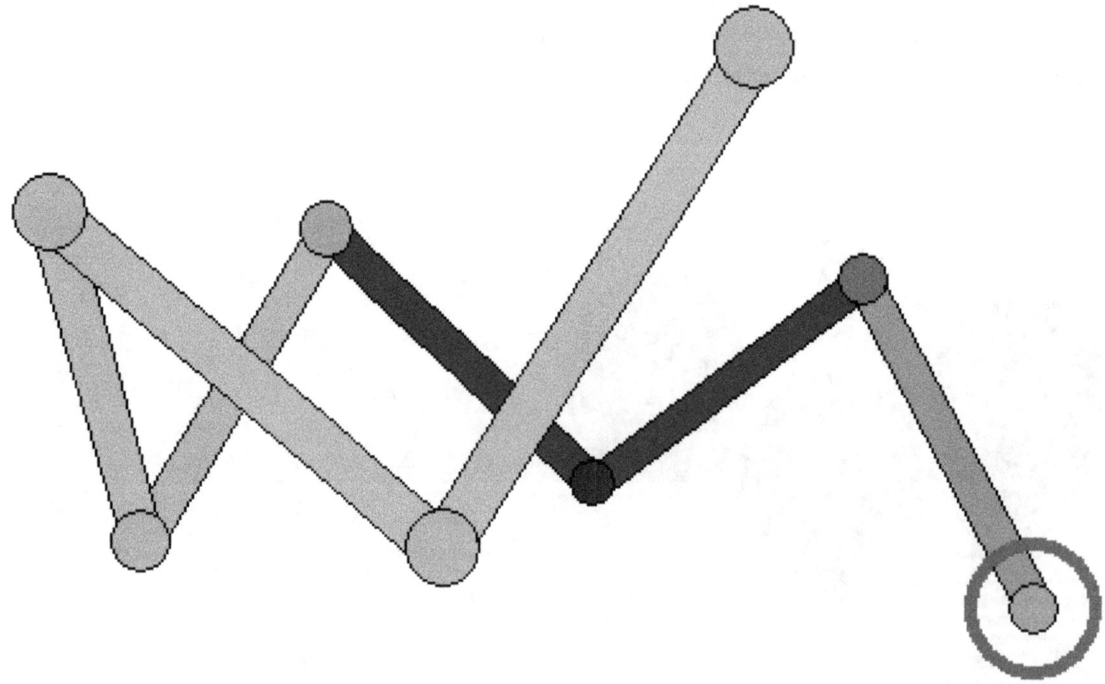

Figure 5. (A) Structure of ring with NH group bonded to OH group, (B) Structure of ring showing hydrogen oxygen bonding

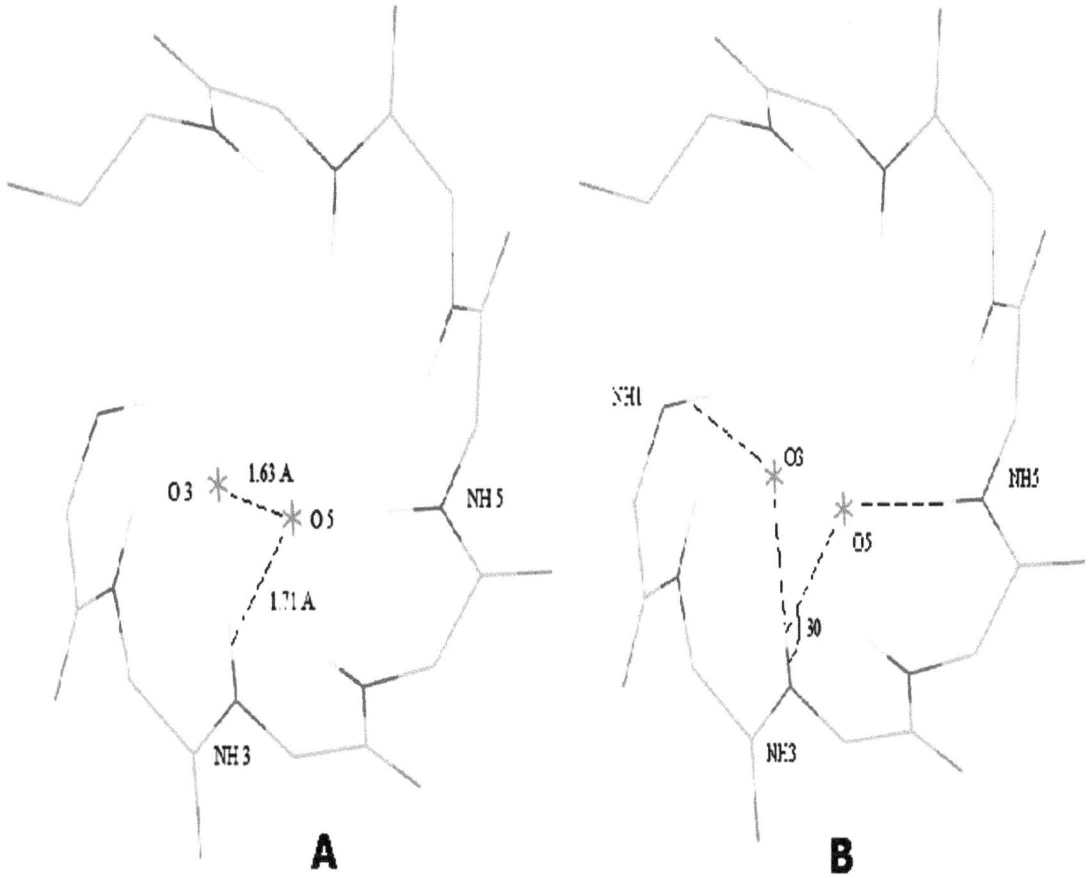

2. Ramachandran Plot Analysis of Polypeptides with Varying φ,ψ Angles

The analysis was done by building many polypeptide models with repeating dipeptides. Two sets of angles φ_A,ψ_A and φ_B,ψ_B. The φ_A,ψ_A angle was kept constant at -90^0,0^0 and φ_B,ψ_B were varied φ_B from 45^0 to 100^0 and ψ_B from -45^0 to +45^0. For each model the distances between oxygen atoms, between hydrogen and oxygen and the NHO bond angle for both RL and LR nests.

2.1 Analysis of OH-OH Distance Atoms in RL Nests

Table 1 gives the data for distances between the two oxygen atoms which are inserted along the NH groups for RL nests. Each distance is calculated by keeping φ_A,ψ_A constant as -90^0,0^0 and varying ψ_B. From the table, if we keep psi B as constant and just vary phi B, the distance decreases gradually from 4.45 A to 2.87 A from 45^0 to 100^0. This means that the distances between the oxygen atoms decreases going from left to right in the Ramachandran plot. Similarly the distance decreases horizontally for all

Table 1. Tabular data of PHI B, PSI B values of RL nests for distances between the oxygen

PHI B	PSI B1	PSI B2	PSI B3	PSI B4	PSI B5	PSI B6	PSI B7	PSI B8	PSI B9	PSI B10	PSI B11	PSI B12	PSI B13
	-20	**-15**	**-10**	**-5**	**0**	**5**	**10**	**15**	**20**	**25**	**30**	**35**	**40**
45	4.45	4.16	3.87	3.59	3.31	3.05	2.8	2.58	2.4	2.26	2.18	2.16	2.2
50	4.33	4.03	3.74	3.45	3.17	2.9	2.65	2.43	2.25	2.12	2.04	2.03	2.1
55	4.21	3.91	3.61	3.32	3.03	2.76	2.51	2.29	2.11	1.98	1.92	1.93	2.01
60	4.08	3.77	3.47	3.17	2.89	2.61	2.36	2.14	1.97	1.86	1.82	1.85	1.96
65	3.94	3.64	3.33	3.03	2.74	2.47	2.22	2.01	1.85	1.76	1.74	1.8	1.93
70	3.8	3.49	3.19	2.89	2.6	2.33	2.09	1.89	1.75	1.68	1.69	1.78	1.94
75	3.65	3.35	3.04	2.74	2.46	2.2	1.97	1.78	1.67	1.63	1.67	1.79	1.98
80	3.5	3.19	2.89	2.6	2.32	2.07	1.86	1.7	1.61	1.6	1.69	1.84	2.05
85	3.35	3.04	2.74	2.45	2.19	1.95	1.76	1.63	1.58	1.61	1.73	1.91	2.14
90	3.19	2.89	2.6	2.32	2.06	1.85	1.68	1.59	1.58	1.66	1.8	2.01	2.26
95	3.03	2.74	2.45	2.18	1.95	1.76	1.63	1.57	1.61	1.72	1.91	2.13	2.39
100	2.87	2.58	2.31	2.05	1.85	1.68	1.6	1.59	1.67	1.82	2.03	2.27	2.54

values of psi B from -20^0 to +40^0. The maximum distance between oxygen atoms is found to be 4.45 A for φ_B,ψ_B as 45^0,-20^0. And the minimum distance is found to be 1.57 A for φ_B,ψ_B as 95^0,15^0 (Fig. 6). The distance between oxygen atoms first decreases from 4.5 A to 2 A, and starts to increase again. The area expected for the nests increases from lower left to upper right as the distance between oxygen atoms decreases. The area occupied by the nests is maximum when the distance is between 1.75 to 2 A and the area is minimum when the distance is minimum. This sudden decrease in area is due to steric hindrance taking place.

2.2 Analysis of O-H Distance and NHO Bond in RL Nests

Table 2 gives the data for distances between the hydrogen and oxygen for RL nests. From the distances calculated by analyzing the models, optimum conformations can be found. As the distances decreases (φ_B constant ψ_B constant) from 45 A to 100 A, the distance decreases going from left to right and bottom to top in the Ramachandran plot. It also shows an increase in the value when minimum value is reached. Fig. 6 shows the contour plot of RL nests showing the distance between hydrogen and oxygen for different models of nests with various degrees of φ_B,ψ_B values. From the plot distance decreases, area increases and the probability of finding optimal conformations also increases. The area for forming optimal nest conformation is more when the distance is 1.75 A and 2 A and the area is less when the distance is 3 A or 1.5 A. Table 3 gives the data for the angle formed by the NHO bond. The angle increases from 45^0 to 100^0 keeping ψ_B constant. Similarly, these angles increase from -20^0 to 40^0 keeping φ_B constant (Fig. 6). The area occupied by the polypeptide models having 60^0 and 70^0 is less than the models with angles greater than 90^0.

Figure 6. Ramachandran plots of showing distances and bond angle for RL nests. (A) O···O distance; (B) (N)H···O distance; (C) angle formed by NHO bond

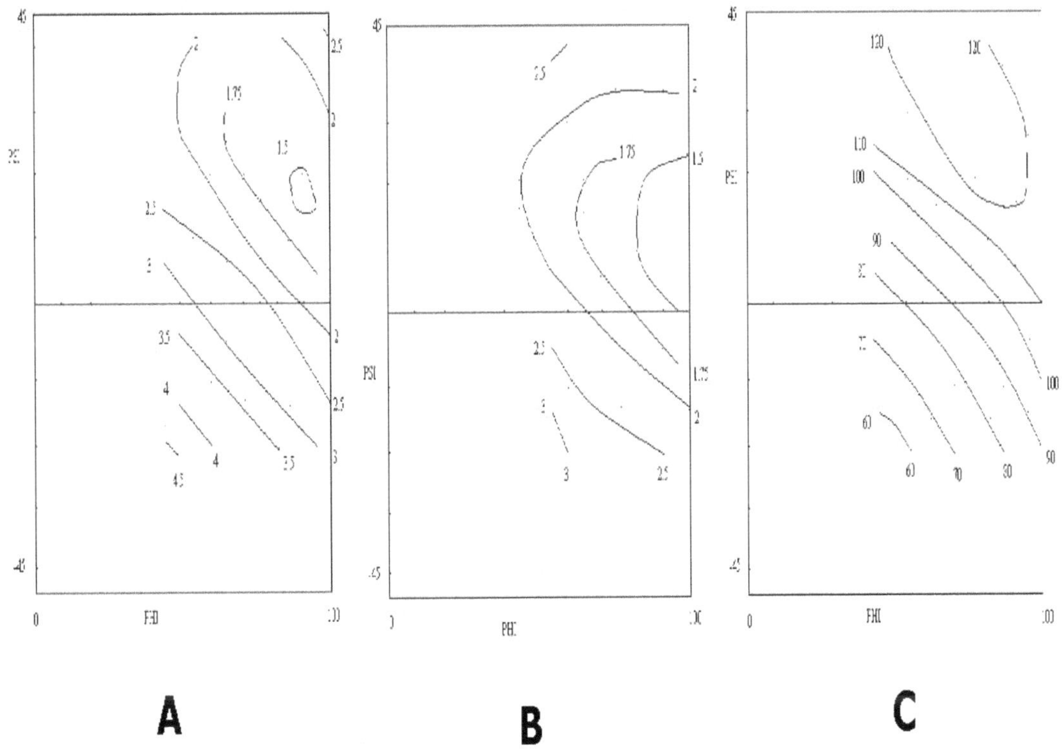

2.3 Analysis of OH-OH Distance and NHO Bond Angle in LR Nests

Table 4 gives the data for distance between oxygen atoms (OH-OH) in LR nests. When φ_B increases, distances decreases (keeping ψ_B constant). If ψ_B increases, distance increases going from left to right in the Ramachandran plot (Fig. 7A).

The data for distance between hydrogen and oxygen for LR nests is given in Table 5 and is plotted on Ramachandran plot in Fig. 7. From Table 5, distance decreases as φB and ψB increases. The data for angle formed by the NHO bond for LR nests is shown in table 6. From the table we see that the data consists of values from 70 to 110 degrees. Angles increases as φ_B and ψ_B increases. The maximum angle the NHO bond can form is 109.31^0. Since the tabular data plotted in Fig. 7C contains all sharp angles, the probability of forming optimal conformation is less.

DISCUSSION

From the results obtained by making models of ring peptides we observe that in nests the main-chain NH groups points away from the center of the ring. This deviation of the NH···O bond forms an angle of approximately 30-40 degrees. This deviation causes O(i+2) to form a hydrogen bond with H(i) of NH(i).

Table 2. Tabular data of PHI B, PSI B values of RL nests for distances between hydrogen and oxygen.

PHI B	PSI B1	PSI B2	PSI B3	PSI B4	PSI B5	PSI B6	PSI B7	PSI B8	PSI B9	PSI B10	PSI B11	PSI B12	PSI B13
	-20	-15	-10	-5	0	5	10	15	20	25	30	35	40
45	3.22	2.98	2.76	2.56	2.38	2.24	2.13	2.07	2.07	2.12	2.22	2.36	2.53
50	3.16	2.92	2.69	2.48	2.3	2.15	2.04	1.98	1.98	2.04	2.14	2.29	2.47
55	3.09	2.84	2.61	2.4	2.21	2.06	1.95	1.89	1.9	1.96	2.07	2.23	2.42
60	3.02	2.77	2.53	2.31	2.12	1.96	1.86	1.81	1.81	1.88	2.01	2.18	2.36
65	2.95	2.69	2.45	2.23	2.03	1.87	1.77	1.72	1.74	1.82	1.95	2.13	2.34
70	2.87	2.61	2.36	2.14	1.94	1.78	1.68	1.64	1.67	1.76	1.91	2.1	2.32
75	2.79	2.53	2.27	2.05	1.85	1.69	1.59	1.56	1.61	1.71	1.88	2.08	2.31
80	2.7	2.44	2.19	1.96	1.76	1.6	1.51	1.5	1.55	1.68	1.85	2.07	2.31
85	2.62	2.35	2.09	1.86	1.67	1.52	1.44	1.44	1.51	1.65	1.85	2.07	2.32
90	2.53	2.26	2	1.77	1.58	1.44	1.37	1.39	1.48	1.64	1.85	2.09	2.35
95	2.44	2.17	1.91	1.68	1.5	1.37	1.32	1.35	1.48	1.64	1.86	2.11	2.38
100	2.35	2.07	1.82	1.6	1.42	1.3	1.27	1.33	1.46	1.65	1.89	2.15	2.42

Table 3. Tabular data of PHI B, PSI B values of RL nests for NHO bond angle.

PHI B	PSI B1	PSI B2	PSI B3	PSI B4	PSI B5	PSI B6	PSI B7	PSI B8	PSI B9	PSI B10	PSI B11	PSI B12	PSI B13
	-20	-15	-10	-5	0	5	10	15	20	25	30	35	40
45	55.71	59.43	63.68	68.69	74.47	81.01	88.1	95.44	102.71	109.17	114.54	118.7	121.46
50	58.83	62.55	66.82	71.86	77.73	84.22	91.42	98.96	106.13	112.15	117.69	121.37	123.6
55	62.02	65.69	69.94	74.97	80.83	87.48	94.8	102.33	109.53	115.79	120.38	123.47	125.07
60	65.27	68.88	78.12	78.21	84.01	90.85	98.2	105.84	112.85	118.61	122.64	125.01	125.96
65	68.46	72.03	76.36	81.4	87.31	94.09	101.58	109.17	115.83	121.04	124.26	125.87	125.97
70	71.72	75.28	79.56	84.58	90.54	97.39	104.9	112.16	118.53	122.89	125.28	125.82	125.21
75	75.07	78.59	82.84	87.9	93.86	100.68	108.16	115.1	120.67	124.03	125.26	125.01	123.63
80	78.39	81.89	86.03	91.1	97.15	103.92	111.12	117.68	122.08	124.21	124.34	123.17	121.39
85	81.76	85.19	89.44	94.47	100.35	107.19	113.91	119.42	122.61	123.49	122.49	120.62	118.39
90	85.16	88.69	92.77	97.81	103.68	110.04	116.21	120.5	122.16	121.64	119.85	117.41	114.84
95	88.64	92	96.13	101.1	106.78	112.68	117.74	120.59	120.73	118.95	116.38	113.64	111.02
100	92.14	95.49	99.54	104.39	109.64	114.93	118.53	119.62	118.14	115.55	112.44	109.52	106.88

It had been supposed that these NH groups would point towards the center of the ring but the observation that they point by 30^0 away from the center and back along the polypeptide is a factor to be considered when bridging of the NH groups in the nest by bound oxygen atoms. Since the project involves finding of optimal conformations of nest-like peptides, three conditions have to be considered. The distance between H(i) and O(i+2), the distance between the oxygen atoms and the angle formed by the NHO bond. If all these conditions are satisfied the optimal conformation is found. From the Tables

Figure 7. Ramachandran plots showing distances and bond angle for LR nests (A) O···O distance; (B) (N)H···O distance; (C) angle formed by NHO bond

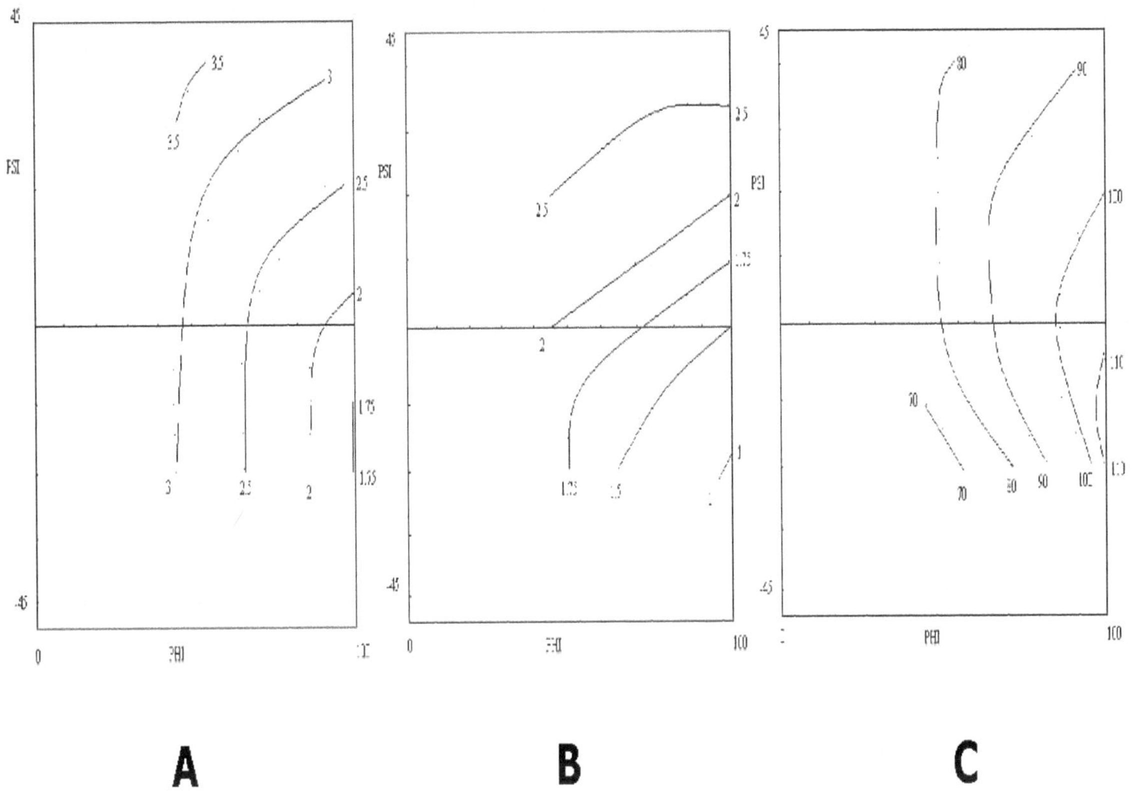

Table 4. Tabular data of PHI B, PSI B values of LR nests for distances between the oxygen.

PHI B	PSI B1	PSI B2	PSI B3	PSI B4	PSI B5	PSI B6	PSI B7	PSI B8	PSI B9	PSI B10	PSI B11	PSI B12	PSI B13
	-20	**-15**	**-10**	**-5**	**0**	**5**	**10**	**15**	**20**	**25**	**30**	**35**	**40**
45	3.01	3.01	3.02	3.05	3.09	3.14	3.2	3.27	3.35	3.45	3.54	3.64	3.74
50	2.89	2.88	2.89	2.92	2.96	3.01	3.08	3.15	3.24	3.33	3.43	3.54	3.65
55	2.77	2.76	2.77	2.79	2.83	2.89	2.95	3.03	3.12	3.22	3.33	3.45	3.56
60	2.65	2.63	2.64	2.66	2.71	2.76	2.84	2.92	3.01	3.12	3.23	3.36	3.48
65	2.52	2.51	2.51	2.54	2.58	2.64	2.72	2.81	2.91	3.03	3.15	3.27	3.41
70	2.4	2.39	2.39	2.42	2.46	2.53	2.61	2.71	2.82	2.94	3.07	3.2	3.34
75	2.28	2.27	2.27	2.3	2.35	2.42	2.51	2.61	2.73	2.86	2.99	3.13	3.28
80	2.17	2.15	2.16	2.19	2.25	2.32	2.42	2.53	2.65	2.79	2.93	3.08	3.23
85	2.06	2.04	2.05	2.09	2.15	2.23	2.34	2.46	2.59	2.73	2.88	3.03	3.19
90	1.95	1.93	1.95	1.99	2.06	2.16	2.27	2.4	2.54	2.69	2.84	3.01	3.17
95	1.85	1.84	1.86	1.91	1.99	2.09	2.22	2.35	2.5	2.66	2.82	2.99	3.16
100	1.76	1.75	1.79	1.85	1.94	2.05	2.18	2.32	2.48	2.64	2.81	2.98	3.16

1, 2, & 3 we found that the area of optimal conformations in the Ramachandran plot for nests is rather limited and also consists of areas where optimal conformations are not found. These are labelled as "bad areas" because the peptides in these areas cannot form bonding because either they are too far away to form a hydrogen bond or are too close generating steric hindrance. The optimal conformations for RL nests are found in the areas coloured yellow on the Ramachandran plot (Fig. 8A). The area below the coloured plot has peptides with large distances and sharp angles. In these areas optimal conformations are not found. The area above the coloured area has peptides with short distances and large angles. In these areas bonding can take place but the bonding generates disruptive steric hindrance. That is why the area in yellow is optimal for RL nests.

From Tables 4, 5, & 6 the area of optimal conformations for LR nests is also limited, as it is in RL nests. Above the colored area the distance between oxygen atoms is too large and the angle formed by the NHO bond is sharp (Fig. 8B). Below the colored area the distance between the oxygen atoms is too small and the NHO bond angle is too large causing steric hindrance. Therefore the area colored in yellow is optimal for LR nests (Fig. 8B). If we take a cut-off of 2 A and compare it with region of optimal conformations, we observe that the area occupied by the cut-off region in RL nests is more than the corresponding area occupied in LR nests (Fig. 8). This means that the chances of forming optimal conformations are more in RL nests than LR nests. If we consider the cut-off of 2 A and subtract it from the cut-off of 1.5 A, we get the area of optimal conformations formed by RL and LR nests.

From tables 1 and 4, 2 and 5, 3 and 6 we see that, at 90^0,0^0 the values are identical for the distances and angles. The peptides above and below the point either have increased or decreased distances and angles. This is because at 90^0,0^0 the distance between oxygen atoms, distance between oxygen and NH group and the NHO bond angle is the same for RL and LR nests. Also from Fig. 8, the 90^0,0^0 is the point where optimal conformation is found in both RL and LR nests. This is a notable feature that has not been described before.

Figure 8. (A) Ramachandran plot showing the area of optimal conformations in RL nests and, (B) Ramachandran plot showing area of optimal conformation in LR nests

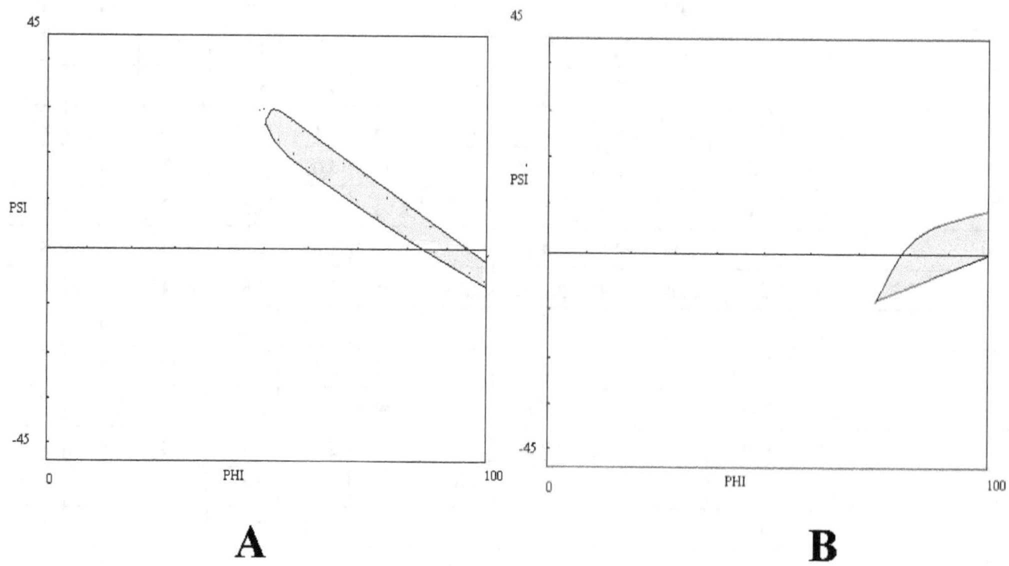

Table 5. Tabular data of PHI B, PSI B values of LR nests for distances between hydrogen and oxygen.

PHI B	PSI B1	PSI B2	PSI B3	PSI B4	PSI B5	PSI B6	PSI B7	PSI B8	PSI B9	PSI B10	PSI B11	PSI B12	PSI B13
	-20	**-15**	**-10**	**-5**	**0**	**5**	**10**	**15**	**20**	**25**	**30**	**35**	**40**
45	1.8	1.86	1.93	2.01	2.1	2.19	2.29	2.39	2.5	2.6	2.71	2.81	2.92
50	1.73	1.79	1.86	1.94	2.03	2.13	2.23	2.33	2.44	2.55	2.66	2.77	2.87
55	1.66	1.72	1.79	1.87	1.97	2.07	2.17	2.28	2.39	2.5	2.61	2.72	2.83
60	1.59	1.65	1.72	1.81	1.9	2	2.11	2.22	2.34	2.45	2.57	2.68	2.8
65	1.51	1.57	1.65	1.74	1.84	1.94	2.05	2.17	2.29	2.41	2.53	2.65	2.76
70	1.44	1.5	1.58	1.68	1.78	1.89	2	2.12	2.24	2.37	2.49	2.61	2.73
75	1.36	1.43	1.52	1.61	1.72	1.84	1.95	2.08	2.2	2.33	2.46	2.58	2.71
80	1.29	1.37	1.45	1.56	1.67	1.79	1.91	2.04	2.17	2.3	2.43	2.56	2.69
85	1.23	1.3	1.4	1.5	1.62	1.75	1.88	2.01	2.14	2.27	2.41	2.54	2.67
90	1.16	1.24	1.34	1.46	1.58	1.71	1.84	1.98	2.12	2.26	2.39	2.53	2.66
95	1.1	1.19	1.3	1.42	1.55	1.68	1.82	1.96	2.1	2.24	2.38	2.52	2.65
100	1.04	1.14	1.26	1.39	1.52	1.66	1.8	1.95	2.09	2.24	2.38	2.52	2.66

Table 6. Tabular data of PHI B, PSI B values of LR nests for NHO bond angle.

PHI B	PSI B1	PSI B2	PSI B3	PSI B4	PSI B5	PSI B6	PSI B7	PSI B8	PSI B9	PSI B10	PSI B11	PSI B12	PSI B13
	-20	**-15**	**-10**	**-5**	**0**	**5**	**10**	**15**	**20**	**25**	**30**	**35**	**40**
45	64.45	67.48	70.03	72.08	73.82	75.18	76.05	76.78	77.16	77.19	77.04	76.67	76.17
50	68	70.99	73.61	75.74	77.34	78.58	79.45	79.96	80.16	80.1	79.78	79.24	78.54
55	71.49	74.62	77.19	72.28	80.93	82.05	82.7	83.1	83.15	82.89	82.3	81.67	80.78
60	75.08	78.34	80.98	82.99	84.39	85.46	85.94	86.17	85.97	82.55	84.81	83.87	82.86
65	78.78	82.07	84.69	86.64	87.95	88.77	89.08	89.06	88.66	87.96	87.08	86.04	84.71
70	82.68	85.94	88.41	90.24	91.37	91.96	92.08	91.82	91.18	90.29	89.17	87.88	86.39
75	86.58	89.91	92.27	93.79	94.73	95.04	94.95	94.37	93.46	92.33	90.99	89.43	87.85
80	90.49	93.56	95.91	97.31	97.92	98	97.52	96.66	95.54	94.2	92.61	90.8	89
85	94.56	97.69	99.57	100.6	101	100.58	99.87	98.7	97.26	95.67	93.86	91.98	89.95
90	98.76	101.55	103.21	103.78	103.63	102.9	101.79	100.82	98.72	96.8	94.81	92.67	90.61
95	102.96	105.31	106.4	106.6	105.9	104.85	103.33	101.7	99.74	97.52	95.48	93.18	90.93
100	107.02	108.84	109.31	108.89	107.74	106.31	104.47	102.45	100.29	97.99	95.68	93.26	90.84

CONCLUSION

With the aim of finding the optimum conformation for nest binding to oxygen atoms, a set of repeating dipeptides with dihedral angles φ_i,ψ_i and φ_{i+1},ψ_{i+1} were modeled on computer using automated tools. These tools Bearing in mind the known conformations of nest in proteins, the φ_i,ψ_i angles were fixed at $-90^0,0^0$ and the φ_{i+1},ψ_{i+1} angles were varied around $90^0,0^0$. The dipeptide (-90^0, 0^0; 90^0, 0^0) is a main chain enantiomer while the others are not.

The results of the investigation showed that RL and LR nests are enantiomers at -90_i, 0_i, 90_{i+1}, 0_{i+1} (where 90_i,0_i is φ_A,ψ_A and 90_{i+1},0_{i+1} is φ_B,ψ_B); where effects such as steric clashes, weak hydrogen bonds, strong hydrogen bonds, etc. are equal in both RL and LR nests. When we fix one set of φ,ψ angles and change the other, differences between RL and LR conformations emerge. As the φ,ψ angles are changed from the 90,0 position, a difference arises between RL and LR nests. This difference is created by the effects (steric clashes, weak hydrogen bonds, strong hydrogen bonds, etc.) which either increases in RL nests and decreases in LR nests if moved from the 90,0 position in either direction. The difference between the models of RL and LR is that RL ones have more conformations optimal for bridging by an oxygen atom and thus RL nests has greater chances than LR nests for optimal conformations. This can now be related to previous observations that, in proteins RL nests occur more commonly than LR nests. Also those LR nests that do occur are less liable to have bound δ- groups or atoms bridging the nest.

In previous work it had been supposed that each NH pointed roughly towards the center of the ring formed by the repeating dipeptide. I have shown that these NH groups (or their projection onto the ring) point at an angle of 40^0 away from the center and towards the N-terminus. When the oxygen atoms are bridged by NH groups, the distance and the angle formed play a major role in determining the optimal conformation in nests. The bridging of oxygen atoms by the NH bonds is favorable in RL nests. If the φ_A,ψ_A value is fixed for RL nests and the φ_B,ψ_B value is changed from 90,0, the structures for RL and LR are no longer elastomeric. If φ_A,ψ_A is at a different value, the φ_B,ψ_B value that produces enantiomers varies.

The φ,ψ values of nests for proteins observed were found to be α_R with φ=-94,ψ=-3 and α_L with φ=76,ψ=22 in RL nests and α_L with φ=73,ψ=22 and α_R with φ=-83,ψ=-22 in LR nests (Watson & Milner-White, 2002a). The average values found from this study were found to be different from the expected values. This investigation of modeled polypeptides has changed the way of observing nests in Ramachandran plot.

ABBREVIATIONS

RL: 1,2-α_R α_L; LR: 1,2-α_L α_R; CO: Carbonyl; R: Right-handed; L: Left-handed; A: Angstroms

REFERENCES

Duddy, W. J., Nissink, J. W., Allen, F. H., & Milner-White, E. J. (2004). Mimicry by asx- and ST-turns of the four main types of {beta}-turn in proteins. *Protein Science*, *13*(11), 3051–3055. doi:10.1110/ps.04920904 PMID:15459339

Hao, X., & Varshney, A. (2006) Geometry guided computation of 3D electrostatics for large biomolecules. Computer *Aided Geometric Design,* 545–557.

Hayward, S., & Milner-White, E. J. (2008). The geometry of alpha-sheet: Implications for its possible function as amyloid precursor in proteins. *Proteins-Structure Function and Bioinformatics*, *71*(1), 415–425. doi:10.1002/prot.21717 PMID:17957773

MathWorks. (2012). Bioinformatics Toolbox: User's Guide (R2012a). Retrieved from www.mathworks.com/help/pdf_doc/bioinfo/bioinfo_ug.pdf

Milner-White, E. J., Nissink, J. W., Allen, F. H., & Duddy, W. J. (2004). Recurring main-chain anion-binding motifs in short polypeptides: Nests. *Acta Crystallographica. Section D, Biological Crystallography*, *60*(11), 1935–1942. doi:10.1107/S0907444904021390 PMID:15502299

Milner-White, E. J., & Russell, M. (2008). Predicting the conformations of peptides and proteins in early evolution. *Biology Direct*, *3*(3), 1–9. PMID:18226248

Milner-White, E. J., & Russell, M. J. (2005). Sites for phosphates and iron-sulfur thiolates in the first membranes: 3 to 6 residue anion-binding motifs (nests). *Origins of Life and Evolution of the Biosphere*, *35*(1), 19–27. doi:10.1007/s11084-005-4582-7 PMID:15889648

Milner-White, E. J., Watson, J. D., Qi, G., & Hayward, S. (2006). Amyloid formation may involve alpha-to beta sheet interconversion via peptide plane flipping. *Structure (London, England)*, *14*(9), 1369–1376. doi:10.1016/j.str.2006.06.016 PMID:16962968

Ochoa, R., Ochoa, J., & Muskus, C. (2011) Automation of bioinformatic tools to detect gene fusion events in the Leishmania braziliensis and Leishmania major genomes *Computing Congress (CCC)*, 6th. doi:10.1109/COLOMCC.2011.5936324

Roterman, I. (1995). The geometric analysis of peptide backbone structure and its local deformations. *Biochimie*, *77*(3), 204–216. doi:10.1016/0300-9084(96)88126-0 PMID:7647113

Taisbak, C. M. (1980). An Archimedean proof of Heron's formula for the area of a triangle; reconstructed. *Centaurus*, *24*(1), 110–116. doi:10.1111/j.1600-0498.1980.tb00368.x

The PyMOL Molecular Graphics System (2010) Version 1.5.0.4 Schrödinger, LLC.

Torrance, G. M., Leader, D. P., Gilbert, D. R., & Milner-White, E. J. (2009). A novel main chain motif in proteins bridged by cationic groups: The niche. *Journal of Molecular Biology*, *385*(4), 1076–1086. doi:10.1016/j.jmb.2008.11.007 PMID:19038265

Watson, J. D., & Milner-White, E. J. (2002). A novel main-chain anion-binding site in proteins: The Nest. A particular combination of phi,psi values in successive residues gives rise to anion-binding sites that occur commonly and are found often at functionally important regions. *Journal of Molecular Biology*, *315*(2), 171–182. doi:10.1006/jmbi.2001.5227 PMID:11779237

Watson, J. D., & Milner-White, E. J. (2002). The conformations of Polypeptide Chains where the main-chain parts of successive residues are enantiomeric. Their occurrence in Cation and Anion-binding regions of proteins. *Journal of Molecular Biology*, *315*(2), 183–191. doi:10.1006/jmbi.2001.5228 PMID:11779238

KEY TERMS AND DEFINITIONS

LR Nests: It is a type of nest forming Left-handed Right-handed helices with L residues having positive phi values and R residues having negative phi values.

Nests: These are anion-binding features of protein and peptide molecules formed by three consecutive amino acids and form a structural motif in proteins.

Polypeptides: Chains of amino acids covalently linked by peptide bonds. Proteins are made up of one or more polypeptide molecules.

Ramachandran plot: The two torsion angles (Ramachandran angles) of a polypeptide chain, describing rotations of the polypeptide backbone around the bonds between N-Cα (Phi/φ) and Cα-C (Psi/ψ). The Indian biophysicist G. N. Ramachandran is credited for this seminal work and the plot is named after him.

RL Nests: It is a type of nest forming Right-handed Left-handed helices with R residues having negative phi values and L residues having positive phi values.

Chapter 8
Strategic Role of Information and Information Technology in Shop Floor Control in Footwear Industry Sector

Sergio Ricardo Mazini
University Center Toledo Araçatuba, Brazil

ABSTRACT

This chapter presents an approach to the strategic role of information and information technology in the shop floor control in footwear industry sector, pointing and tracking through the various stages of the production process. Discusses the importance of industries perform monitoring of production processes, with the goal of identifying information needs, actions and solutions that will contribute to the improvement and efficiency of the production process. The chapter also discusses the contribution of information technology to the information systems of companies, through the resources and solutions available today, such as Enterprise Resource Planning - ERP, Manufacturing Resource Planning - MRP and Shop Floor Control - SFC. The research method is the case study conducted in firm located in an industrial Brazilian footwear. This study examines the use of a solution called GradeSFC tracking and pointing of the production process.

INTRODUCTION

We are in the XXI century and living in a knowledge society, where business is changing rapidly and where success and survival depend on the organization's ability to adjust to dynamic business environment. In this context, some questions arise: How can we develop the best strategy for business? How information systems can contribute to the strategy? What is the contribution of information technology to the existing information systems in business?

The purpose of this chapter is to seek to understand the importance and strategic role of information and information technology in the shop floor control in the footwear industry sector. After the main themes

DOI: 10.4018/978-1-4666-8693-9.ch008

of literature related to the subject as: systems and enterprise production process, shop floor control, the strategic role of information and the contribution of information technology for information systems, we defined the following research questions, which will be key to the development of this research:

- How to identify the need for information, solutions and actions needed that will contribute to the improvement and efficiency of the production process?
- How to identify the need and contribution of resources and solutions currently available through information technology.

The research will be conducted, with company of an industrial Brazilian footwear, aims to raise and analyze the degree of maturity of these companies in tracking and pointing the various stages of the production process and identify how these industries identify needs information, actions and solutions that will contribute to improvement and efficiency of the production process. The importance of effective control of production processes and the use of information technology industries in Brazil have been meeting the expectation of growth that emerging countries like Brazil, has shown in recent years. In the case study, we will analyze a solution called GradeSFC tracking and pointing of the production process.

BACKGROUND

The Strategic Role of Information and Information Technology

Systems and Enterprises

The concept of systems lead to understanding the complexity of modern business as a whole. It is considered system a set of interdependent elements or an organized or interacting parts forming a unitary whole and complex. However, we must distinguish closed systems, such as machines and watch open systems, such as biological and social systems: the man, the organization and society (Laudon & Laundon, 2007). A system is closed if no material enters or leaves it is open if there is import and export and, consequently, changing components. The open system can be understood as a set of parts in constant interaction, constitute a whole geared for certain purposes and in permanent interdependent relationship with the external environment. A system may be composed successively of subsystems (also set of interdependent parts) that relate to each other, forming the larger system. Open systems involve the idea that certain entries are entered into the system and processed, generate certain outputs. The company draws on material resources, human and technological, whose processing resulting goods or services to be supplied to the market, as illustrated in Figure 1.

Another important concept is the information and information systems. Information can be considered as the result of data handling and provided with relevant purpose. Good information has significant value for organizations. Information systems are defined as a set of interrelated components that collect (retrieve), process, store and distribute information to support decision making and organizational control. According to Oliveira (2011), can approach an information system with a subsystem of the company, which owns the activities of inputs, processing and outputs as shown in Figure 2.

The components of an information system may be defined, according to O'Brien (2004):

Figure 1. The company seen as an open system (Author Self, 2010)

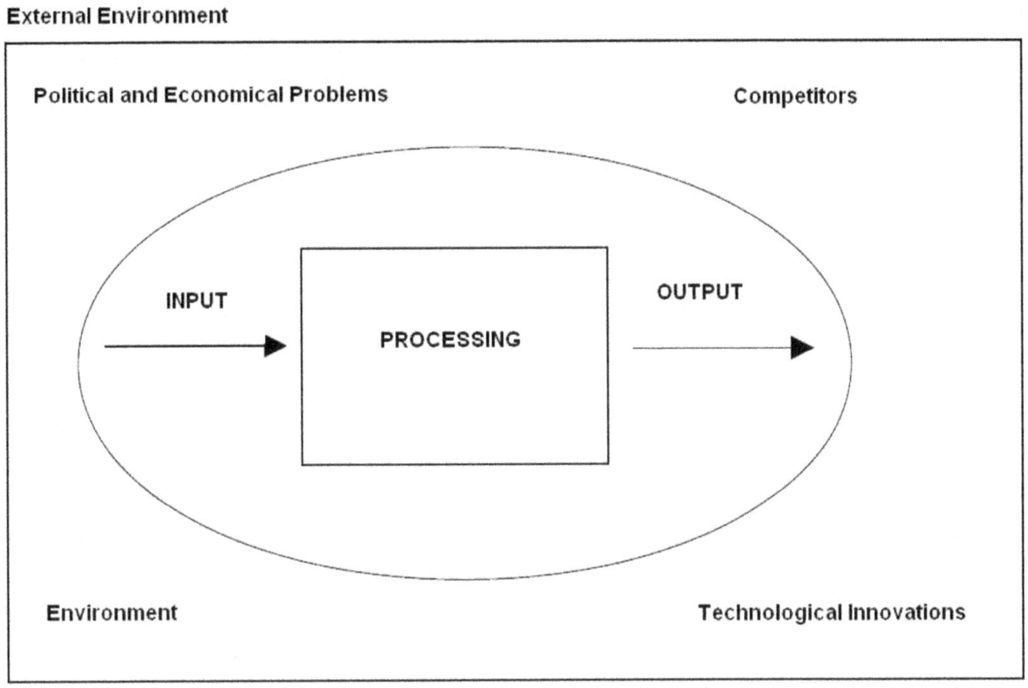

Figure 2. Activities of the Information Systems (Oliveira, 2011)

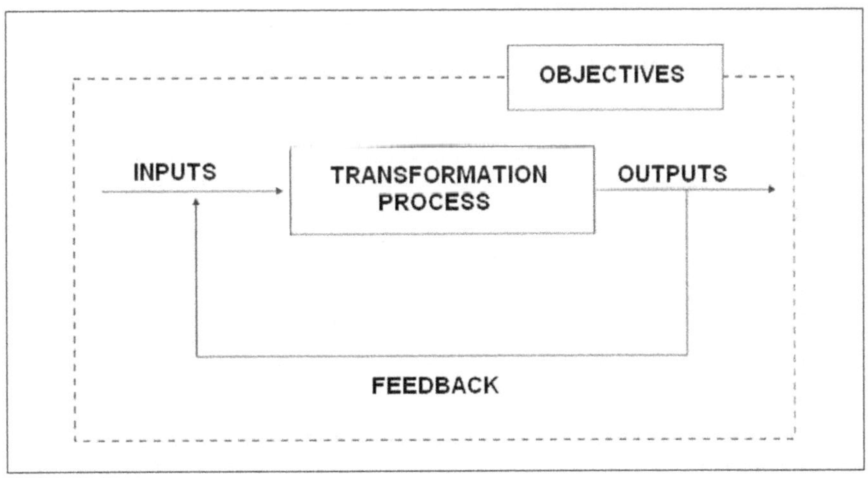

- **Organizations:** each organization has a specific culture, values and a way to make their processes and have been accepted by company employees. Some procedures, such as how to fill out a purchase order, or to correct an invoice is added to information systems.
- **People:** people use information from computer-based systems in their work, integrating them in the workplace. They are required to enter data into the system, placing them directly. Employees need special training to perform their tasks efficiently or using information systems.

- **Technology:** the technology and the means by which data is transformed and organized (generating information) for use in people. An information system can be a manual system, using only the technology of pencil and paper and then make use of technological resources and computer information technology, such as hardware, software, telecommunications.

Every system, using no resources or information technology, which generates and manipulates information can be considered an information system, which is designed primarily to assist the processes of decision-making and control processes in organizations.

Strategic Planning, Information and Information Technology

The term strategy is - from the Greek strategos - initially referred to the position of general in command of an army, later came to designate the art of the general, meaning the application of skills in the general exercise of its military function (Chiavenato & Sapiro, 2003). Strategy can be understood as a pattern or plan that integrates the major policies, objectives, goals and actions of the organization. A good strategy can ensure the best allocation of resources in anticipation of movement, planned or not, the competitors or the circumstances of the environment.

Organizational strategy is the basic action structured and developed by the company to achieve adequately and preferentially differentiated objectives envisioned for the future, to better position the company towards its environment and strategic planning is a process of formulating organizational strategies in which seeks the inclusion of the organization and its mission in the environment in which it is acting.

According to Beal (2008), information is an essential element in the creation, implementation and evaluation of any strategy and strategic planning of information is the management of information resources of an organization from a strategic and planning must be derived from strategic organization. Also according to Beal (2008), the strategic value of information can be represented by:

- **Strategic resource:** the basic purpose of the information is to enable the company to achieve its strategic objectives, the efficient use of its resources.
- **Factor supporting the decision:** the various hierarchical levels in the organization have different information needs. The information must be reliable and be available at the right time and must be linked to the company's strategic planning.
- **Factor of production:** the information is important to create and market products (goods and services) of greater value to customers.
- **Factor synergy:** the organization's performance is directly related to the quality of organizational relationships and this will directly depend on the existing information flow.
- **Determinant of behavior:** information influences the behavior of individuals and groups within and outside organizations.

Information technology (IT) is an integral and fundamental to support, sustain and grow a business (Mohamed & Singh, 2012). Several perspectives are defined for the governance of information technology, among them the activities and structure to align business and information technology that targets principles to organize activities, use of resources, risk management, governance structures and performance measurement. The relationship between business and information technology should be made through strategic alignment, value delivery to the business through information technology, performance management, risk management, policies and procedures and control and accountability (Ward & Peppard, 1996, Webb, Pollard & Ridley, 2006).

The relationship between information technology and business management has changed over time. In the early stages of automation, the primary objective of information technology in relation to the business was the establishment and improvement of organizational efficiency and effectiveness and in a second stage companies realized that information technology plays an important and crucial to organizational success in support of the business and management strategies and emphasis on (Klouwenberg et. al., 1995).

The Contribution of Information Technology to Production Systems

From the late 1960s, the production function has to be perceived as a strategic asset and new approaches to better understand its strategic role began to emerge (Gianese & Correa, 1993). The content of manufacturing strategy consists of a pattern of decisions regarding the structure and infrastructure of a manufacturing organization. The decisions include the following: product scope, process technology, manufacturing alliance and competence production. The strategic alignment between manufacturing strategy and IT strategy must be considered by industrial firms (Ho, 1996).

The highly dynamic and complex environment that currently manufacturing companies face, make these companies have to react quickly to these changes daily. The implementation of technological processes in production systems that are flexible in terms of quantities and variations are particularly interesting. These production systems consist of a number of sequences which are necessary to the successful manufacture of the products, where each sequence is a compatible combination using different manufacturing technologies (Grienitz & Hausicke, 2013).

Currently in competitive markets, companies that develop and industrialize new products need to sustain their competitive advantage by reducing the time and cost in the development of new products, without compromising the performance of these products (Boudouh, Boxberger & Gomes, 2013). With increasing global competition, reducing time and cost favor the competition and creating competitive advantage, in addition to the major costs in the development of new products are assigned during the design phase, where this regard, the use of prototyping tool digital is a solution to test and validate a new product early in its life cycle (Affonso, Cheutet, Ayadi & Haddar, 2013).

One of the most important tools for the industrial sector is information technology, since they help in the automation and control of operations, play an important role in reducing costs while adding value to the production process, enabling manufacturing companies to obtain and sustain competitive advantage (Abu-Shanab, E., & Al-Tarawneh, H., 2013).

The proposal of computer integrated manufacturing (CIM) can help manufacturing companies to quickly respond to changes in product design, demand and product variety, and enable cost reduction, improved quality and reduced production lead time (Marri, H. B., Gunasekaran, A., Irani, Z., & Putnik, G., 2008).

The governance of information technology can generate better business results, leading to better decision making by firms (Teo, W. L., & Tan, K. S., 2013). In this context, the use of information technology arises as a response to assist in various business processes, data quality, information and decision making. Some concepts and solutions are presented below:

Enterprise Resource Planning (ERP)

The beginning of the concept of enterprise management systems (ERP) can be identified in the early 1960s, where this time the first solutions - software - were responsible for inventory control for production systems. In the following decade, 1970, the production systems demanded more refined controls, then the emerging systems of material requirements planning (MRP) that helped control the master production plans and needs for raw materials. In the 1980s, emerged the manufacturing resource planning (MRP II) that in addition to controlling the production of master plans and needs for raw materials also allow the calculation of equipment needs and labor-intensive. Subsequently, other organizational areas such as purchasing, finance, sales, among others, were added to MRP II, thus creating the concept of enterprise business systems - ERP (Yen, Chou & Chang, 2001). The need to improve the organization, control and management of data and information in its internal processes, motivated many companies to implement systems for enterprise resource planning - ERP (Davenport, 1998, 2000).

The ERP systems allow an organization to automate and integrate business processes, share common data and practices across the enterprise and can produce and access information in a real time environment. The ultimate goal of an ERP system is that the information should only be entered once (Marnewick & Labuschagne, 2005). Furthermore, they need to simplify, streamline, integrate and boost productivity in the various organizational processes. Some studies seek to analyze the selection of ERP systems for companies or even analyze success by business managers and information technology (Ifinedo, 2007; Ziaee & Sadjadi, 2006)

A proposed conceptual framework for the understanding of ERP systems is defined by Marnewick and Labuschagne (2005), which are based on the 4 Ps of marketing (people, product, promotion and price), defining the 4 Ps of the conceptual model of ERP (persons product, process and performance), as shown in Figure 3.

The four components of the conceptual model are described below:

- **Software:** is the most visible component for users and then seen as a product. It consists of several integrated modules designed to meet various organizational processes, simplifying and automating them.
- **Process flow:** is the component that handles the flow of information going through the various modules of ERP. Before an ERP system can be implemented in an organization, business processes need to be modeled and if necessary, redesigned to allow for a smoother implementation.
- **Customer mindset:** is the component that handles the resistance to ERP, which can be a critical factor in the successful implementation and use. The involvement and participation of clients in the implementation project is extremely important because it can help to break paradigms and changing mentality of it.
- **Change management:** is the component that plays an important role in the implementation of ERP systems, since the change should happen at all levels: operational, managerial and strategic support from top management is a decisive factor for the success of the project.

Figure 3. Components of the conceptual ERP (Marnewick & Labuschagne, 2005).

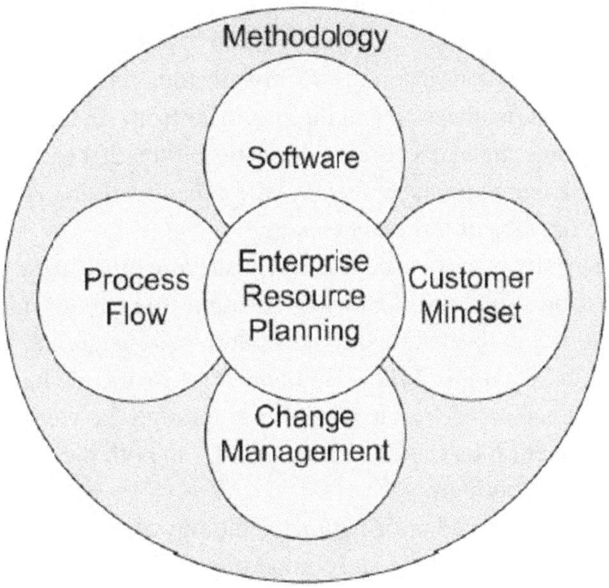

Material Requirements Planning (MRP) and Manufacturing Resource Planning (MRP II)

The MRP (Material Requirements Planning) and MRP II (Manufacturing Resources Planning) systems are the administration of large production which has been used by companies and have as main objective the calculation of resource requirements productive order to meet delivery deadlines of customer orders, with minimum stockpiling, production planning and purchasing components for items that occur only at such times and quantities required (Gianese & Correa, 1993).

The MRP / MRP II is an IT-based approach to deal with uncertainty and complexity that arise in the manufacturing environment. It is a highly sophisticated information system which controls the complexity and uncertainty that occur in a production environment. Uses the schedule of planned production to design requirements for individual parts or subassemblies. These requirements are compared with the levels of inventory and scheduled receipts for the batches can be scheduled to be produced or received as needed (Ho, 1996). It has five modules, as follows:

- Module production planning (production planning): Helps decision to aggregate levels of inventory and production period to period, also based on forecasts of aggregate demand.
- Module master production planning (master production schedule): It is a plan for the production of end products items, period to period.
- Calculation module need materials (materials requirements planning): Based on the list of materials (sheet) is made the explosion of materials needed for the production.
- Calculation module capacity requirements (capacity requirements planning): Locates infeasibility of a given master production plan in relation to productive capacity.

- Control module factory (shop floor control): Responsible for production control at the factory level, making the production of notes held.

Another contribution of information technology for organizations is related to organizational resilience. The organizational environment is always changing originate from several sources such as customers, suppliers, competitors and government. According Ates and Bititci (2011), Managing change to adapt to an uncertain future is a challenge requires resilience - the capacity of an organization to survive, adapt and sustain the business in the face of turbulent change.

Resilience is seen as a key organizational capability for sustainability in the current turbulent environment. Also according Ates and Bititci (2011), for organizations to be more sustainable and resilient, the delivery of innovative responses to the market through continuous change and improvement is necessary.

According to Burnard and Bhamra (2011), the concept of resilience has a firm grounding within the realm of ecology and gained considerable recognition through the work of Hollings (1973, 2001), Walker et al. (2002) and Walker (2004) and resilience resides in both the individual and organizational responses to turbulence and discontinuities.

The strategic role of information and information technology are also identified by enterprises as an important factor for organizational resilience and competitiveness. Enterprise Resource Planning – ERP systems - provide companies among other benefits the organization of the flow of data and information between various organizational processes, which contributes greatly to the speed, efficiency and elimination of waste proposed by Lean Six Sigma. Among some important to monitor the production process and that ERP systems can contribute information, we can highlight:

- The appointment of the planned manufacturing orders versus the pointing of manufacturing orders made.
- The appointment of inputs provided versus the inputs used to supply the production process.
- Production lots that are in arrears in the production process.

Compared to Six Sigma, ERP systems can also contribute to the continuous improvement process that reduces the variability by identifying and pointing rework caused during the production process. Another important factor for the development of Lean Six Sigma aspect is the establishment and monitoring of a set of metrics - performance indicators (KPIs), which guide the efforts and resources used by companies as well as the adoption of corrective or preventive measures . In this sense the BI solutions help to monitor these performance indicators that can be used at strategic, managerial or operational level.

The solution GradeSFC, presented later in this chapter, fits like a software with a focus on the control module factory described above.

MAIN FOCUS OF THE CHAPTER

Shop Floor Control Solution: GradeSFC

The aim of the solution GradeSFC developed by Orion Management Solutions (http://www.oriongestao.com.br) is to provide an innovative solution to control and pointing of the factory floor, regardless of the physical location of the factories. In summary the GradeSFC proposes:

- Capture production information held and published on a web portal.
- Take information from production planning to the shop floor.
- Allows viewing of the production information in real time.

As shown in Figure 4, the GradeSFC solution consists of the following modules:

- **Control:** Responsible for capturing the occurrences of a control point on the factory floor. Aims to ensure that the lower orders of manufacturing production is consistent and executed even if the ERP backend is not available. This module also allows the operator himself checkpoint track the progress and make an analysis of the evolution of its activities. Figure 5 illustrates the control module GradeSFC solution.
- **Monitor:** Responsible for showing the current situation in real time from one or more plants, functioning as a sort of control panel (dashboard) that shows graphically how is the progress of the activities of the entire production. Figure 6 illustrates the monitor module GradeSFC solution.
- **Production Back End Web:** Responsible for synchronizing the information generated at checkpoints and update them on the company's ERP. This module will run persistently on the application server and will need only one point of access to the database of the ERP.

Solution Features

- **Simplicity and Agility:** The basis for the creation of the prototype of this solution is the simplicity of its operation and the speed with which information must be updated to allow for making safe decisions.
- **Open Platform:** Developed in Java with MySql-database solution has the concept of multi-platform, which allows the application to run on any operating system installed on the factory floor. The MySQL database is a database robust, agile and cost without license;
- **Free Software:** Developed with open technologies that do not require an additional payment of the cost of third party license.
- **Technology Education:** From the beginning of the development of the system, it was possible to incorporate this solution development best practices and technologies widely recognized by the software market, which shows the robustness of the application and ensures their rapid evolution.

Figure 4. Modules of the solution GradeSFC (Author Self, 2013)

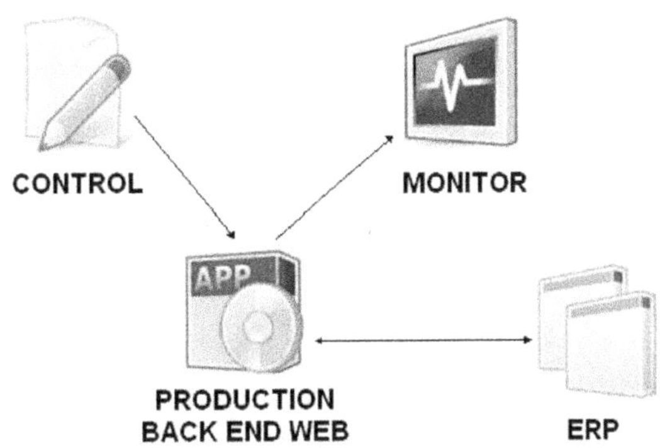

Figure 5. Control Module Solution GradeSFC (Author Self, 2013)

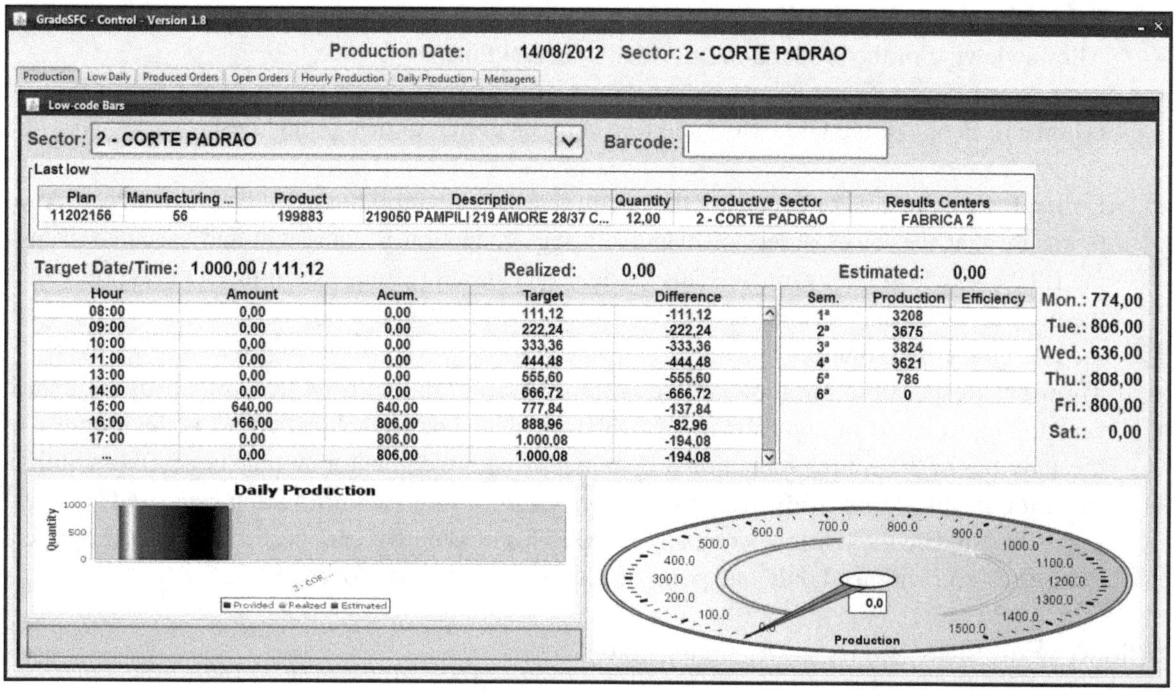

Figure 6. Monitor module solution GradeSFC (Author Self, 2013)

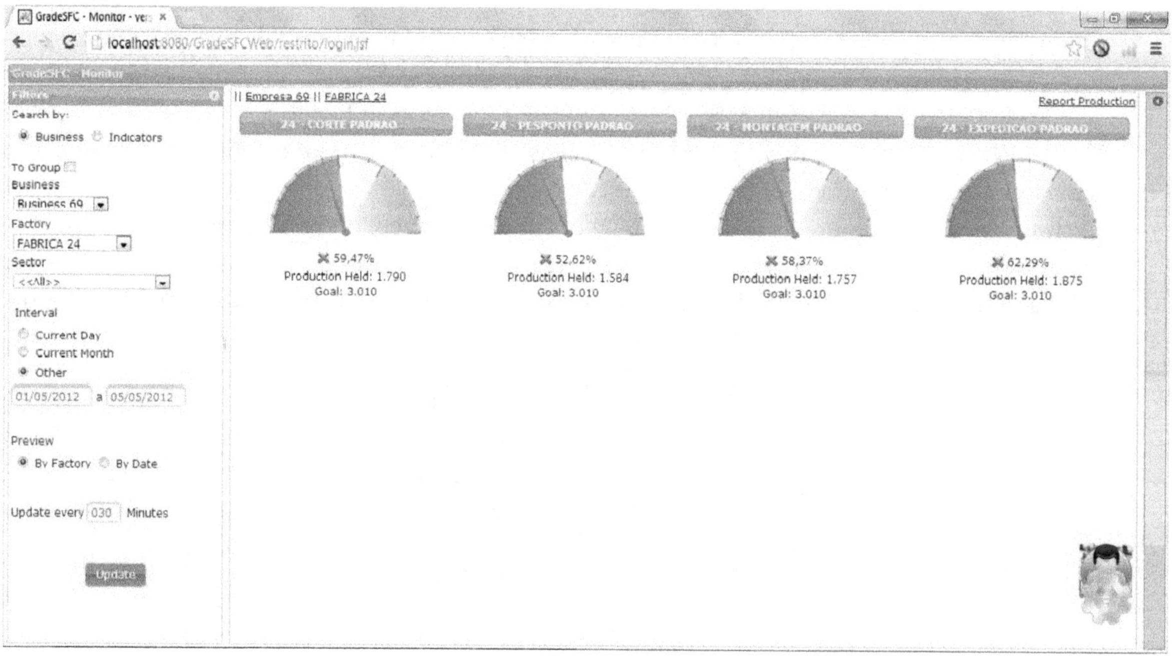

Benefits GradeSFC

- **Reduction in Paper Consumption:** Using this solution, the manufacturing order can have a unique bar code for all stages, since the control point that is running the control module has the information of the step that is being singled out, decreasing the size required for the issuance of work order and, consequently, spending on paper.
- **Low License Cost:** because of the extensive use of free software and open platforms.
- **Robust and Dynamic Applications Running in Production:** Compared to a simple collector, using a thin-client - workstation lean - allows applications more robust, dynamic and easy to use to run in production and to provide information to the agile and safe operators graphically and dynamic;
- **Independence of ERP:** The system is designed to be integrated with any ERP that uses a relational database. The production back end web module was built for interfaces with other ERP systems can be built.
- **Total Cost of Ownership:** The cost of the solution - hardware and software - is less than the cost of the collectors and the possibilities for its use are much broader and more flexible.

Research Method

With the aim of taking information about the follow-up appointment and the various stages of the production process and identify information needs, actions and solutions that will contribute to the improvement and efficiency of the production process, an exploratory research was carried out through a qualitative/case study approach. The exploratory character of research is justified, because even with the existence of publications about the use of information technology on business, there is the lack of research that considers the use of IT resources applied to the information systems of the company - especially when it comes to Brazilian scenario.

The qualitative research was used, as this approach makes possible the comprehension of people's opinions about the phenomenon studied, which promotes the study of an exploratory research. The survey unit for the qualitative study/case research was a company of the Brazilian footwear industry that already performs the controls to notes in its production process. To collect the information, primary data sources were used that according to Yin (2005) allows the researcher to dedicate a huge diversity of questions, such as: historical, behavioral and attitudinal.

The collection of primary information from companies was made through interviews:

- **Interviews with Company Personnel:** Interviews were conducted with the production managers of companies that were using the GradeSFC solution. Interviews were not structured ones and conducted with a set of questions (that was recorded to raise the reliability on the study). Interviews collected the data about the origins, necessities, structures and variables that influenced the use of GradeSFC. A non-structured interview was also conducted (that was recorded to raise the reliability on the study) with the persons responsible for Information Technology area of user companies and data was collected about the organizational structure of the project, the organizational process and the implementation of variable that influences the project.

On the stage of analyzing information, interviews were transcribed and the relevant information, as per aim of the research, was organized in a way to get all the main references to the GradeSFC project. There was also a confrontation with the theory and the analysis of the case study according to the GradeSFC solution. The construct validity of the research happened through the using of multiple sources of information and the interaction with the involved constructs since the literature.

Case Studies

The company Industry and Trade Pampili Shoes, located in the shoe industry for Birigüi, state of São Paulo, Brazil and operates in the children's shoes since 1987. Produces around 20,000 pairs / day in 3 plants themselves (Birigüi-SP, Vicentinópolis-SP and Paranaiba-MS) and 3 suppliers Birigüi, with about 37 lines per collection and about 3000 direct employees. The company has several product lines and factories in different physical locations. The production process of the footwear industry has some characteristics, described below:

- Customer orders are usually grouped into batches of production, in order to optimize the production process.
- The productive sectors most common are: cutting, stitching, preparation, assembly and shipping.
- Some components of the finished product, they are made in the footwear industry outsources, are named beneficiaries and components need to be available in inventory in the industry at the time of supply of the production plans of the finished products.

The monitoring and control of production plans need to be made to each work order generated by production plans in order to minimize unforeseen delays in the production process and that may impact the delivery date agreed with the customer.

Figure 7 illustrates how the infrastructure is used in the solution GradeSFC, where there is a main server in the industry to receive the notes of the production of several existing factories and industries and also a secondary server located at manufacturing unit Paranaíba, Mato Grosso do Sul, Brazil.

The solution provides leading GradeSFC production, important information for the daily management of production plans, enabling better decision making on the necessary actions and solutions that will contribute to the improvement and efficiency of the production process. Some information presented by management solution and used in the production process, are presented below:

- **Daily low:** low ratio of production made the day, enabling the monitoring of the implementation of the planned manufacturing orders for the day.
- **Orders produced:** list of manufacturing orders produced in the reporting period.
- **Open orders:** list of manufacturing orders open awaiting completion and fulfillment of notes and low production.
- **Production per hour:** graph shows the production of the current time, thus tracking and monitoramente of unexpected failures and the production process.
- **Production per day:** chart that shows the daily production, enabling the tracking and monitoring of actions to be performed in order to recover and finish on the planned production.

Figure 7. Infrastructure Solution GradeSFC Industry and Trade Footwear Pampili (Author Self, 2013).

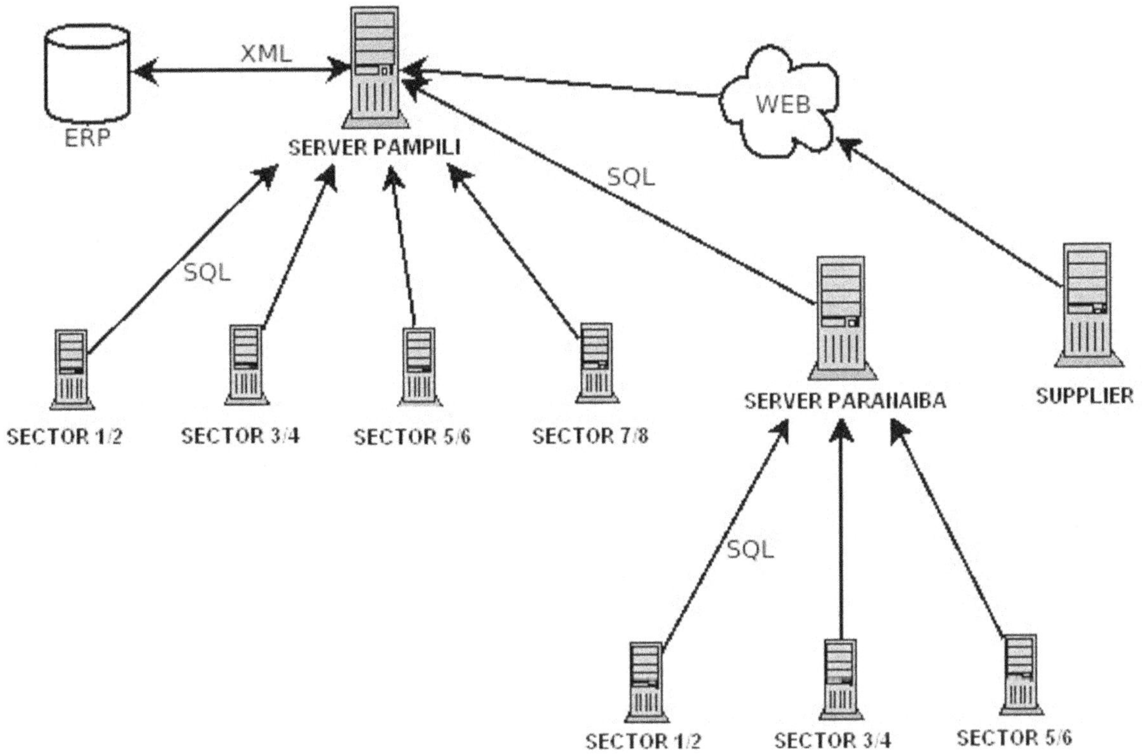

- **Posts:** exchange of messages between the leading manufacturing and productive sectors of the existing factories, allowing the communication of important information for tracking and monitoring production.

Description and Analysis of Data Obtained in Field Research

The case studies reported GradeSFC demonstrates that the solution is practical example of the use of information technology in tracking and monitoring the production process of a footwear industry. In the specific case study, the use of the solution possible at first, the consolidation of information-floor factory in a web portal, which greatly facilitated the management of individual factories that the company owns, in addition to allow direct communication between the leaders and the production-floor factory in order to direct the actions needed and solutions that contribute to the improvement and efficiency of the production process. Another benefit generated by the use of the solution is the provision of real time information about the various plans and production stages of the production process, allowing the area of planning, scheduling and control of production greater assertiveness of actions and provision of information to other areas organizational, eg, the commercial and shopping.

The availability of on line information on the progress of production plans in-floor makes possible the monitoring of any problems that arise during the production process, resulting in the solution of the same agility and reducing the number of production plans in arrears. This directly impacting deliveries to the customers, because the number deliveries delayed due to issues and problems in the production process decreased significantly.

Other issues can be analyzed based on a case study:

- **Decision making:** with the available data and information relating to the production process and better control of the factory floor during the various steps of the manufacturing process it is possible that industrial managers have a better quality in the process of decision making, directing questions and problems that may arise during this process.
- **Disadvantages:** current dependence on information technology by businesses in general, can generate conflicts by not using these resources, thus causing problems in obtaining data and information from various organizational processes.
- **Advantages:** good use of information technology resources available, may lead companies to a significant improvement of their business processes, because the flow of data and information will be better organized, thus avoiding delay in decision making.
- **Convenience:** we are currently living in a highly competitive environment, where companies need to search competitive and differential advantages perceived by customers. The understanding of the strategic role of information and information technology helps to create an environment conducive to search for differences and competitive advantages.
- **Benefits:** benefits as we have the best organization of the flow of data and information, creating simplicity, agility, integration and increased productivity in the various organizational processes.

FUTURE RESEARCH DIRECTIONS

The use of information technology in the business environment is already a reality in many industrial companies. The improvement of controls in various production processes, contributing to the increase in operational efficiency and the quality of products produced, thus avoiding delivery delays, quality problems, among other issues that may affect the company's relationship with its customers. This chapter contributed in this direction, but some suggestions are summarized below:

- Research focusing on the various processes and their relationship with the environment.
- Research focusing on the influence processes of companies' relations with the entire production chain.
- Research focusing on the contribution of information technology in managing the life cycle of products.

CONCLUSION

Increasingly, organizations are seeking productive forms of management and control for use in their various organizational processes, including the production process. In this context, information systems and information technology stand out as key pieces for the management and control of the production process.

Through the proposed use of the solution GradeSFC, managed to identify factors essential to the management and control of the production process in several existing factories in the company, regardless of its physical location, thus contributing to improving the efficiency and effectiveness of the production process.

Resuming the research questions proposed, was identified by analyzing the case study, the literature review and the solution used by the company, that the identification of the need for information, actions and solutions needed that will contribute to the improvement and effectiveness the production process is an amalgamation of factors such as organizational processes, people and technology used in this case clearly notice the use of information technology. Also identified is the importance of the use and contribution of information technology to the information systems of the company, through the use of various solutions with ERP, MRP II and GradeSFC solution presented in this chapter.

In a development country like Brazil or some other emergent economic, a solution like GradeSFC, presented by this chapter, can provide competitive advantages to companies which use the solution, achieving like this increasing the control of production processes, improves the flow of data and information on the factory floor and aid in decision making.

For industrial companies, the management of some important processes such as planning, scheduling and production control, manufacturing orders on the shop floor, calculation of productive resources, shopping and manufacturing costs, are extremely critical and should be using the resources available information technology to assist in automation, productivity, simplicity and quality of data and information.

REFERENCES

Abu-Shanab, E., & Al-Tarawneh, H. (2013). Production Information Systems Usability in Jordan. In I. Management Association (Ed.), Industrial Engineering: Concepts, Methodologies, Tools, and Applications (pp. 975-989). Hershey, PA: Engineering Science Reference. doi:10.4018/978-1-4666-1945-6.ch053

Affonso, R. C., Cheutet, V., Ayadi, M., & Haddar, M. (2013). Simulation in Product Lifecycle: Towards a better information management for design projects. *The Journal of Modern Project Management*, *1*(1), 112–119.

Ates, A., & Bititci, U. (2011). Change process: A key enabler for building resilient SMEs. *International Journal of Production Research*, *49*(18), 5601–5618. doi:10.1080/00207543.2011.563825

Boudouh, T., Boxberger, J., & Gomes, S. (2013). Project Management and Lean Engineering: An Industrial Application. *The Journal of Modern Project Management*, *1*(1), 50–55.

Beal, A. (2008). *Gestão Estratégica da informação: como transformar a informação e a tecnologia da informação em fatores de crescimento e alto desempenho nas organizações*. São Paulo: Atlas.

Burnard, K., & Bhamra, R. (2011). Organisational resilience: Development of a conceptual framework for organisational responses. *International Journal of Production Research*, *49*(18), 5581–5599. doi:10.1080/00207543.2011.563827

Chiavenato, I., & Sapiro, A. (2003). *Planejamento Estratégico. Fundamentos e Aplicações*. Rio de Janeiro: Elsevier.

Correa, H. L., & Gianese, I. G. N. (1993). *Just in time, MRPII e OPT: um enfoque estratégico*. São Paulo: Atlas.

Davenport, T. (1998). Putting the enterprise into the enterprise system. *Harvard Business Review*, *76*(4), 121–131. PMID:10181586

Davenport, T. (2000). *Mission Critical*. Boston, MA: Harvard Business School Press.

Ho, C.-F. (1996). Information technology implementation strategies for manufacturing organizations: A strategic alignment approach. *International Journal of Operations & Production Management*, *16*(7), 77–100. doi:10.1108/01443579610119171

Holling, C. S. (1973). Resilience and stability of ecological systems. *Annual Review of Ecology and Systematics*, *4*(1), 1–23. doi:10.1146/annurev.es.04.110173.000245

Holling, C. S. (2001). Understanding the complexity of economic, ecological, and social systems. *Ecosystems*, *4*(5), 390–405. doi:10.1007/s10021-001-0101-5

Ifinedo, P. (2007). An empirical study of ERP success evaluations by business and IT managers. *Information Management & Computer Security*, *15*(4), 270–282. doi:10.1108/09685220710817798

Laudon, K. C., & Laudon, J. P. (2007). *Sistemas de informação gerenciais*. *7*. São Paulo: Prentice-Hall.

Klouwenberg, M.K., KOO, W.J.D., & van Schaik, J. A. M. (. (1995). Establishing business strategy with information technology. *Information Management & Computer Security*, *3*(5), 8–20. doi:10.1108/09685229510104945

Marri, H. B., Gunasekaran, A., Irani, Z., & Putnik, G. (2008). Collaborative Alliance for the Implementation of Computer Integrated Manufacturing in Small and Medium Enterprises. In G. Putnik & M. Cruz-Cunha (Eds.), *Encyclopedia of Networked and Virtual Organizations* (pp. 216–224). Hershey, PA: Information Science Reference; doi:10.4018/978-1-59904-885-7.ch029

Marnewick, C., & Labuschagne, L. (2005). A conceptual model for enterprise resource planning (ERP). *Information Management & Computer Security*, *13*(2), 144–155. doi:10.1108/09685220510589325

Mohamed, N., & Singh, J. K. G. (2012). A conceptual framework for information technology governance effectiveness in private organizations. *Information Management & Computer Security*, *20*(2), 88–106. doi:10.1108/09685221211235616

O'Brien, J. A. (2004). *Sistemas de Informação e as decisões gerenciais na era da Internet*. São Paulo: Saraiva.

Oliveira, D. P. R. (2011). *Sistemas de Informações Gerenciais*. São Paulo: Atlas.

Grienitz, V., & Hausicke, M. (2013). Strategic Optimization of Future Manufacturing Process with Grafem, Technology Roadmaps and Scenarios Technique. *Proceedings of the 22th International Conference on Production Research, Foz do Iguaçu, Brazil.*

Teo, W. L., & Tan, K. S. (2013). Adoption of Information Technology Governance in the Electronics Manufacturing Sector in Malaysia. In I. Management Association (Ed.), *Industrial Engineering: Concepts, Methodologies, Tools, and Applications* (pp. 887-906). Hershey, PA: Engineering Science Reference. doi:10.4018/978-1-4666-1945-6.ch049

Ziaee, M., Fathian, M., & Sadjadi, S. J. (2006). A modular approach to ERP system selection A case study. *Information Management & Computer Security*, *14*(5), 485–495. doi:10.1108/09685220610717772

Walker, B., Carpenter, S., Anderies, J., Abel, N., Cumming, G. S., & Janssen, M. et al. (2002). Resilience management in social-ecological systems: A working hypothesis for a participatory approach. *Conservation Ecology*, *6*(1), 14–35.

Walker, B., Holling, C. S., Carpenter, S., & Kinziq, A. (2004). Resilience, adaptability and transformability in social – ecological systems. *Ecology and Society*, *9*(2), 5. Retrieved from http://www.ecologyandsociety.org/vol9/iss2/art5

Ward, J., & Peppard, J. (1996). Reconciling the IT/business relationship: A troubled marriage in need of guidance. *The Journal of Strategic Information Systems*, *5*(1), 37–65. doi:10.1016/S0963-8687(96)80022-9

Webb, P., Pollard, C., & Ridley, G. (2006). Attempting to define IT governance: wisdom or folly? *Proceedings of the 39th Hawaii International Conference on System Sciences, Kauai, HI*, 1-10. doi:10.1109/HICSS.2006.68

Yen, D. C., Chou, D. C., & Chang, J. (2001). A synergic analysis for web-based enterprise resources planning systems. *Computer Standards & Interfaces*, *24*(4), 337–346. doi:10.1016/S0920-5489(01)00105-2

Yin, R. (2005). *Case Study Research: Design and Methods*. London: Sage.

KEY TERMS AND DEFINITIONS

Computer Integrated Manufacturing: The aim of CIM is to produce the required amount of the product of acceptable quality at the right time.

Enterprise Resource Planning: ERP is a software-driven business management system that integrates all facets of the business, including planning, manufacturing, sales, and marketing. ERP systems can be used to manage operational business information for corporate resource planning.

Information Systems: are defined as a set of interrelated components that collect (retrieve), process, store and distribute information to support decision making and organizational control.

Information Technology: Technological resources and computational generation and use of information. Its main components are: hardware, software, telecommunications and data management.

Information Technology: is the application of computers and telecommunications equipment to store, retrieve, transmit and manipulate data, often in the context of a business or other enterprise. The term is commonly used as a synonym for computers and computer networks, but it also encompasses other information distribution technologies such as television and telephones.

Organizational Strategy: the basic action is structured and developed by the company to achieve adequately and preferentially differentiated objectives envisioned for the future, to better position the company towards its environment and strategic planning is a process of formulating organizational strategies in which one seeks the inclusion of the organization and its mission in the environment in which it is acting.

Manufacturing Resources Planning: are the systems of administration of large production which has been used by companies and have as main objective to calculate the needs of productive resources in order to care for delivery of customer orders, with minimal inventory building, planning purchases and production of components for items that occur only at such times and quantities required.

Shop Floor Control: is concerned with the detailed management of activities and the flow of materials inside the plant - including the workers, materials, machines and time utilized in production.

Chapter 9
An Intuitive Teleoperation of Industrial Robots:
Approach Manipulators by Using Visual Tracking Over a Distributed System

Andrea Bisson
University of Padova, Italy

Stefano Michieletto
University of Padova, Italy

Valentina Ferrara
Comau S.p.A., Italy

Fabrizio Romanelli
Comau S.p.A., Italy

Emanuele Menegatti
University of Padova, Italy

ABSTRACT

Teleoperation of manipulator robots with RGB-D sensors is now mainly done using inverse kinematics techniques. In this chapter, we describe an intuitive way to teleoperate an industrial manipulator through vision sensors by directly controlling manipulator joints retargeting specific human motion. In this way the human operator has the full control of robot movements with practically no training, because of the intuitivity of this teleoperation method. The remapping into the robot joints is done by computing angles between vectors built from positions of human joints, tracked by the selected vision sensor. The obtained system is very modular which allows to change either the tracking sensor or the robot model with some small changes. Finally, the developed teleoperation system has been successfully tested on two real Comau robots, revealing to be fast and strongly reliable.

INTRODUCTION

Teleoperation is a technique that allows human operator to move and to program a robot by simply controlling it from a certain distance. This technique is used to move robots particularly in dangerous environments and tasks, such as bombs dismantling, exploring human inaccessible sites, maintenance of nuclear facilities, underwater operations, etc. Moreover, teleoperation can be also used for robot offline programming, allowing a human operator to save a lot of time.

DOI: 10.4018/978-1-4666-8693-9.ch009

Teleoperation tasks can use both trivial (e.g. joysticks, keyboards, etc.), but also more complicated human machine interfaces, such as vision-based interfaces. The first ones are less intuitive to use with respect to the second ones, in fact they require some training to the human operator in order to be used properly and efficiently. The vision-based teleoperation, instead, is generally more intuitive and easier to use. In fact, it finds applications in particular for the programming of more complex robots with an high number of degrees of freedom (DoFs), such as humanoid robots (Dariush et al., 2009). Nevertheless, recently, these techniques have been applied also to industrial manipulator robots, in order to move and program them more easily and intuitively (Kofman, Wu, Luu, & Verma, 2005) (Marinho, Geraldes, Bó, & Borges, 2012). Vision-based teleoperation can be of two types, marker-based or markerless: the first is more uncomfortable because it requires to the operator to wear additional clothes, while the second is more complex to develop, because human keypoints must be estimated via software.

In this Chapter, we will present a technique that allows to teleoperate a manipulator robot with a markerless vision-based system, by using in particular a RGB-D sensor. The sensor chosen for this work is a Microsoft Kinect, that is a very cheap and powerful RGB-D device, which allows to track human movements without using any kind of uncomfortable marker. Moreover, Microsoft Kinect has a large developer community, which implies the existence of many already implemented software packages to work with.

The implementation of the whole system has been made the more modular as possible. At this purpose, Robot Operating System (ROS) middleware has been used as framework to connect the developed modules one to each other.

ROS middleware is spreading even more in both academic and industrial environments. While for the first, ROS has a great potential allowing to test new algorithms on robot on high level programming; for the second it is seen as a platform that allows to program industrial manipulator robots in order to create complex applications for their customers in less time.

THE DEVELOPED TELEOPERATION SYSTEM

In this Section we will describe how the developed teleoperation system works. First of all we will describe how the physical system used is composed and interconnected, then we will focus on how the developed algorithm works.

The System Architecture

The developed system architecture is composed by two computers:

- A common Ubuntu Linux computer connected to the Microsoft Kinect sensor, on which is installed ROS middleware.
- A real-time Linux computer connected both to the computer previously described and also to the robot controller.

The Algorithm Pipeline

The developed teleoperation algorithm can be partitioned in three macro-blocks (Figure 1):

Figure 1. The yellow blocks represent the steps of the algorithm pipeline: the first two blocks work within ROS framework on a Standard PC, while the last block runs on a Real-Time PC

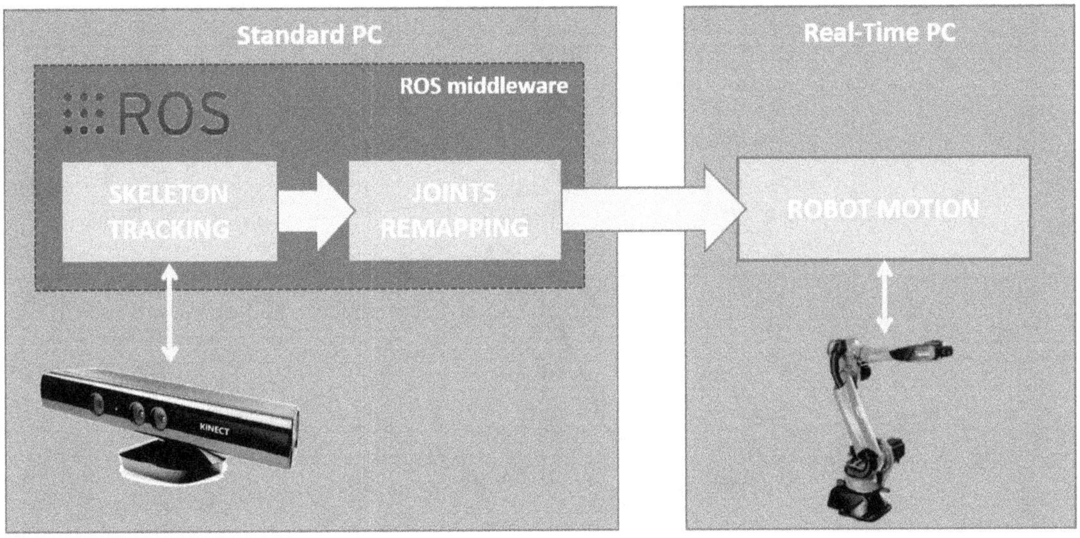

1. Skeleton Tracking
2. Joints Remapping
3. Robot Motion

The first two blocks run on the first computer on the ROS middleware and they are responsible of the human tracking and the joints remapping. The last block, instead, runs on the real-time Linux PC and it is responsible of sending the motion to the real robot. Now, we will describe these three blocks more in detail.

Skeleton Tracking

This block of the algorithm is the responsible of the skeleton tracking of a user in front of the Kinect sensor. This part of the algorithm is done by some ROS nodes which goal is to track every motion of the human joints by using OpenNI and NiTE libraries in order to achieve this goal. These nodes publish the position and the orientation of the human joints as TFs disposed as in Figure 2. The reference coordinate system for all the TFs is called *camera_rgb_optical_frame* and it is located on the RGB camera of the Kinect sensor.

In order to acquire the precise orientation of the user's hands, requested for a precise remapping of human hands motion onto the robot wrist, the NiTE tracker has been extended

Because NiTE library does not provide a complete tracking of the human hands, we extended the human tracker and add other custom TFs to the already existing *left_hand* and *right_hand*, which positions correspond to the centroid of the respective hands. This is because the standard information provided by the NiTE tracker were not enough to acquire a precise orientation of the hand.

Therefore, our NiTE human tracker extension adds two more TFs to each arm, one on the wrist and one on the middle fingertip.

Figure 2. On this image are represented the joints and the links of the human model tracked by Kinect during a frontal acquisition with the standard NiTE skeleton tracker. Notice that this model is flipped horizontally, in fact the TFs of the left side of the body are on the right and vice versa

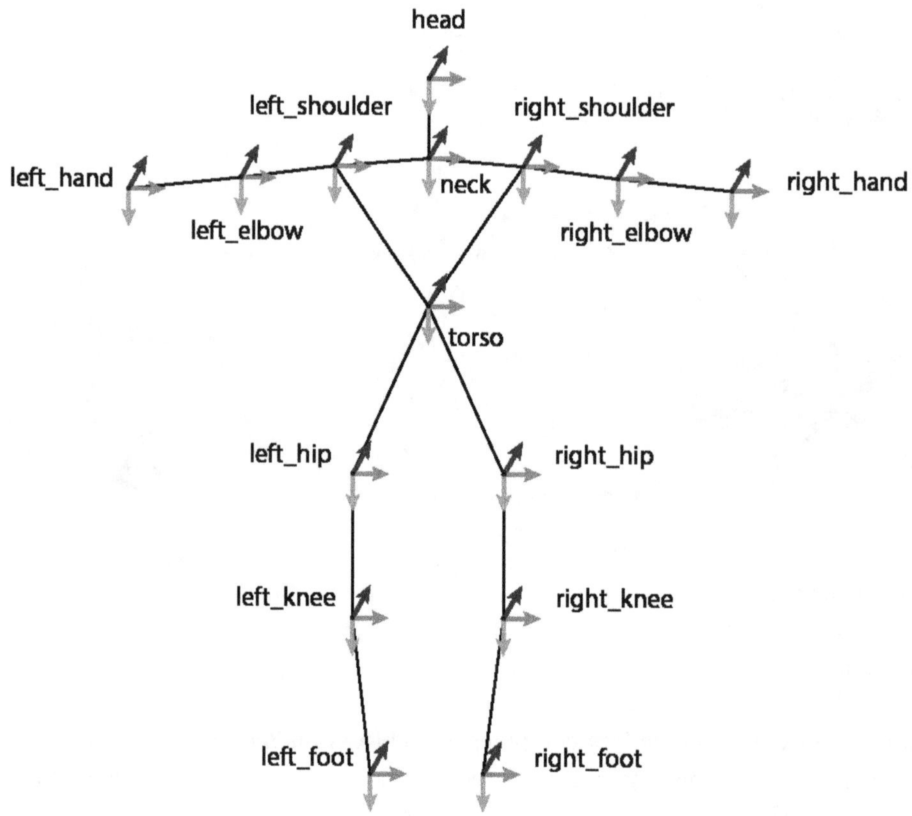

But, in order to do this, we had to work with the human point cloud given by OpenNI, which is built from depth information acquired by Microsoft Kinect.

Once acquired the human point cloud, the first thing to do is to segment the point clouds of each hand, from fingertips to wrist. This segmentation is computed after estimating the positions of both hands centroids, allowing to obtain two clusters of points representing the two human hands.

Now, in order to extrapolate the orientation of both hands, we exploit some results taken from Principal Component Analysis (PCA). Given a set of points distributed in a three dimensional space, using PCA theory, we can find principal directions of the points cluster (Jeong, Ziemkiewicz, Ribarsky, & Chang, 2009), which consists in finding the vectors (also called principal components) along which the variance of the points cluster is maximum (Figure 3).

Formally, let $X = \{x_1, x_2, \ldots, x_n\}$ be a set of points $\mathbf{x}_i \in \mathbb{R}^3$ representing the hand point cloud. The first step to do is to calculate the normalized covariance matrix Σ:

$$\Sigma = \begin{pmatrix} \sigma_{xx} & \sigma_{xy} & \sigma_{xz} \\ \sigma_{yx} & \sigma_{yy} & \sigma_{yz} \\ \sigma_{zx} & \sigma_{zy} & \sigma_{zz} \end{pmatrix}$$

Figure 3. This is a graphical example of a bi-dimensional Principal Component Analysis, in which it has been computed the principal components, $\mathbf{u}$ *and* $\mathbf{v}$*, of the blue points cluster on the XY plane*

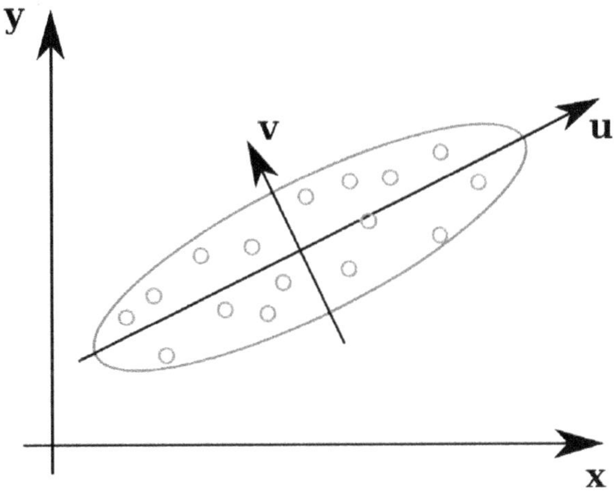

where:

- $\Sigma_{ij} = \Sigma_{ji} = \frac{1}{n-1}\sum_{k=1}^{n}(\mathbf{x}_{\mathbf{k}i} - \mu_i)(\mathbf{x}_{\mathbf{k}j} - \mu_j)^T$
- $\mu = \frac{1}{n}\sum_{k=1}^{n}\mathbf{x}_{\mathbf{k}}$, instead, is the mean point in $\mathbb{R}^3$ of the $\mathbf{X}$ vector, i.e. the centroid of the hand point cloud.

The next step is to compute the 3 eigenvectors of the Σ matrix. These vectors will be the principal components of our hand point cloud, allowing us to obtain all the information needed in order to achieve the hand orientation. So, computing roots of the characteristic equation:

$$\det(\boldsymbol{\Sigma} - \lambda \boldsymbol{I})$$

We obtain the three eigenvalues λ. Once obtained also the associated eigenvectors, we take the one associated with the greatest eigenvalue: this eigenvector will be the component vector v_{WF} going from wrist to fingertips.

Now, before publishing our new TFs, we must first compute their origin position and orientation. Taking the direction given by v_{WF} we add and subtract half of the hand length from the hand centroid, obtaining the positions respectively of the hand wrist and the middle fingertip.

For the TF orientation, instead, we calculate the needed quaternion by using the angle between the Microsoft Kinect z axis and the v_{WF} vector.

In synthesis, the algorithm for the generation of custom TFs is the next:

1. Segment the hand point cloud starting from the user point cloud

2. Compute the hand point cloud centroid $\mu = \frac{1}{n}\sum_{k=1}^{n} \mathbf{x}_k$
3. Calculate the normalized covariance matrix Σ of the hand point cloud, given its centroid μ
4. Compute eigenvalues and eigenvectors of matrix Σ and get the eigenvector associated with the greatest eigenvalue. It will be called v_{WF} and it is a vector going from wrist to fingertip
5. Take the centroid μ and add and subtract half the hand length along the direction given by vector v_{WF}, obtaining the origin position of custom TFs
6. Get TFs quaternions by taking the angle between the z axis and v_{WF} vector.

After applying this procedure for both left and right hands point clouds, we can publish our new custom TFs, which names are respectively: *left_wrist* and *left_fingertip* for the right hand, and *right_wrist* and *right_fingertip* for the left hand. The new human model obtained with the modified skeleton tracker can be seen on Figure 4.

These changes, applied to the skeleton tracking node, allow us to remap easily the human movements on the wrist joints of the manipulator robot.

Figure 4. This image represents the new human model (in a frontal view) published by the modified skeleton tracker. Respect to the original tracker, the total number of published TFs is now increased by four (two new TFs for each hand)

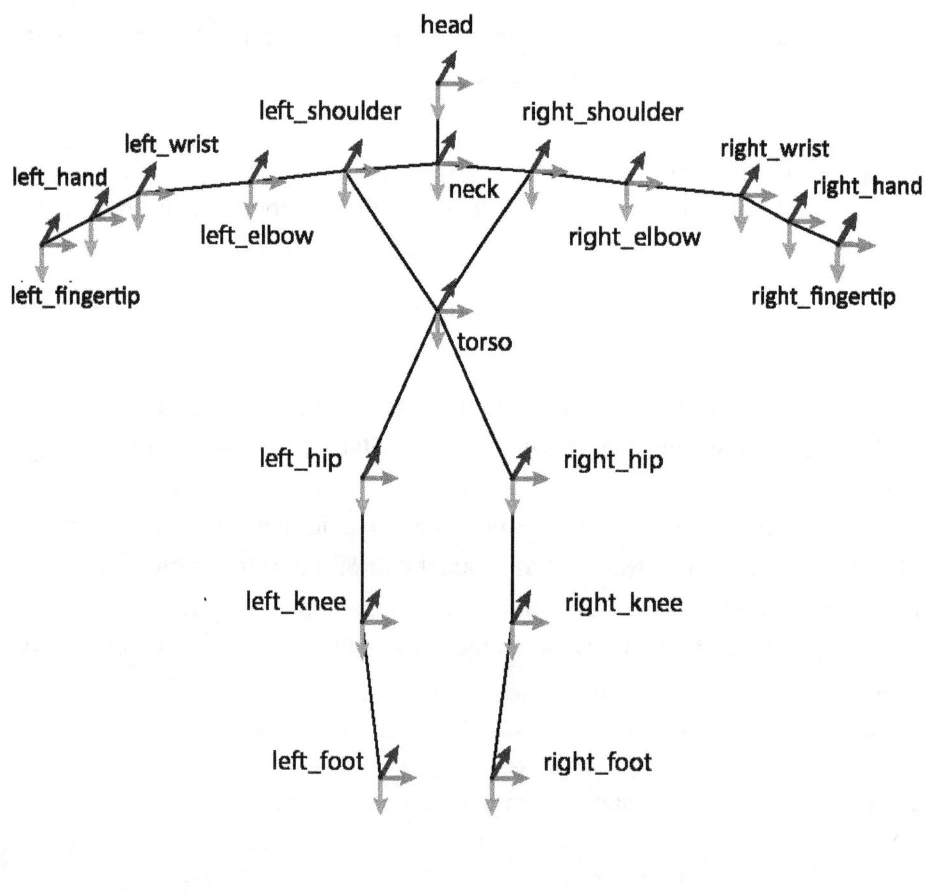

Before considering joints remapping, we first explain basic concepts about the virtual robot model used within ROS and Gazebo, in order to better understand the implemented remapping functions.

Robot Model

For testing purposes, we implemented also a robot model in a simulated environment within ROS framework (Figure 5). This virtual robot has been developed using Gazebo simulator, which can simulate both the physics of a complex tridimensional environment and naturally also the kinematics and the dynamics of any kind of robots (Figure 6).

In our case, the creation of a virtual robot arm has been useful in order to test the intuitiveness and the correctness of the remapping of each joint without using a real robot in the test phase.

Figure 5. This image shows the TFs position on the implemented virtual robot model

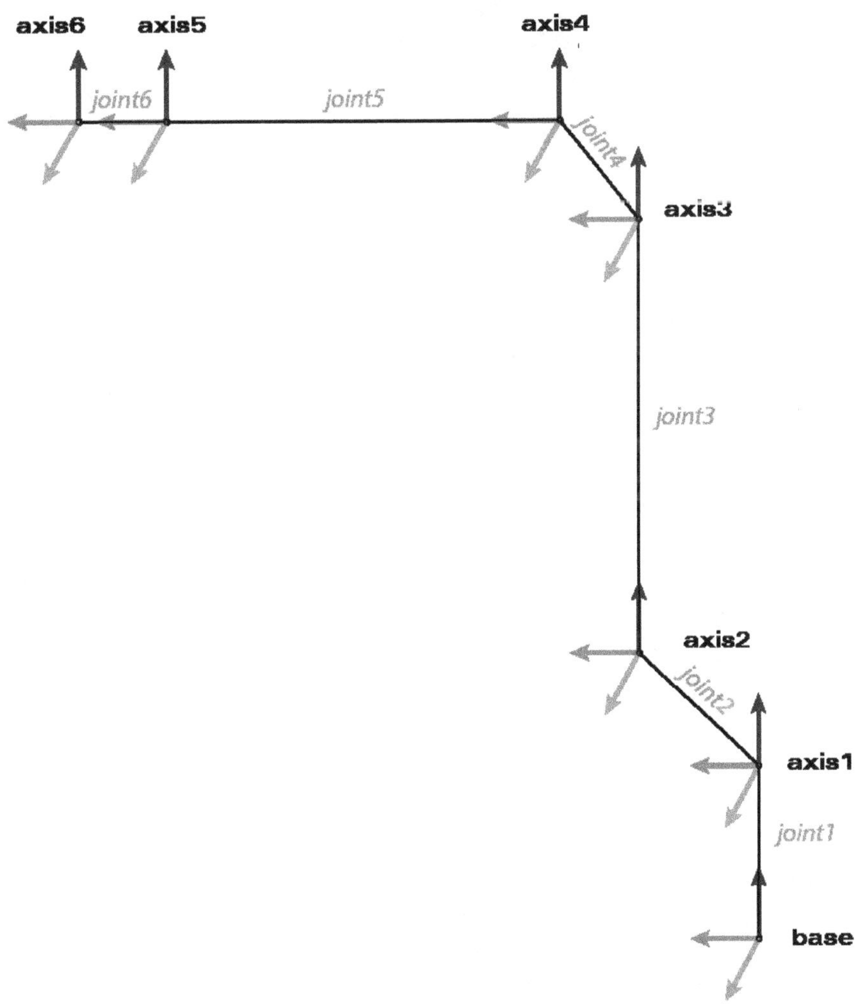

Figure 6. This is a screen shot of the virtual robot model implemented within Gazebo simulator

Joints Remapping

Now we can talk about the ROS nodes implemented in order to remap, in real-time, the human joints motion into the robot joints. The nodes responsible of this task manage the remapping algorithms in order to transform the joints motion coming from the TFs, received from the skeleton tracker nodes, into joints angles of the virtual robot model.

Two ways has been implemented to remap the human movements into the robot using a Microsoft Kinect sensor:

1. **Direct Joints Remapping**: this technique is aimed to remap the motion of the human body parts into the robot links in the most intuitive way for the human operator. This is the remapping technique developed in this work.
2. **Theory of Inverse Kinematics**: this technique is widely used in literature and it has been implemented for comparison purposes with our direct remapping method. The theory of inverse kinematics allows to calculate the robot joints angles by acquiring only the hand Cartesian position of the human operator.

Now, we describe in details our Direct Joints Remapping teleoperation technique.

One of the main goals of this work has been to find an intuitive way to move an industrial manipulator robot by simply acquiring the motion of the person, which moves its body parts as if they were robot links (Michieletto, Chessa, & Menegatti, 2013). In order to accomplish this objective, it has been created a remapping function for each of the first five robot joints, i.e. joints 1, 2, 3, 4 and 5. Movements for joint 6 has not been considered because its remapping would have not been intuitive to implement, and also because it would have required the acquisition of finer movements by the sensor, such as complex hands movements, which a Microsoft Kinect sensor cannot supply.

For the remapping of all the joints we must first consider a change of Cartesian coordinate system, because human and robot TFs have different coordinate frames. So to pass from the human tracked model to the robot virtual model, we have to do the next transformation:

$$x \leftarrow -z$$

$$y \leftarrow \mathrm{x}$$

$$z \leftarrow -y\,.$$

From now on, all the coordinates frames of the human skeleton will be considered already transformed in this way (Figure 7).

Joint 1

Joint 1 allows robot to rotate about its axis z. For moving this joint using the information acquired by Microsoft Kinect, it has been used the orientation of the human actor in front of the sensor.

In order to do this, first we must take the TFs positions of the neck $\mathbf{w}_n$ and of the left hand $\mathbf{u}_{rh}$. Formally:

$$\mathbf{w}_n = (x_n, y_n, z_n) \qquad \mathbf{u}_{rh} = (x_{rh}, y_{rh}, z_{rh})$$

and now from these two points we calculate the vector going from neck to left hand

$$\mathbf{v} = \mathbf{u}_{rh} - \mathbf{w}_n$$

Figure 7. On the left the Cartesian coordinate system of the human model and on the right the coordinate system of the robot model

From this vector we can now get the orientation of $\mathbf{v}$ along z axis, by calculating the angle α between the vector projection on XY plane and the x axis, using the next formula:

$$\alpha = \begin{cases} \arccos\left(\dfrac{x_v}{\rho_{XY}}\right) & \text{if} \quad y_v \geq 0 \\ -\arccos\left(\dfrac{x_v}{\rho_{XY}}\right) & \text{if} \quad y_v < 0 \end{cases}$$

where x_v and y_v are respectively the x and y components of the vector $\mathbf{v}$, while $\rho_{XY} = \sqrt{x_v^2 + y_v^2}$ is the length of its projection on the XY plane. α corresponds to the joint 1 rotational angle (Figure 8).

Joint 2

This joint allows to rotate the robot about the y axis. This type of movement allows the robot to make a longitudinal advance along the direction of joint 1. The angle needed for the remapping of joint 2 is obtained from the relative position of the human demonstrator respect of its initial tracking position.

Before calculating this angle, that we call γ, we have to get the direction of the person in front of the sensor by using the orientation of the human shoulders. Let

$$\mathbf{v}_{ls} = (x_{ls}, y_{ls}, z_{ls}) \qquad \mathbf{u}_{rs} = (x_{rs}, y_{rs}, z_{rs})$$

be the vector representing the positions respectively of right and left shoulders. Now, we can take the vector

Figure 8. Graphical representation of angle α, used for the remapping of joint 1

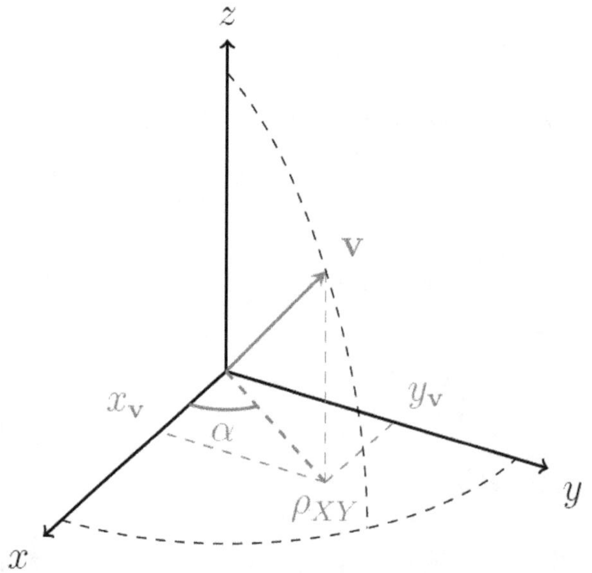

$$\mathbf{w} = \mathbf{v}_{ls} - \mathbf{u}_{rs}$$

that represents the initial direction of the person respect to the z axis. In order to change the coordinate system from the initial to the actual $\mathbf{w}$ orientation about the z axis, we use the next rotation matrix:

$$R_z = \begin{pmatrix} \beta & \alpha & 0 \\ -\alpha & \beta & 0 \\ 0 & 0 & 1 \end{pmatrix}$$

where $\alpha = \frac{x_w}{\|\mathbf{w}\|}$ and $\beta = \frac{y_w}{\|\mathbf{w}\|}$, with x_w and y_w and y components of vector $\mathbf{w}$ and $\|\mathbf{w}\|$ its L^2 norm.

Inverting the matrix R_z we get the matrix

$$R_z^{-1} = \begin{pmatrix} \frac{\beta}{\alpha^2+\beta^2} & -\frac{\alpha}{\alpha^2+\beta^2} & 0 \\ \frac{\alpha}{\alpha^2+\beta^2} & \frac{\beta}{\alpha^2+\beta^2} & 0 \\ 0 & 0 & 1 \end{pmatrix}$$

which can transform the coordinate system given by the vector $\mathbf{w}$, representing the actual orientation of the shoulders, into the coordinate system of their starting orientation.

Once known the shoulder orientation, we must compute the shift of the person respect to its initial position. To do this we take a shift vector $\mathbf{s}$ going from the initial position of the neck to its actual position:

$$\mathbf{s}' = \mathbf{r}_{n,a} - \mathbf{r}_{n,i}$$

where $\mathbf{r}_{n,a}$ and $\mathbf{r}_{n,i}$ are respectively the actual and the initial position of the neck. Now, we have to rotate $\mathbf{s}'$ using rotation matrix R_z^{-1} obtaining a vector $\mathbf{s}$ referred to the starting coordinate system (Figure 9):

$$\mathbf{s} = R_z^{-1}\mathbf{s}'$$

To convert this vector into an angle for joint 2, we calculate the arccosine of the ratio given by the projection of $\mathbf{s}$ with the length of axis 2 of the industrial manipulator robot used. To this quantity we have then to add a $\frac{\pi}{2}$ corrective factor due to the robot model, giving us the final rotation angle γ:

Figure 9. Graphical representation of vectors **s** *and* **s**$'$ *with their respective coordinate systems*

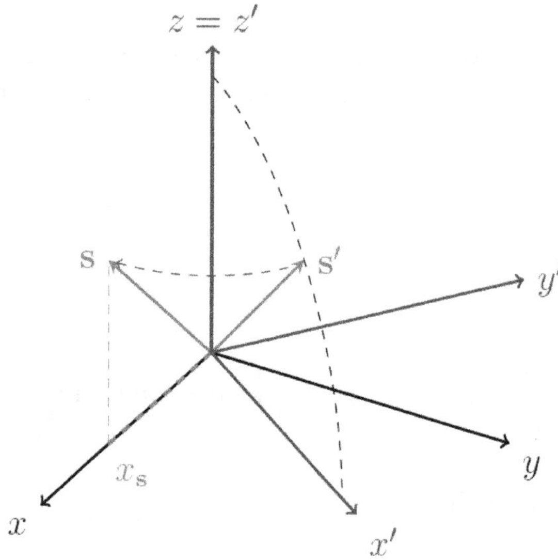

$$\gamma = \frac{\pi}{2} - \arccos\left(\frac{x_s}{\text{axis 2 length}}\right)$$

Joint 3

This joint rotates about axis y and is responsible of moving up and down the robot forearm.

The angle for this joint has been computed looking the direction of the left arm of human actor respect to his torso. Moreover, because the rotational axes of joints 2 and 3 are the same we have to consider also the inclination γ for joint 2, in order to guarantee a correct angle remapping for joint 3.

Let's take the vector $\mathbf{v}$ already used for joint 1:

$$\mathbf{v} = \mathbf{u}_{rh} - \mathbf{w}_n$$

that represents the vector going from neck to left hand.

Calculating the vertical inclination of this vector respect to the XY plane we get the angle φ' (Figure 10):

$$\varphi' = \begin{cases} \arccos\left(\dfrac{\rho_{XY}}{\|\mathbf{v}\|}\right) & \text{if} \quad z_v \geq 0 \\ -\arccos\left(\dfrac{\rho_{XY}}{\|\mathbf{v}\|}\right) & \text{if} \quad z_v < 0 \end{cases}$$

Figure 10. Graphical representation of the φ' angle used for the remapping of joint 3

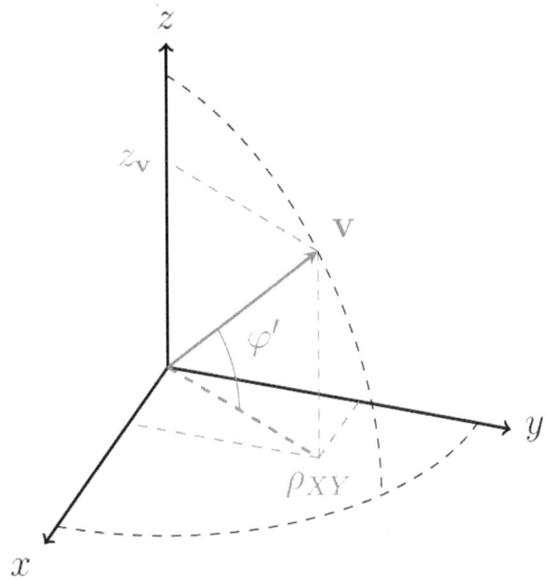

where ρ_{XY} is the length of **v** projection on XY plane, $\|\mathbf{v}\|$ is the vector L^2 norm and z_v its component along z axis.

Taking into account the previous considerations about the inclination of joint 2, the final angle will be:

$$\varphi = \gamma - \frac{\pi}{2} - \varphi' = \begin{cases} \gamma - \frac{\pi}{2} - \arccos\left(\frac{\rho_{XY}}{\|\mathbf{v}\|}\right) & \text{if} \quad z_v \geq 0 \\ \gamma - \frac{\pi}{2} + \arccos\left(\frac{\rho_{XY}}{\|\mathbf{v}\|}\right) & \text{if} \quad z_v < 0 \end{cases}$$

where $\gamma - \frac{\pi}{2}$ is the angle computed for joint 2 without the corrective factor.

Joint 4

The joint 4 allows to rotate clockwise or counterclockwise the robot forearm. To rotate this joint like a human forearm, we get the required information by tracking the human hand rotation about the axis given by the forearm direction. To do this, we can exploit the quaternion contained in the TFs of the hand wrist, but first we have to convert the quaternion to Euler angles, using the Z-Y-X convention. Given the quaternion $\mathbf{q}$ of the left hand wrist:

$$\mathbf{q} = (q_x, q_y, q_z, q_w)$$

we convert it to Euler angles:

$$\begin{cases} \varphi_q & = \arctan 2(2(q_z q_w + q_x q_y), (q_x^2 - q_y^2 - q_z^2 + q_w^2)) \\ \theta_q & = \arcsin(-2(q_y q_w - q_x q_z)) \\ \psi_q & = \arctan 2(2(q_y q_z + q_x q_w), (q_x^2 + q_y^2 - q_z^2 - q_w^2)) \end{cases}$$

where $-\frac{\pi}{2} \leq \theta_q \leq \frac{\pi}{2}$.

The final angle θ for joint 4 has been obtained using the next equations:

$$\theta = \begin{cases} \theta_q & \text{if} \quad \psi_q \geq 0 \\ -\theta_q & \text{if} \quad \psi_q < 0 \end{cases}$$

This two cases are necessary in order to correct the direction of the joint rotation (clockwise or counterclockwise) respectively when the hand is pointing up or down.

Joint 5

Joint 5 is the one that moves the wrist of the robot. It can rotate about the y axis and its motion is really similar to the one of human wrist. For the calculation of the angle for this joint, in fact, we will use the inclination of the hand respect to the wrist.

Formally we take:

$$\mathbf{t}_{rs} = (x_{rs}, y_{rs}, z_{rs}) \qquad \mathbf{u}_{ls} = (x_{ls}, y_{ls}, z_{ls})$$

Figure 11. This plot represents the angles used for the remapping of the joint 4. The red arrow represents the rotational axis (arm) about which the hand can rotate. The blue vector, instead, represents the hand

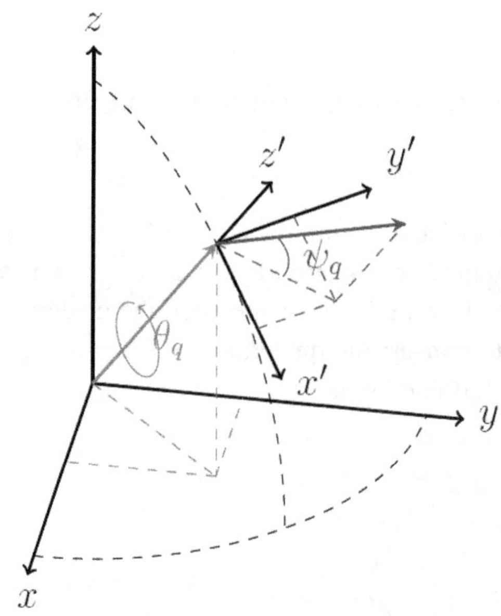

$$\mathbf{v}_{rw} = (x_{rw}, y_{rw}, z_{rw}) \qquad \mathbf{w}_{rf} = (x_{rf}, y_{rf}, z_{rf})$$

which are respectively the vectors of the left and right shoulder, the left wrist and the left fingertips. From these we calculate:

$$\mathbf{a} = \mathbf{u}_{rw} - \mathbf{t}_{rs} \qquad \mathbf{s} = \mathbf{t}_{rs} - \mathbf{u}_{ls} \qquad \mathbf{h} = \mathbf{w}_{rf} - \mathbf{v}_{rw}$$

where $\mathbf{a}$, $\mathbf{s}$ and $\mathbf{h}$ are vectors representing respectively the directions of the left arm (from shoulder to wrist), the shoulders (from the left to the right) and the left hand (from fingertips to wrist).

In order to obtain the orientation of the hand respect to the arm, we have to calculate the angle ψ' between the two respective vectors in the next way:

$$\psi' = \arccos\left(\frac{\mathbf{a} \cdot \mathbf{h}}{\|\mathbf{a}\|\|\mathbf{h}\|}\right)$$

where $\cdot$ is the dot product, and $\|\mathbf{a}\|$ and $\|\mathbf{h}\|$ are the L^2 norms of the respective vectors.

Now we have to get the direction for ψ' angle, in order to remap it to the joint 5. For this purpose, we take the vector obtained from the cross product between $\mathbf{a}$ and $\mathbf{s}$. This vector, in fact, is pointing to the side of the plane given from $\mathbf{a}$ and $\mathbf{s}$, in which the angle ψ' should be negative. In order to add the information of the direction, we do:

$$\psi = \begin{cases} \psi' & \text{if} \quad \mathbf{h} \cdot (\mathbf{a} \times \mathbf{s}) \geq 0 \\ -\psi' & \text{if} \quad \mathbf{h} \cdot (\mathbf{a} \times \mathbf{s}) < 0 \end{cases}$$

where $\times$ is the vector product.

In other words, the equation above is correct the sign of ψ' angle, by looking if the hand vector $\mathbf{h}$ has the same direction of vectors given by $\mathbf{a} \times \mathbf{s}$. If their direction are the same (the dot product is positive), the joint angle ψ is positive, otherwise it is negative (Figure 12).

Now the final angle ψ can be sent to the robot.

Before sending these 5 computed angles to the robot, we have first to filter the output values in order to smooth them and to eliminate all the noise coming from the human tracker and from tremors of human movements. In the next Section we will describe the technique adopted for values filtering.

Values Filtering

In order to eliminate the noise coming from the tracker and from natural tremors of human movements, a smoothing filter has been developed. This filter takes the angles calculated with the direct joints re-mapping method, computing the arithmetic mean of their last n values. Formally, the equation used in the implemented smoothing filter is the next:

Figure 12. This plot represents the angles used for the remapping of joint 5. We can see the vectors **a** *,* **h** *and* **s** *representing respectively the arm, the hand and the shoulders*

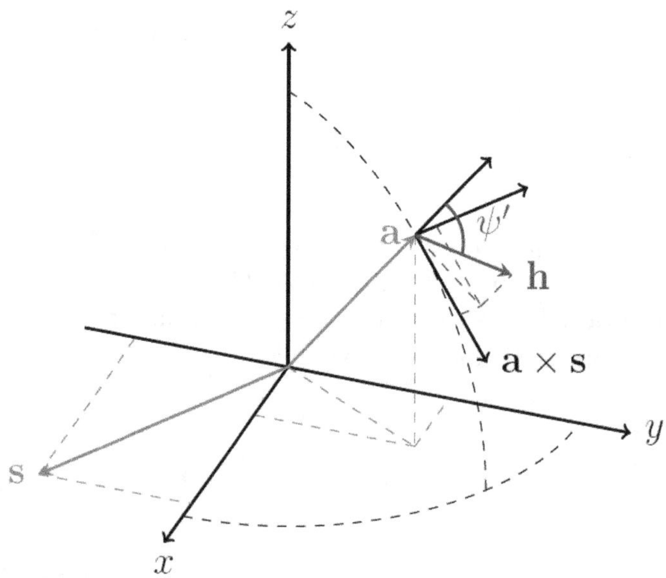

$$\bar{\xi}_k = \frac{1}{n}\sum_{i=0}^{n-1}\xi_{i,k}$$

Where:

- n is the number of values of a single joint angle to be smoothed by the filter;
- k is the number representing the remapped joint index (from 1 to 5)

Using the output values $\bar{\xi}_k$ of this filter instead of using directly the remapped angle values, we can notice a smoother and a less noisy motion on the real robot.

the best value of n has been found experimentally by testing different values, in the final version of the code we fixed the value to $n = 20$. This value revealed to be an optimal compromise between a well smoothed remapped motion and a less responsive robot control.

After that robot joints values have been computed and filtered, they are sent to the real-time computer, via a TCP connection.

Robot Motion

In this work, we applied the developed retargeting algorithm to two real models of Comau manipulator robots: a Smart5-Six and a 7 axes prototype. In order to move the robot we used the Comau C5G Open architecture, which allows developers to move any Comau robot via a B&R real-time computer connected to the controller via a Powerlink Ethernet (Romanelli, & Ferrara, 2014).

The task of moving the Comau robot is entirely done by an application that runs on the real-time PC. As previously said, this application receives the new joints from the computer via a TCP connection and then it forwards them to the robot controller, which will move the robot according to the target computed from the other PC.

The implemented software allows to use the described retargeting techniques with all the Comau robot models. In fact, it is enough to change the values of maximum velocities and maximum accelerations of the axes of the robot model in order to use it with other robot models.

Comparison between Direct Joints and Inverse Kinematics Remapping Techniques

It has been done some experimental comparison between the two remapping techniques about usability and intuitiveness. During this comparison tests it has been noticed that:

- The inverse kinematics has more limited movements respect to the direct joints remapping method, because its working area must be bounded in order to avoid singularities;
- With the direct joint remapping method, sometimes it can be difficult to control the end-effector position with a certain accuracy, because the operator has not the direct control of the position of the robot tool;
- With the direct joint remapping method, the human operator can control the angle of each robot joint in a natural and intuitive way: this freedom of control can be particularly useful in crowded industrial environments;
- With the inverse kinematics method, the human operator can control only the end-effector position. This could be a problem in some crowded industrial environments.

CONCLUSION

The programming of industrial manipulator robots can be currently done using only robot controller terminals or computers. Nevertheless this robot programming technique can take a long time to move a robot in specific positions. For this purpose, the teleoperation technique presented in this work allows to move an industrial manipulator robot, controlling directly the joints motion, just using a markerless based vision sensor that does not constrain the human operator to hold or to wear any physical devices.

In this work, in particular, we have implemented a system which allows to teleoperate a Comau industrial robot in an intuitive and in a real-time way using a RGB-D sensor, which in our case is a Microsoft Kinect sensor. The implemented software packages exploit the potentialities and the modularity of both ROS and the C5G Open architecture, allowing to apply the developed retargeting algorithm on each model of the Comau robots family, and potentially also to other brands of industrial manipulator robots.

The implemented system revealed to be strongly reliable and fast on the two Comau robot models tested.

Moreover, thanks to this modular architecture, developers can easily modify for example the retargeting algorithms or the tracking sensor.

ACKNOWLEDGMENT

Part of the software developed during this work has been used as base in some recently published works (Munaro et al., 2014) (Michieletto, Tosello, Romanelli, Ferrara, & Menegatti, 2014) (Tosello, Michieletto, Bisson, Pagello, & Menegatti, 2014).

REFERENCES

Dariush, B., Gienger, M., Arumbakkam, A., Zhu, Y., Jian, B., Fujimura, K., & Goerick, C. (2009). Online transfer of human motion to humanoids. *International Journal of Humanoid Robotics*, *6*(02), 265–289. doi:10.1142/S021984360900170X

Jeong, D. H., Ziemkiewicz, C., Ribarsky, W., & Chang, R. (2009). *Understanding principal component analysis using a visual analytics tool. Charlotte Visualization Center*. UNC Charlotte; doi:10.1109/SBR-LARS.2012.59

Kofman, J., Wu, X., Luu, T. J., & Verma, S. (2005). Teleoperation of a robot manipulator using a vision-based human-robot interface. *Industrial Electronics. IEEE Transactions*, *52*(5), 1206–1219.

Marinho, M. M., Geraldes, A. A., Bó, A. P., & Borges, G. A. (2012, October). Manipulator control based on the dual quaternion framework for intuitive teleoperation using Kinect. In *Robotics Symposium and Latin American Robotics Symposium (SBR-LARS), 2012 Brazilian,* (pp. 319-324). IEEE.

Michieletto, S., Chessa, N., & Menegatti, E. (2013, November). Learning how to approach industrial robot tasks from natural demonstrations. In *Advanced Robotics and its Social Impacts (ARSO), 2013 IEEE Workshop on* (pp. 255-260). IEEE. doi:10.1109/ARSO.2013.6705538

Michieletto, S., Tosello, E., Romanelli, F., Ferrara, V., & Menegatti, E. (2014). ROS-I Interface for COMAU Robots. In Simulation, Modeling, and Programming for Autonomous Robots (pp. 243-254). Springer International Publishing. doi:10.1007/978-3-319-11900-7_21

Munaro, M., Antonello, M., Moro, M., Ferrari, C., Clemente, G., Pagello, E., & Menegatti, E. (2014). FibreMap: Automatic Mapping of Fibre Orientation for Draping of Carbon Fibre Parts. In *ROS-Industrial in European Research Projects, IAS-13 Workshop on* (pp. 272–275).

Romanelli, F., & Ferrara, V. (2014). *C5GOpen: The latest generation of the Industrial Robots Open Control System for University and SMEs*. Comau.

Tosello, E., Michieletto, S., Bisson, A., Pagello, E., & Menegatti, E. (2014, June). A learning from demonstration framework for manipulation tasks. *Proceedings of ISR/Robotik 2014; 41st International Symposium on Robotics,* (pp. 1–7). VDE.

KEY TERMS AND DEFINITIONS

Inverse Kinematics: The use of kinematics equations to compute the joint configuration of a robot able to provide a certain position and orientation of the end-effector.

Joint Remapping: A map used to translate the motion a joint to another one with different characteristics, usually belonging to different entities.

Manipulator: robot: An arm-like device consisting of a series of automatically actuated segments linked together through rotational or translational joints; it is used to manipulate and move objects in a fast and reliable manner.

RGB-D Sensor: A device able to provide registered information regarding color and depth of the surrounding environment in order to obtain a digital map of colored points in the 3D space.

Skeletal Tracking: An algorithm used to track a certain amount of joints composing the human body; each joint is provided with spatial position and an orientation in order to obtain a good representation of the human motion.

Teleoperation: Also known as remote control, it refers to operation by a person of a device or machine, usually a robot, at a distance.

Transformation Frame: The relationship between two coordinate frames over time; it can provide transform matrices between the considered coordinate frames at any desired point in time.

Section 4

Service Applications of Robotics, Automation and Control

Chapter 10

A Gamification Mechanism for Advertising in Mobile Cloud

Zongwei Luo
South University of Science and Technology of China, China

Tao Jiang
The University of Hong Kong, Hong Kong

Qixing Zhuang
The University of Hong Kong, Hong Kong

Yang Liu
The University of Hong Kong, Hong Kong

Feng Yi
The University of Hong Kong, Hong Kong

ABSTRACT

In this chapter, we introduce a gamification mechanism for advertising in mobile cloud. Gamification for advertising uses game thinking and mechanism in non-game contexts to engage users in developing and deliver advertising content suitable for mobile devices. To support this gamification advertising mechanism, we develop a cloud based service platform for media integration and distribution, supporting flexible interactions and collaboration among media content providers, advertisers, and developers. Contribution of this chapter is it introduces game theory and mechanism design into gamification for advertising which is demonstrated as feasible and just in time. And the gamifiation for advertising is the first in the literature ever discussed as we know in the context of mechanism design. A layering solution with introduction of an advertising layer for developing gamified applications for mobile devices is also the first ever in the literature as we know.

1. INTRODUCTION

Today intelligent mobile phones can be seen every-where, overtaking PC in terms of users. Mobile applications have become a new focus of attention for advertising. However, the mobile ecosystem has encountered difficulties more than what has been there for PCs. That is, applications are free to use desirably in the spirit of Internet. Leaders, e.g. Apple, in some of the mobile ecosystems offer application stores where applications can be downloaded with compensation. Figure shows that more than 50 billion applications have been downloaded from Apple's App Store up to May 2013 (Lowensohn, 2013)

DOI: 10.4018/978-1-4666-8693-9.ch010

compared with 25 billion downloads in 2012 (Miller & Monaghan, 2012) and 15 billion downloads in 2011 (Pope & Muller, 2011). But a major number of mobile applications are still offered free. This is because often only top one or two applications in their application category can succeed. Another way to compensate for the application cost is to offer it for free but get return from advertisement. This strategy seems very promising as market leaders including Apple and Google are actively leading mobile operating system development (e.g. iOS and Android) with advertising supporting infrastructures built-in. The problem of this strategy is that the screen size of mobile phones is so small, any display of advertising messages would be rather annoying, instead of encouraging to reach and affect potential advertising targeted buyers.

In this chapter, we propose a gamification mechanism for supporting advertising in mobile devices. Gamification for advertising uses game thinking and mechanism in non-game contexts to engage users in developing advertising content and delivering it into mobile devices. To support this gamification advertising mechanism, we develop a cloud based service platform for media integration and distribution, supporting flexible interactions and collaboration among media content providers, advertisers, and developers. Media content providers supply the advertising resources to the cloud. Advertisers provide requirements to customize the media contents for advertising. Developers will offer systems and tools to assemble advertising resources and integrate them with game content.

Contribution of this chapter is it introduces game theory and mechanism design into gamification for advertising which is demonstrated as feasible and just in time. And the gamifiation for advertising is the first in the literature ever discussed as we know in the context of mechanism design. A layering solution with introduction of an advertising layer for developing gamified applications for mobile devices is also the first ever in the literature as we know.

Organization of this chapter is as follows. Section 2 presents the motivation of this chapter and the problems faced in current mobile advertising. In Section 3, mobile gamification strategies are discussed. Section 4 presents gamification development with a POI advertising method introduced. In Section 5, a city touring game is introduced to illustrate gamification advertising. In Section 6, implementation of the gamified application for city touring is presented to demonstrate the feasibility of mobile gamification for advertising. In Section 7, we discuss about digital advertising identity integration for gamification. In Section 8, related work is reviewed and compared. Evaluation of this gamification mechanism is discussed in Section 9 to highlight the research significance, and Section 10 concludes this chapter.

2. MOTIVATION

2.1 Mobile Advertising

Mobile media compete with traditional media like press, broadcast, television with time and functional displacement. Time displacement refers to that mobile media will compete with traditional media in terms of audiences' time (Qi, 2011). Commonly speaking, in terms of time displacement, mobile media is negative correlated with transitional media, i.e. the more mobile media consumption will lead to the less traditional media consumption. Functional displacement often means to satisfy different requirements with non-reciprocal functions which have the complementary relations (Qi, 2011). Comparing with other traditional media, mobile media have incomparable unique advantages. It can integrate image, audio, and movie. Mobile media's mobility, availability and interactivity makes it a promising candidate for carrying advertising messages.

2.2 Problem Statement

Mobile advertising, at the first beginning, in fact borrows directly what is already successful in the PC era. But in the case of mobile phone, the tiny screen affects considerably about the acceptance of mobile advertising messages displayed. Mobile users often will just delete those applications with advertising and install another "clean" one with similar functions. Clicking through the advertising message is not as promising as in PCs.

Resulted reality is that many mobile portals claim a huge volume of user base. But many of them are still searching for a good business model to turn this huge mobile user stream into revenue stream. In the case of mobile advertising, advertisers are hesitating because the effectiveness is in double and is difficult to measure.

Even Google's successful business model in PC based Internet is very difficult to replicate in mobile advertising. The keyword searching based business model is no longer considered the key to gold reservoir in a mobile world.

2.3 Advertising Strategy

One possible way out for this tiny screen problem is to deliver advertising messages embedded in answering user's queries. For example, when a user queries about a hotel nearby, advertising messages of a particular hotel fitting with the query can be returned. This type of advertising needs a huge database and a good ecosystem with active participation of a large number of advertising owners. The database will be used to understand the user and in a good running ecosystem, advertising messages can be effectively linked to what the users is seeking. However, all the set up depends on the availability of good quality data. Otherwise, without good data support, user experiences will be negative- affected significantly.

An alternative is that the advertising message can be embedded in the mobile application itself. In this way, the whole application is developed based on certain advertising needs. When carefully designed, acceptance of this application advertising is certainly better than those schemes by displaying advertising messages directly on the tiny screens. Toward this direction, we propose a gamification mechanism to embed and deliver advertising messages into mobile phones.

3. GAMIFICATION FOR ADVERTISING

3.1 Mobile Gamification Mechanism

Gamification is adopted to implement such an advertising strategy to embed advertising messages in mobile applications. The idea of gamification has been visible in human-computer interaction (HCI), dating back to early 1980s (Deterding, et al., 2011). Early research result shows that a system adopting gamification will offer inquiry in variety, scale and data quality and significantly increase user experiences (Deterding, et al., 2011). Mobile gamification for advertising proposed in this chapter is a strategy to apply game design and mechanism to mobile advertising, aiming to improve user experience and user engagement. It targets to encourage mobile users to accept gamified adverting messages, by turning the boring advertising messages into something interesting and interactive.

In gamification for advertising, there are players like advertiser, media content provider, developer, and user. Game theory (Osborne, 2003), in fact, provides a mathematical tool to model and optimize the decision-making when their strategies intervene with each other. To reach the best choices for all, information, for example, users' profile and advertising media, has to be shared. Action of each gamification participant can be either in or out. That is, each can decide to join the gamification for advertising or to withdraw from it. A set of action chosen by each player is an action profile. Each participant can also have preferences which can be defined upon action profiles as the decision of one participant will affect those of the others. Usually the preferences are defined using a utility function. The higher the utility value, the more preferred will be the profile.

This game theoretic approach to gamification allows negotiation among participants to make optimal trade-off with the ultimate purpose of improving mobile media as an ideal carrier for advertising. It can be easily observed from the above setup that equilibrium state will be either all in or all out.

3.2 Leader-Follower Game

In fact, in gamification for advertising, the developer, among all the participants, is the one most interested in an equilibrium state of all in. In order to convince other participants, developer will have to play a leading role to understand user preferences, provide tools to assist media content provider to develop mobile media to attract advertiser. The bargaining power of developer will increase if good user profiles are established and good tools are available. In the end, developer will have to negotiate with advertiser in adopting gamification for advertising. Thus, this gamification can be reduced into a leader-follower game between developer and advertiser in which advertiser is in a leading role with more power in negotiation.

In (Camuffo, Furlan, & Rettore, 2007), a leader-follower game is modeled in which the buyer is leading with willingness to absorb risks in sharing sensitive information. In the above discussed gamification, developer is willing to absorb risk in sharing sensitive information, but advertiser is leading the game.

The willingness to share sensitive information can be modeled and advertiser has more power than developer in negotiation. Both of them will have privacy concerns but at the same time try to share as much information as they can to reduce the impact of information asymmetry. However, in gamification, advertiser will not necessarily compensate developer for its extra information provided. But the uncertainties can be reduced through information exchange such that developer can convince advertiser to stay in the game.

3.3 Gamification Strategy

The willingness for information sharing can be mod-eled as information sharing level, i.e. information shared v.s. total information available. Information sharing level can always be measured in the boundary between 0 and 1. A diagram of information sharing level for such a leader-follower game can be drawn in Figure 1 (Zhang, et al., 2010). The green dotted line is the developer's information sharing level. And the blue dotted line is the advertiser's information sharing level. The solid dark blue line is the total sharing level of both parties. And information sharing ratio is measured as either advertiser's information level over developer's information sharing level or vice verse.

From Figure 1, it can be concluded that willingness to share information of both parties will increase at the beginning. But the willingness to share information will eventually drop even when the other party is still willing to share more. This shows that advertiser may not want all the information from developer. It is a warning sign when advertiser starts to withhold from information sharing. While exact information level to share will vary from case to case and can be further studied, in the gamification discussed in this chapter, in order to attract advertiser to stay in the game, a good strategy will be for developer to first demonstrate its core competence to understand user and provide good tools for gamification.

4. GAMIFICATION DEVELOPMENT

4.1 Point of Interest for Gamification

Point of Interest (POI) often refers to a specific location point which attracts people's interest in location based services. In advertising, this POI is useful to attract people's attention. In the case of gamification for advertising, this term of POI is reused as a virtual placeholder attaching to the mobile media content for advertising. POI is used to embed advertising messages. Identification of POI in mobile gamification is more complex than in location based services. The simplest form of POI is a property of mobile media. The whole mobile application can be a POI. For the sake of easy explanation and comparing with physical advertising carriers, for example, a location of an image displayed in a mobile application (see Figure 2) can be viewed as POI.

POI will be the focus for interaction among advertising participants in a gamified application. First of all, the media content provider carefully designs POIs in its media resources. Then, the advertiser browses the media contents and POIs and selects POIs candidates for putting advertising messages. Lastly,

Figure 1. Information sharing strategy

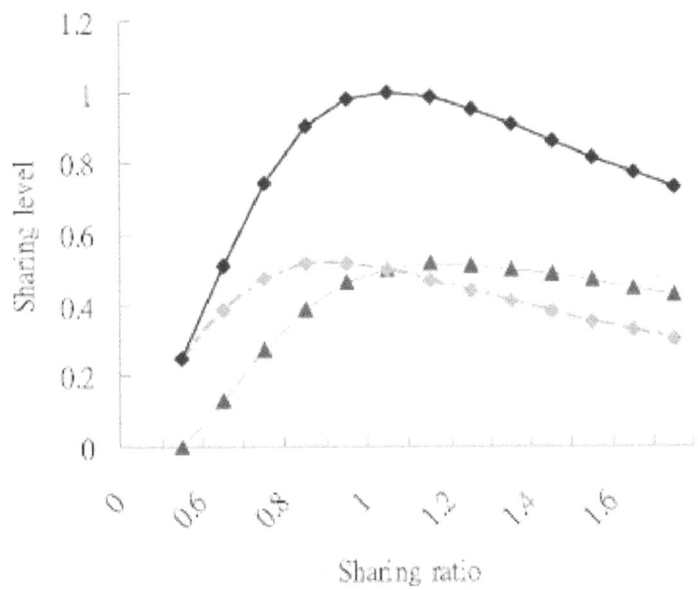

Figure 2. Mobile touring game

mobile user interacts with this gamified application through this POI. Meanwhile, the user preference is collected as feedback for POI.

User preference will signal how good POI is designed. An advertiser, for example promoting new products, on the one hand, can rent the POIs for a time period. On the other hand, POI can be bought permanently. In this way, the correspondence between this POI and advertising messages is recognized. The resulted filled POI will rep-resent a digital advertising identity which will be discussed further in later sections of this chapter. Further, in a gamified application, all the POIs in it can be bundled and offered for a single advertising task.

4.2 POI Preference

POI leads to revenue stream. Value of each POI may differ, so it may be priced differently. In terms of POI pricing, we can collect the statistics to determine the value of each POI. Such statistics can include user preference in a straightforward way to justify supply with demand. But at the beginning of POI evaluation, POIs can be ranked according to certain attributes.

As shown in Figure 2, each of these attributes will influence effectiveness of POI (taken by the ads displayed, e.g. HSBC). POI at International Financial Center (IFC) of Hong Kong is obviously eye-catching. This is to say advertising at this POI is easier to be noticed. As for POI frequency of occurrence, it means more often POI is used to push advertising messages to the game players, more valuable the POI is. Also, POI's appearance will influence effectiveness of advertising messages, such as POI's size, orientation, and animation (e.g. to make advertising messages more colorful and attractive). The media form of POI is not limited to static background. Image, game property and game characters can also be POI. To sum up, we need to evaluate POI with corresponding weight given to each of these ranking attributes. The weight will have to be updated from time to time For example according to Table 1, we calculate the ranking of POI 1 and 2 like below. The ranking of POI 2 is higher than that of POI 1.

$$Ranking\ (POI\ 1) = 3*10\% + 2*20\% + 2*30\% + 1*40\% = 1.7$$

$$Ranking\ (POI\ 2) = 5*10\% + 4*20\% + 2*30\% + 5*40\% = 3.9$$

Of course, ranking will not be limited to those four at-tributes. After a gamified application is released into the market, user preference is a good way to update POI ranking. For example in an application, the mobile user may often choose (user preference) to go sightseeing at Victoria Harbor in Hong Kong, and such statistics will give the Victoria Harbor (POI) a higher value. This evaluation method is reasonable, since a higher value indicates a higher degree of attractiveness of POI.

4.3 Layering Principle

Layering is a quite general concept to present media in terms of layers. Layers resemble sheets of stacked acetate (Kelby, 2015). For example, a picture can be viewed as a combination of multiple sub-images stacking on top of each other. As each sub-image has a portion visible from the top, the complete picture will contain all the features that all the sub-images collectively will display.

For example, in a typical 2D game (Strougo & Wenderlich, 2011), there are three layers in a game scene. Background Layer usually contains one single sprite, which is often a static image. Main layer is the one responsible for character movement and interactions with enemies and bonus. It is always consisted of multiple sprites. Control layer allow users to steer and interact with a game.

It is obvious that in this 2D game play scene, layers stack with a spatial order for convenience of game development and play. For example, the control layer always resides upon the main layer and main layer is always on top of the background layer. Elements in one layer will not be mixed with ones in the other layers. This basic layering principle allows developers to focus on their own layers and enables them full control of adding elements to their needs in each layer.

4.4 Advertising Layer

This layering concept is applied in media integration in gamification, that is, to embed advertising messages in POIs. Adopting layering in gamification can assist developers to separate one individual component from another. It would help developers focus on solving a particular advertising problem without getting lost in a much bigger context. Adding advertising messages in a separate layer enables

Table 1. POI ranking

POI	V (10%)	F (20%)	A(30%)	M(40%)
1	3	2	2	1
2	5	4	2	5

game designers to easily manage advertising resources just like updating or modifying other game resources. By working with other layers like the background, it helps integrate game environment and enable seamless gaming experiences.

For example, in Figure 3, when the background is moving backward as player moving forward, position of advertising content, i.e. the images of company icons, needs to be aligned with building blocks as a whole. Since the background and advertising images reside in different layers, it is necessary to update pixel locations of those building blocks frequently so advertising layer is aware of such changes. One alternative to this is to apply the same moving pattern to the both layers so that players will not notice any gap between two layers.

Media integration is more difficult in a 3D game, for example, the background layer is replaced with with another 3D city block like what Apple Maps will provide. The layering principle still works as an ideal solution to add an advertising layer in an existing game, and the advertising media will be resized and transformed correspondingly in its own layer. To simplify explanation, we will use Figure 3 as an example. You will find that the far way objects tend to be smaller than those closer ones. By adding a new advertising layer between the main layer and controls layer, it is much easier to control shape, position and size of each advertising messge element.

5. GAME OVERVIEW

This gamified application shown in Figure 2 is a digital city touring game adding amusing features to enhance user experience and engagement. It is a surfing and shooting game with music integration for fun. Meanwhile it is also a digital city touring game with advertising message integrated for sightseeing.

Surfing is a kind of sport game that a surfer (mobile user) carries a surfboard and surfs on the wave surface. The surfer goes up and down according to the fluctuation of the wave. Perhaps, he can jump

Figure 3. Processed output result

higher on the peak of the wave surface or fall into the sea and die. Only if the surfer finishes the whole course safely, will he win the game and get a reward. The wave curve in the game is generated based on the music volume.

As for shooting, the operation is similar to the well-known mobile game "Angry Birds". A mobile user throws weapons to the enemy in a parabolic curve. But the difference is that both the user and enemy are moving on both vertical and horizontal direction at every moment, in order to cater for digital touring. When the enemy is shot by the user, the user gets a score. The enemy moves synchronized with the music.

The game can automatically detect the current location (city) of the user. Then, the user selects which street he wants to travel virtually as the user may want to get familiar with some touring attractions before the real travel. Within the game, such city street image is set as game background image and will be scrolled according to the game role position. Besides, the advertising messages are linked to buildings in background image. Its size, shape and orientation will be changed in real time along with the buildings seen in the game.

As depicted in Figure 4, the system architecture of this gamified application is developed based on a cloud based resource exchange platform. First of all, primary target user is tourists who want to play the game for fun and to get familiar with a city at the same time. The interactions among four participants in the cloud are shown in Figure 4. It can be divided into two phases, the preprocessing development phase and real time game play phase. The preprocessing development phase will make the gamified

Figure 4. Interactions in the cloud platform

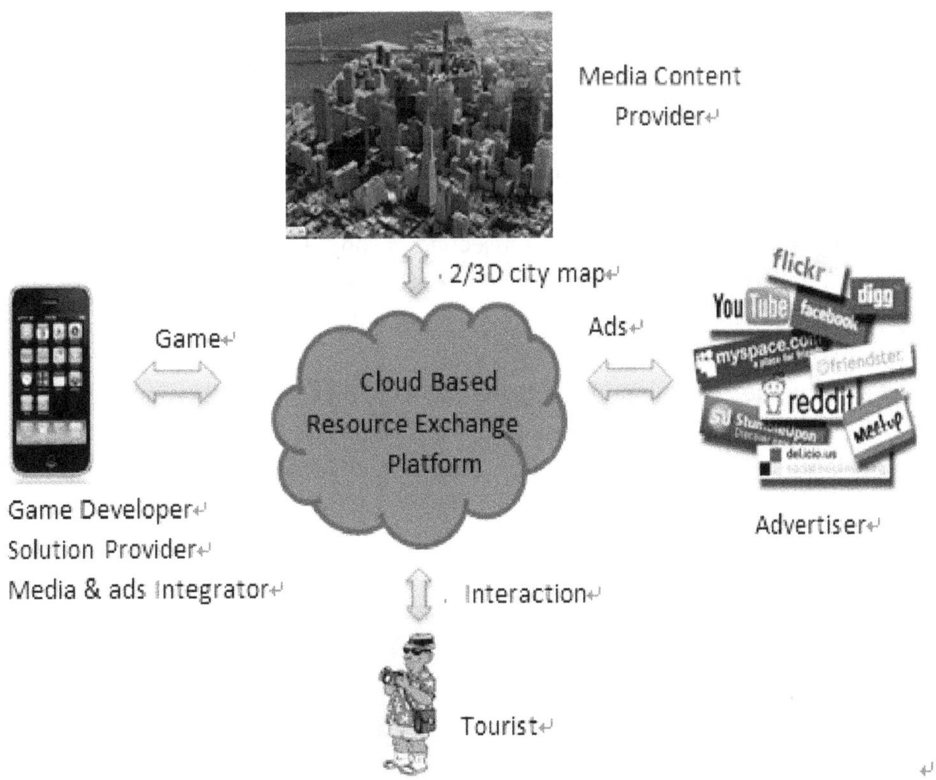

application ready before launching the game, while the game play phase begins from the launch to the close of the game. Note that the cloud based platform is used in both preprocessing development phase and the real time game play phase.

6. GAMIFICATION APPLICATION DEVELOPMENT

6.1 Music Preparation

We need two properties from a music file: length of the music and decibels at each specific time. Length is a trivial property to obtain. As to obtaining decibels, the first problem is how often to obtain a decibel value. The more frequent we ask for a value, the more accurate the final wave will be, compared to detailed original music content.

An array of data will be drawn on the screen. And sine waves will be constructed based on these dots. If intervals between those dots were wide, it would decrease difficulty level. It would disappoint player if the music picked were a very challenging one. If intervals were too narrow, there would be no much room for sine wave to flow naturally. With testing and experiments, a pleasant time interval to pick is one second. It proves that it is a good balance between curve generation and user experience.

6.2 Wave Curve Generation

In this game, specific music properties are used to generate game elements which will affect user experience. The key element in this game is the wave curve for surfing from the starting point, chosen by user, of the landscape to the end point of the landscape. After parsing the music about the music volume, it produces a set of points. The vertical coordinate of every point indicates the height of the wave surface in the game, which is the most significant property in the game. When we draw a curve passing through all the given set of points, it somehow reconstructs the music volume along the time (see Figure 2).

If the music volume sampling rate is 1 point/second, a song, lasting 3 minutes for example, would produce 180 points. It is hard to find a function which corresponding curve passes through such a large amount of points. As a consequence, we use divide-and-conquer (i.e., piecewise) to divide the complex curve into a number of curve segments.

Then, it poses a question how to connect these curve segments smoothly in order to form a whole curve. Practically, we can consider three types of local properties and decide which properties we should satisfy. These are continuity (position), slope (direction) and curvature (speed), which corresponds to the curve connection in terms of zeroth derivative (C0), first derivative (C1) and second derivative (C2) respectively. If C0 is not met, the curve is break. If C1 is not met, there is a sharp change between two connecting points. If C2 is not met, the speed change between two connecting points will be very different. A summary of these three local properties of curve is listed in Table 2.

6.3 Polynomials for Curve Smoothing

Then, we use Hermite cubic polynomials (Shirley, & Marschner, 2009) to define a curve function for every curve segment. Matrix expression of cubic polynomials can also be produced, P= CA, where C is called constraint matrix. In order to solve A, we want to find an inverse matrix of C. The Hermite cubic

Table 2. Three local properties of curve

Property	Dissatisfied	Requirement
Continuity	Curve break	Zeroth derivative
Slop	Sharp change	First derivative
Curvature	Sharp change speed	Second derivative

polynomials allow us to specify the positions and first derivatives C1 of the start and end point. If we specify the same values (position and C1) for the end point of the current segment and the start point of the next segment, we can guarantee C1 continuity. Apart from that, Hermite cubic polynomials provide local control, namely, a change in one segment would not lead to changes in other segments. Thirdly, they guarantee that the curve will pass through those control points (i.e., the start point and the end point of the segment), and hence they can accurately reflect the fluctuation of music volume at every given point. One disadvantage of Hermite cubic polynomials is that it cannot ensure C2 continuity. Therefore, it is a trade-off among local control, C2 continuity and passing through the control points. C2 continuity problem seems not too obvious in the game shown in Figure 2.

The idea of applying Hermite cubic polynomials is that two positions of points are provided indicating the start point and the end point, and then the slope (i.e., first derivatives C1) of such two points is provide and the curve segment will be like the one depicted in Figure 5. The positions of the start and end points can be obtained from the music volume parser. The question is that how to specify the slopes of the start and end points. Frankly, if the slope of the start point differs from the slope of the end point, C1 continuity cannot be ensured, and there would be a sharp change in the connection. We can choose both slopes of two points equal to a value, e.g. 1 (i.e., 45 degree) to make the curve more similar to a sine wave. Certainly, a higher value of slope means a more volatile wave, and then a more difficult game.

There could be a problem in the fixed increment of x coordination. The interval between the start point and the end point is obtained from music parser, and it may be a float value in pixel (e.g., 32.4 pixels). Therefore, if we use a fixed increment of 3 pixels, for example, eventually the curve generated will not reconcile with the background music. If the chosen increment is too small, the system performance will drop significantly. Thus, we have to balance this increment with how smooth the game can be played.

Figure 5. Hermite cubic polynomials for curve reconstruction

6.4 Movement Synchronization

Since the wave curve is generated based on the music volume, user will feel that the wave is fluctuating according to the current background music. The speed of the movement in the horizontal direction is totally up to the user, that is the user can surf faster if he is more familiar with this game, or slower if he makes a mistake. Thus, the question is that how to keep the user movement roughly at a predefined speed all the time. One way to solve this is to provide an enemy or target for user to follow. If the user is left behind, the surfing speed is increased to let the user catch up with the enemy as well as the background music. If the user is moving closer to the enemy, we will slow down the surfing speed for the movement synchronization. The next question is how to let the enemy move along the wave surface at the correct position at a predefined speed.

The enemy or target in the game moves along the wave surface with a fixed offset in the horizontal direction, and the game finishes when the background music stops. We have a set of points to depict the whole wave curve, instead of the mathematics function of the whole curve based on image pixel. The curve is calculated based on an increment of how many pixels per second. It becomes the focus of how to get the y coordination according to the given x coordination becomes the focus during the move of the enemy.

The solution to this problem is that we use a vector to store every discrete point for the whole curve with an interval equal or less than increment. Note that $x_j = x_i$ + increment (or less than increment) and we have $x_i < x_j$, if $i < j$. Thus, given a coordination x_k, we can use two available points (x_i, y_i) and (x_j, y_j) to estimate the point (x_k, y_k) where $x_i \leq x_k < x_j$, that is y_k is the desired coordination with respect to x_k.

With an ascending order in the x coordination, it is not necessary to traverse the vector for every given x_k. Instead, we use an index, called last access index, to record the element in the vector which has been accessed in the last search. Therefore, to find a given x_k, we start the searching from the last access index rather than the first element of the vector, and hence after one-pass traverse, we have obtained all the desired points (x_i, y_i).

To move in synchronization, we should firstly know how many pixels corresponding to one second denoted by p pixel/second, and then the update frequency of the game denoted by t time/second. Then, we can get how many pixels the enemy should move in horizontal direction per update. Finally, we can use the coordination acquisition method discussed above to get the corresponding y coordination.

7. DIGITAL IDENTIFY FOR ADVERTISING

7.1 Advertising Identity

In gamified advertising, advertising messages shall be easily recognizable. They are subtly noticed without any negative influence to users desirably. This is the goal of gamification advertising. In a well-designed gamified application, advertising messages are presented and meaningfully embedded in the overall game application.

Since the advertising messages are embedded, how these messages can be effectively linked with an advertising theme become a critical issue. The concept of digital identity for advertising (DIA) is used for addressing this issue. DIA can be considered a condensed version of normal advertising messages for it is delivered in mobile devices. A company logo could be a good DIA. An intangible DIA could also exist. But if a company wants to promote more than just its company, but a product or service, it has to be reflected in a new DIA. This is, the DIA, when perceived, will be easily linked to an advertising theme of a product or service of a company.

When a single DIA cannot convey the advertising messages, more than one DIA will be used, collectively. In many cases, the whole applications can be used to construct DIA to reach and affect users.

7.2 Advertising Media Adaptation

Let's use the city touring game again to illustrate DIA construction, that is, to integrate advertising messages with POI. A 3D game scene shown in Figure 2 could be a good game background for discussion. Let's assume the gold bag is a DIA to be presented in this game.

In this touring game, expected game background will be city streets. We will need to deal with problems to let advertising media (e.g. company icons) adapt to changing game environment, i.e. integrating DIA with the game. One way to accomplish the adaptation is to apply image processing techniques to produce an image that the game engine can support.

In a game, the game engine will track four dots to construct a perspective view of each scene in real time. Fixing lower and upper boundaries of streets and buildings can locate those four dots for each scene. For each image with four boundary dots denoted, the next step is to compose an image for further processing.

7.3 DIA Integration

To integrate DIA images, we need to analyze POI regions denoted by those four dots in each building. Sometimes, the POI locations for embedding advertising DIA images are not too rigorous. We can, based on a building's shape, scale DIA images down to fit in the corresponding POI region. In many cases, there is a need to register a DIA image with POI location. Luckily poses of POI and DIA media for registration can be known before.

Another issue is to deal with movements of buildings. Views can be updated when camera pose or eye location changes. When buildings move backward or forward relatively to a player, views of DIA images should change as well in a natural manner. This is achievable if we know the POI locations in those buildings, because relative relationship between POI in those building blocks and advertising DIAs is already predefined. What we need to do is get the properties of viewing camera and display the content accordingly. We can apply the same movement pattern to both background layer and advertising layer to simply image processing in those layers.

It is also an issue to fit the DIA images with the surrounding environment. For instance, a DIA image may have higher resolution compared to skin images of nearby buildings. Under this scenario, a blurring image filter can be applied to ease unnatural effect of the DIA images when embedded into the building blocks.

8. COMPARSION AND RELATED WORK

8.1 Advertising Media Adaptation

There are several mobile advertising platform operators (Cody,Cosmas, & Tsekleves, 2009). Among them, MobiSage (AdSage, 2013) is a relative successful case for the mobile device application advertising platform. It is very representative as it owns lots of resource in both advertiser side and mobile application developer side and provides good links between them. However, there are still many limitations for those platforms.

Firstly, the end user experiences and acceptance need to be improved. The advertising messages embedded are "hard" advertising messages. That means that the platforms are doing hard selling to users. Besides that, the advertising messages are not really embedded in the mobile applications. It occupies a part of screen like what is shown in a PC. In this situation, acceptance rate of most of the users is low. In other words, it is lack of interaction and entertainment. Many researchers have claimed that entertainment is a crucial factor for mobile advertising (Zhang, & Xiong, 2012, Shuang, 2011). Haghirian and Madlberger (2007) showed that perceived entertainment and user's attitude were closely related. In fact in those platforms, the click rate for the advertising is not reaching the expectation of both mobile application developer alliance and advertisers.

Secondly, those platforms are more suitable for PCs and tablets with larger size of screens. They are not really targeting mobile phones with smaller screens. Therefore there is a long way to go to develop the mobile phone market because the mobile phone screens are too small to let users accept those hard selling advertising messages provided by those platforms.

8.2 Mobile Media for Advertising

There are various kinds of mobile media for advertising, like SMS, mobile newspaper, mobile video/ TV and so on. Some of them are quite successful. For example iZigg focuses on SMS for advertising (Clay, 2010). But the major problem of them is it is lack of interaction. Those media just present normal display with adverting messages. In many cases, they evolve into a form of spamming as the advertising content is not requested by users, but enforced by those media advertising operators.

This is especially true for SMS media which are re-quested to avoid sending out hard-selling advertising text messages. Mobile TV and mobile newspaper has limitation in attracting subscribers for a long period of time because of the tiny screen size (BBC, 2007).

Gamification provides a way to increase the attractiveness and interaction of those mobile media for advertising. However, gamification strategy varies with applied cases.

8.3 Gamification Mechanism

In terms of gamification mechanism, there are a good number of gamified applications developed. Earlier in this chapter, gamification strategy is discussed in the con-text of gamification processes to develop gamified applications. Evaluation of the developed gamified applications has not been touched upon yet. In fact, game theory can also be applied to test whether the gamified applications would yield good results as expected.

For example, a gamified traffic sharing application is introduced in (Law, Kasirun, & Gan, 2011) encouraging interaction among users to solve real life urgent problem, namely, to avoid the traffic congestion by observing the traffic condition in real time. However, user can only be helped if other users upload relevant traffic accident photos which happened at those locations. Since the online photos cannot cover all places people can reach, there will be certain blind spots where users cannot get assistance from the others. Moreover, the update frequency of traffic condition is uncertain depending on the other users. Users, in fact, cannot solve urgent or real time problem (i.e., find the best route) when there is no available resources at that specific area recently. Further, when the update frequency is low, the previous traffic condition may not be able to reflect current traffic condition. Lastly, it is rather difficult to validate the correctness of those photos uploaded. For instance, if a user wants to select a route A for traveling, then he may post fake photos on route A, indicating that there is an traffic accident there by gaining benefit in the hope to lower the chance for people to select route A. Since there are a large amount of photos updated every day, validation of these photos seems impractical by the operator.

In order to help resolve those problems, a voting mechanism can be introduced to build up credit for users to encourage them to upload correct photos. Similarly a rewarding mechanism can also be introduced as to encourage users to upload more photos. The timing is also critical to release the application out for users to try.

9. RESEARCH SIGNIFICANCE

Research significance of this chapter can be characterized in three aspects.

First, it provides a promising direction to develop mobile advising toward a future of good potentials. According to the statistics from Limelight Networks, mobile media industry output presented breathtaking development in 2011 (Limelight, 2014). Based on the data for the first five months of 2011, output of mobile media industry all over the world presented a 600% increase compared with the same period in 2010 (Limelight, 2014). But majority of those mobile figures were achieved in bigger screens on the move. Up to now, there is no significant progress regarding to business models made for the smaller screens of mobile phones. The gamification for mobile advertising provides a meaningful case to pave for future development for advertising leveraging mobile phones.

Second, current market development seems echoing what is proposed in this chapter for gamified advertising applications. It has been observed that the market size of mobile searching will reach 5 Billion RMB in 2014 in China alone (Qianlima, 2014). Focus is now shifted and placed on mobile application and online/offline integration instead of keyword based search business model (Qianlima, 2014), which is consistent with what we have observed and reported about mobile advertising gamification for enriching user experience and interaction.

Third, it provides a systematic analysis of advertising gamification from the perspective of game theory to guide mobile advertising gamified application development. The layering method with advertising layer introduced is demonstrated to be feasible and practical.

10. CONCLUSION

In this chapter, we introduce a gamification mechanism for supporting advertising in mobile devices. Gamification for advertising uses game thinking and mechanism in non-game contexts to engage users in developing advertising content and delivering it into mobile devices.To support this gamification advertising mechanism, we develop a cloud based service platform for media integration and distribution to collaboratively develop gamified applications for advertising.

Contribution of this chapter is it introduces game theory and mechanism design into gamification for advertising which is demonstrated as feasible and just in time. A layering solution with introduction of an advertising layer for developing gamified applications for mobile devices is also the first ever in the literature as we know.

This gamification mechanism is discussed in the context of game theory for mechanism design, first ever in the literature as we know. The game theory can also be used to validate a gamified application. The gamified application in this chapter illustrates a meaningful touring assistance application to show case how to develop mobile applications to enrich user experiences and interaction.

Firstly the location based display is helpful and very effective as local image can grasp users' attention especially with well-planned local adverting messages. The embedded DIA advertising messages on the local build-ings of digital counterpart, for example, can be fun to accept. When a user changes to another place, street scene will also change which will bring freshness to the game players.

Secondly, gamification is used instead of pushing out "hard-selling" advertising images. As shown in Figure 2, the advertisement images are put on the buildings in the background image. It is well integrated just like what is presenting in the Victoria Habour in Hong Kong. Such advertising messages embedding method can improve user acceptance and entertainment.

Further, we do not require users to click through the advertising messages. We rely on the whole gamified application to attract users to come back again. The feedback from the users' street selection will be a good indicator to show how successful the gamification is. Feedback is collected without notice of users. That means that the users will perceive advertised messages without interruption, thus minimizing any effect to affect game playing experience.

Future direction of this work will be big data enabled mobile advertising gamification. Difficulties in obtaining and utilizing data to support gamification analysis will be addressed.

ACKNOWLEDGMENT

We thank all the volunteers, and all publications support and staff, who wrote and provided helpful comments on previous versions of this document. We also thank the startup fund of SUSTC (Y01236115/Y01236215) for its support.

REFERENCES

AdSage. (2013). *Introduction to mobisage*, Retrieved from http://mobi.adsage.cn/about

Camuffo, A., Furlan, A., & Rettore, E. (2007). Risk Sharing in Supplier Relations: An Agency Model for the Italian Air-conditioning Industry. *Strategic Management Journal*, *28*(12), 1257–1266. doi:10.1002/smj.635

Clay (2010). *iZigg, SMS Mobile Media 90210*. Retrieved from http://www.mobilemediaforbusiness.com/14/izigg-mobile-media-marketing-90210/

Cody, R. L., Cosmas, J., & Tsekleves, E. (2009). Open-standards Rich Media Mobile Platform & Rapid Service Creation Tool. *IEEE Global Mobile Congress*, (pp. 1-6). doi:10.1109/GMC.2009.5295896

Deterding, S., Sicart, M., Nacke, L., O'Hara, K., & Dixon, D. (2011). *Gamification, Using Game-design Elements in Non-gaming Contexts. Proceedings of the 2011 annual conference extended abstracts on human factors in computing systems*, (pp. 2425-2428). doi:10.1145/1979742.1979575

Haghirian, P., & Madlberger, M. (2007). Consumer Attitude towards Advertising via Mobile Devices – An Empirical Investigation among Austrian Users. *International Journal of Mobile Communications*, 48–67. doi:10.1504/IJMC.2007.011489

Kelby, S. (2015) *Adobe Photoshop Layer basics*, Retrieved from http://helpx.adobe.com/en/photoshop/using/layer-basics.html

Law, F. L., Kasirun, Z. M., & Gan, C. K. (2011). Gamification towards Sustainable Mobile Application. *Proceedings of IEEE Software Engineering 2011 5th Malaysian Conference*, (pp.349-353). doi:10.1109/MySEC.2011.6140696

Lowensohn, J. (2013). Apple's App Store downloads hit 50 billion. *CNET*. Retrieved from http://www.cnet.com/news/apples-app-store-downloads-hit-50-billion/

Miller, T., & Monaghan, C. (2012). Apple Press Release Library, Apple's App Store Downloads, Top 25 Billion. *Apple.com.* Retrieved from http://www.apple.com/pr/library/2012/03/05Apples-App-Store-Downloads-Top-25-Billion.html

Mobile Media Consumption. (2014). *Limelight.com.* Retrieved from http://blog.limelight.com/2011/06/a-look-at-the-numbers-mobile-media-consumption-2010-2011/

Mobile TV Predicted To Be A Hit. (2007). *BBC News.* Retrieved from http://news.bbc.co.uk/2/hi/technology/6639249.stm

Osborne, M. J. (2003). *An Introduction to GameTheory*. Oxford, UK: Oxford University Press.

Pope, S., & Muller, T. (2011). Apple Press Release Library, Apple's App Store Downloads, Top 15 Billion. *Apple.com.* Retrieved from https://www.apple.com/pr/library/2011/07/07Apples-App-Store-Downloads-Top-15-Billion.html

Qi, Y. (2011). Risk of Social Disruption and Integration Function Development of News Products of Traditional Media in the Digital Age. *Proceedings of 6th International Conference*, (pp.70-74). IEEE.

Qianlima (2014). *Mobile Market Reaches 5 Billion*, Retrieved from http://digi.163.com/14/0521/20/9SQ0GDET001618JV.html

Shirley, P. & Marschner, S. (2009). *Fundamentals of Computer Graphics*, Wellesley, Massachusetts: A K Peters, Ltd.

Shuang, X. (2011). Consumer Adoption of Mobile Advertising: An Empirical Investigation among China Users. *Proceedings of IEEE Product Innovation Management (ICPIM), 2011 6th International Conference*, (pp.461-464). doi:10.1109/ICPIM.2011.5983702

Strougo, R., & Wenderlich, R. (2011). *Learning Cocos2D: A Hands-On Guide to Building iOS Games with Cocos2D, Box2D, and Chipmunk.* Boston, Massachusetts: Addison-Wesley.

Zhang, X., Cheung, W. K., Luo, Z., & Tong, F. (2010). A Game Theoretic Approach for Sensitive Information Sharing in Supply Chain. *International Journal of Applied Logistics*, *1*(4), 1–12. doi:10.4018/jal.2010100101

Zhang, X., & Xiong, K. (2012). A Conceptual Model of User Adoption of Mobile Advertising, *Proceedings of IEEE Computer Science and Electronics Engineering 2012 International Conference*, (pp.124-127). doi:10.1109/ICCSEE.2012.454

KEY TERMS AND DEFINITIONS

DIA: Since the advertising messages are embedded, how these messages can be effectively linked with an advertising theme become a critical issue. The concept of digital identity for advertising (DIA) is used for addressing this issue. DIA can be considered a condensed version of normal advertising messages for it is delivered in mobile devices.

Game Theory: According to Wikipedia, game theory is the study of strategic decision making, i.e., the study of mathematical models of conflict and cooperation between intelligent rational decision-makers.

Gamification: Gamification refers to use game elements and game design techniques for non-game context. It is often used to derive motivational strategies.

Mobile Advertising: Mobile advertising refers advertising where advertising messages are carried on mobile devices.

Mobile Cloud: It refers to a cloud system where cloud computing, mobile computing and wireless computing are organically mixed.

POI: Point of Interest (POI) often refers to a specific location point which attracts people's interest in location based services.

Service Platform: It refers a computer information system providing tools and functions where services could be hosted, orchestrated, managed and offered.

Chapter 11
Robotic Transformation and its Business Applications in Food Industry

Anas Mathath
Universiti Sains Malaysia, Malaysia

Yudi Fernando
Universiti Sains Malaysia, Malaysia

ABSTRACT

The role of robots is becoming substantial for industrial applications and business competitiveness. The robot transformation in food industry has increased business productivity, reduced cost and enhanced customer experiences. The usage scale of robots has an increasing trend globally when industries modernize and increase the production capacities with ability in handling complex tasks. The objective of this chapter is to explore robotic transformation in literature and to investigate its business applications in food industry. There are two points raised in the discussion, would the robot technology which has been developed only capable owned by large scale food companies and the experiences gained in the restaurant which serves by robots can replace the human touch. At the end of this chapter, some solutions are given to shed light on the application of robot in food industry and deepen critical analysis for researchers, technocrats and business practitioners.

1. INTRODUCTION

In the modern competitive business the role of robots is becoming significant for industrial applications. The important factor for using robots in the industry aims at reducing human inference and to increase the productivity. According to Chiu-Chi (1995), the shortage of manpower led Taiwan industry to use more robots and it enhanced the annual growth rate of robots globally. Chiu-Chi (1995) also states that the robotic research have expanded the domestic industry, promoted the technological standard and the development of the small and medium manufacturers. The robot transformation in service sector has significantly increased business productivity and customer experiences.

DOI: 10.4018/978-1-4666-8693-9.ch011

Manshi and Shukla (2012) analyzed the market growth of industrial and service robots, mainly involving the growth of the two categories of robots in 21st century. The growth of service robots seems to be higher since they own a huge market share, out of which personal robots contributing the major share of growth. Manshi and Shukla (2012) compared the growth of robots in the industry in terms of sales and volume. The growth of service robots showed a higher percentage than the industrial robots depending on the areas of application. Manshi and Shukla (2012) concluded that the technological transformation and the huge requirement in various industries have raised the need for robots at a big scale there by developing the robotics industry during the early 21st century.

Robots are used in promoting product quality and reducing costs (Chiu-Chi, 1995). According to Gorle and Clive (2011) reported that there is a great potential for new job creation in the years up to 2016. According to Moran (2011) Shenzhen –Foxconn, the largest manufacturers of Apple products plan to replace workers with nearly one million robots in recent years. Low cost maintenance with high productivity in the business is considered as the prime advantage for the robots. Baxter, a new inexpensive robot is an example, which provides user friendly operations and performs various tasks in small manufacturing companies (Ben, 2012).

When looking into the food industry, the robots serve for much purpose, mainly in production systems for material handling and packaging operations (Wallin, 1997). The evolution of robotic applications in food industry evolved in 1980s and the growth was not effective due to high technology related costs. After realizing the consistency of robotic jobs, with improved efficiency and reduced work space, food industry expanded the use of robots for various applications which resulted in high productivity.

Although the investment required in the robotics was an obstacle to the development of the robotic application, Rene et al. (2010) outlined the design feature for low cost robots for the production in the food industry. Robots should fulfill the basic requirements like hygiene, speed and safety, cost, operational speed and ease of programming (Rene et al. 2010). Rene et al (2010) also suggested that low cost robots in the food industry will pose developments in the industry with reduced costs for the production. The objective of this chapter is to explore robotic transformation in literature and to investigate its business applications in food industry. This is can be seen than many restaurants have extended robot applications to increase business productivity, keep the loyal ones and to attract new customers.

2. LITERATURE REVIEW

A robot is an electro mechanical machine operated with the computer manipulating various functions. According to International Standard Organization (1997), a robot can be automatically controlled, reprogrammable multipurpose manipulator programmable in three or more directions. Basically robots can be categorized on the basis of degree of autonomy, industry or field where they are used and goals they are fulfilled. On the basis of degree of autonomy, it can be categorized as stationery, ground, underwater, and aerial. Robots under the category of goals fulfillment includes contest, personal enrichment, manufacturing and entertainment.

Based on the industry, robots can be divided into categories- Industrial, service and military robots, agricultural, mobile and telerobots. Engelberger integrated the first industrial Robot (Unimate) in 1962 at New Jersey General Motor's automobile factory. It was an automated die-casting mold, used to avoid human interference in holding car parts made of molten steel. Handling hazardous job is one of the most advantages of robots in the industry. The success of the first industrial robot, Unimate, shed light on the robot welders in the auto industry in Japan; influence them to make use of robots for their operations in the future (Mickle, 2014). Robot automation is introduced in the industry to optimize efficiency and to reduce the costs (Bogue, 2009).

On the other side, service robots, which operates semi or fully autonomously are mostly used to aid human activities. Service robots are further divided into professional and personal service robots. The application of professional service robots is concentrated in the areas of logistics, transport, medical and health, education, security and surveillance and agriculture, whereas personal robots are used for domestic uses like daily chores, assistance jobs and for entertainment. The application of robots can be again categorized into many types. For example, professional service robots are used as waiter robots, guide robots and tour robots. The market research report on service robotics shows that service robotics segments the largest portion in the market by value. On the other side, personal service robots cover the maximum market in terms of volume (San Jos, 2010).

2.1 Robot Transformation in Business

According to Moore's (1970) law, today the computer devices are more powerful than it was 20 years ago, eventually with the increase in performance level. The observation made by Gordon Moore, founder of Intel, stated that the number of components used in the circuits increased at regular intervals of time and would increase at a higher pace in the future (Schaller, 1997). It is forecasted that the performance and applications for such devices cannot be predicted for the future. Unlike the computer devices, robots are also supposed to advance in the future as computers are today. Starting from the manufacturing operation, robots have now entered all the industries for numerous applications, even in helping humans at home to perform daily chore activities are some of the novel applications. Business owners and entrepreneurs were able to bulge their pockets by saving costs with the introduction of automated business operations. Where economies struggle against labor shortage and high labor costs, this electro mechanical machine (robots) were able to exterminate those issues rising to growth and profits everywhere.

"I cannot define a robot, but I know one when I see one". The remark was raised by the father of robots, Joseph Engelberger who founded the world's first robotics company (Kilbane, 2008), when robots represent remarkable inputs irrespective of the sectors in business. It even made an uncertainty in excerpting widely used definition of robots indicates that robotics evolved an enormous augmentation with advancements. While analyzing the transition made by the robots, it is essential to look at which areas they act as a cusp for the business growth. Relentless transformation is ongoing from the early 1960s until now, with the introduction of stunning innovations based on the needs. Whether it is for manufacturing or space or the nuclear industry, the applications of robots are inevitable. New market place have outpaced the use of robots, as the applicability has changed from caged to adjacent, static to portable, connected to autonomous and easily trainable device (Tobe, 2012).

Recently, industrial robots were put into more use and it is apparent that the manufacturing, repair and maintenance, makes the best use of the robots helps to transform growth level of the business. The usage scale of robots has an increasing trend globally when industries modernize and increase the production capacities. In addition, industries have realized the complexity of the tasks performed, which can be effectively enhanced by robots. Automotive industry increased the use of robots when the robot manufacturers come up with cutting edge technology robots. When looking into the annual shipping of industrial robots worldwide, it gives a clear cut for the significance of robotic application in the industry (Figure 1).

Figure 1. Robotic application in the industry Source: IFR Statistical Department (2012)

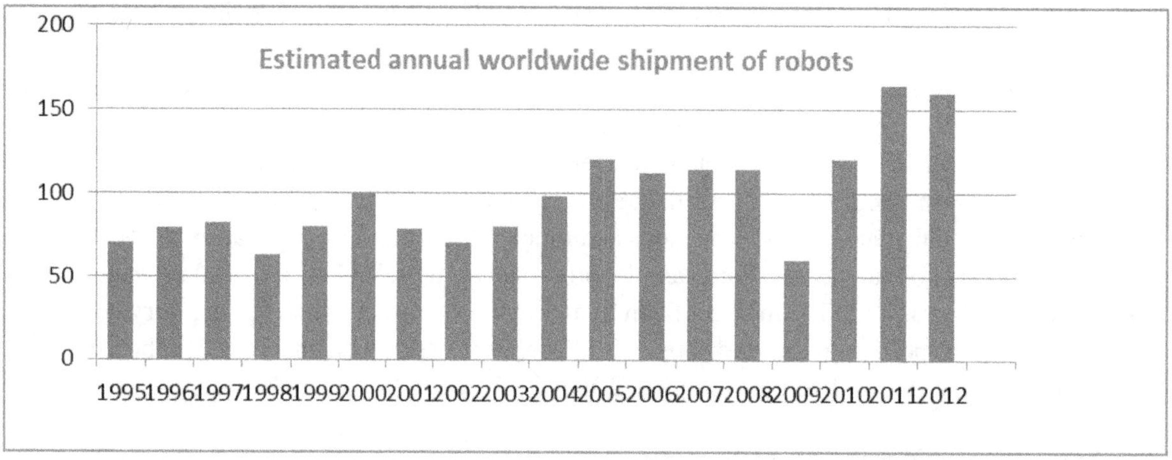

The global shipments of robots were dominated by Japan (28700 units) followed by US (22414 units), Asia and Australia (84645 units) and Europe (41200 units). Between 2007 and 2012, Asia records the maximum Compound annual growth of robots (CAGR) with an average of 8% while Europe has the least growth with 3% (World Robotics 2013). However the current business trend aims at providing flexibility and versatility which could not be implemented well by the traditional ones. Thus the need for reinvention was a necessity for the industry, which witnessed the growth of industrial robots presently by fulfilling the current innovative demands. Uncountable contributions have been dominated by the service industry with the passage of time.

For instance, Google has introduced various types of robots from seven tech companies that provide various applications ranging from supply chain to the distribution channels to the consumer's front door thereby creating a substantial breakthrough.

Several types of robots were put into use from those seven high tech companies. It is apparent that robotics will lead the giant companies like Google which aims at spurring innovation and investment in the future. In addition, Apple with an investment of 10.5 billion, in supply chain robots and automation equipment and Amazon (750 million) for delivering goods to a picker/packer robotically. Thus, the transformations followed by these giant companies will make the rest of the global companies to follow the same suit. When industries were confronted with rising labor costs, low production levels, outdated management methods and changing demographics, robotics acquisition helps to uncover all these challenges with a robust support by providing financial savings, time and space by ensuring top quality and production level. However the consistence and quality in the operation achieves efficiency, reduced injuries, waste reduction and maximum utilization of space.

Almeida et al. (2001) analyzed the significance of robots in e-business, by developing an analytical model for the interaction between robots and websites. With the increased dependence on websites for the business, the role of robots is momentous, as they perform task like mirroring, data mining and link maintenance. In e-business, robots form as an interface between the customers to find out the requested information. Crawlers, shop bots, price bots and autonomous software agents are the main categories of robots used in the interaction between the web and e-business, for providing top quality of service.

Crawler helps to generate the requests with the use of robots, pricebots set prices and shopbots seek prices that minimize costs for customers (Almeida et al., 2001). The impact of robots on websites was significant and it impacts the quality of service provided by the websites (Almeida et al., 2001).

Automotive industry was the largest to use robots in their operation (IFR statistical department, 2014). Robots have been utilized in the automotive industry used for welding, foundry operations, laser applications, painting and final assembly operations. Robotics has a significant role behind the continuous growth in the industry, with the advancements made by the Kuka Robotics, German manufacturer of industrial robotics. Automotive industry has transitioned to a great extent with the robotic advancement. For instance, robots in Tesla motors do multiple tasks such as welding riveting, bonding and installing a component in its assembly line, whereas the normal robots can perform single function only (Markoff, 2012)

With the flexibility attribute, robotization has extended its use over the automation industry. With the implementation of tricept robots, BMW utilized its flexibility for its innovative machining in the axle design (Kochan, 2004). Flexibility always costs money, *"but, you will get a return on your money if you take a long term view, and if you place a high value on innovation rate,"* says Konrad Pagenstert, Director of axle production, BMW (Kochan, 2004). With the intention of raising the productivity, General Motors (GM) has increased the rate of automation and efficient ways of parts handling (Mortimer, 2009). By investing to acquire Fanuc robots, the leading supplier of robots in North and South America, General Motors were successful in achieving the target of building 1,000,000 units of Van. Fanuc robots, with its high reliability and innovative and high quality solutions were used for welding operation in General Motors and it has raised the number of welds, leaving the traditional ones with less usage (Kochan, 2004).

However the robot utilization never tends to be ceased in the industry rather it comes up with the new innovative technologies a faster rate. Let's take the example of auto drive car, Robots have taken over the driver seat, with the introduction of new Google auto drive car auto industry will be revolutionized in the near future with this breakthrough. It is supposed to be the dominant contribution laid by the robotics involvement in the auto industry. When looking into the manufacturing industry robotic automation started in 1962 with the introduction of Unimate, a robotic arm attached to a steel drum used for pouring liquid metal into die casts, auto welding and controlling heavy loads (Lamb, 2013). Handling tedious jobs with consistent speed and precision was one of the main features of this robotic arm. Perhaps robots were successful in handling complex tasks; low productivity was a threat to the industry in the earlier decades since the operation was based on the human intervention and decision making.

Flexibility and growth were the main issue faced by the manufacturing firms, as a result of fierce competition and advanced technology by the global firms but it can be overcome by the incorporating automation that considers various disciplines in the manufacturing decisions. Although robots are being used in the manufacturing industry, the initial investment made for the robotics was high made a challenge for the small scale manufacturers. However small and medium factories and business was an unexploited area of robotic utilization, and looking forward to implement the automation at its fullest to remain competitive in the global business environment. With the innovations made by the robotics industry, the dilemma to create inexpensive robots was a major threat in the recent times. But rethink robotics, a splendid robotics company paved the way for robotics with the introduction of the new robot Baxter, cheap and user friendly robot with ensured safety for the manufacturers. In fact it made a revolution for the small and medium manufacturers who were in search for such an affordable product to fight with the low cost manufacturers in the world. It operates with presence of custom software or robotic experts, in the assembly lines to perform menial tasks (Duhamel et al., 2013). Baxter, with its affordability helps to reduce the outsourcing for the small U.S manufacturers (Wang, 2013). Its application line has crossed more than a traditional robot with some of its pioneering features. Common sense, safety, user friendly, versatility are some of its features that even can complete any task in short time irrespective of the size of the manufacturers. Unlike the traditional robots, Baxter does not require any additional capital investment and custom software and ensuring a long term endure in the industry.

Manufacturing industry find it productive when they have incorporated robotic operation in lean manufacturing. According to International federation of robotics (IFR), certain characteristics of robots can enhance lean manufacturing, listed as follows

- Save Costs for operation
- Product quality and reliability
- Enhancing work environment quality
- Raise the production output level
- Increase product manufacturing flexibility
- Intensify the yield and waste reduction
- Ensures safety at the work place
- Employee turnover can be reduced
- Reduce capital costs
- Standard size for operation with maximum utilization

While scrutinizing the robotic impact in food industry, studies has been focused on the role of robots in palletizing, material handling and tertiary packaging (Wallin, 1997). When robotics opened the doors for food industry applications in 1980's, operations were limited to packaging assorted chocolates into trays and gradually evolved into pizza assembly, meat processing, cake decoration and cake and pie handling (Wallin, 1997). But companies could not encourage the high scale use of robotics as a result of lacking investment. According to Wallin, (1997), proposes that packaging, palletizing, depalletizing and labeling, carton loading will be the common areas covered by robots, and the complex applications can be handled with the introduction of sophisticated computerized systems in the future.

Now the current trends in food industry has witnessed the robotic operations including primary packaging and processing such as curd slicing, sausage packing, ultra-sonic cheese slicing and cheese multiplexing etc. However, with the recent developments in the food industry, different types of robots were put into operation for several purposes. The old model SCARA (selective compliance assembly robot arm) robots for pick and place, spider robots for high speed picking and placing of light weight objects are the recent examples of robots used in the industry. SCARA robots are one of the types of stationery robots also known as horizontal articulated arm robots, with motions same as human arm. Its reliability for fast and repeatable movements make it fit for packaging palletizing, loading and unloading purposes (Brumson, 2011).

Nevertheless, these robots have some kind of downsides in terms of efficiency as it was prone to heavy load pick and place operation. In 2006, the invention of five axis robots by Fanuc robotics, focused on increasing flexibility in pick and palletizing. The robotic operation in end of line packaging in the food industry can be differentiated with the traditional ones in terms of efficiency, safety, hygiene, risk consistency and reliability (Table 1).

Table 1. The robotic operation (Source: Fanuc Robotics America, 2013)

Traditional End-of-Line Methods	End-of-Line Methods Utilizing Robots
Manual Pick, Place, and Palletizing	Robots can pick primary and secondary product, place into cartoners, and palletize product efficiently and effectively
Unsafe process or a process that is at a significant risk for repetitive stress injury	Robots are precise, clean, and repetitive and can run without failure for several cycles
Hard automation is inflexible	Sudden product or packaging changes are handled with ease
Risk contamination	Are hygienic and sterilized
Risk product damage during packaging	Highly sophisticated sensors monitor pressure, and grippers that have been developed specifically for food processing can eliminate product or package damage
Tedious, repetitive, and heavy-handling tasks	Consistent movements and high payload capacity perform tasks with ease
Imprecise equipment handling can damage equipment	Precise, consistent handling can greatly increase the service life of the equipment.
Unreliable equipment	Reliable with Mean Time Between Failure rates > 90,000 hours
Large single-task equipment requirements	Small work cell footprints with robots that can handle multiple operations

According to Wilson (2010), the real business growth of the food industry lies at the implication of the concept- *"untouched by human hand* "will become the significant marketing differentiator for the food products in the future. Moreover, the robotic usage tends to be escalated in the future as a result of its revolution in the industry irrespective of its applications. Based on the global statistics in 2012, Andreas Bauer, chairman of IFR industrial robots supplier group propose that there is forecast of 6% future growth with three main sources (World robotics). Huge potentials growth in future growth for many sectors, growth based on innovative technological developments and usage scale escalation in emerging countries.

2.2 Business Productivity

Business operations are more productive when they produce more output from a given set of inputs (Brown & Dev, 2000), simply called the efficiency in production (Syverson, 2011). It is a notion that is easy to define but difficult to measure (Djellal, 2008). In a nutshell, it can be measured based on the relationship between the production of goods/services and the factors of production. However measuring productivity is crucial for the organization and it's a complex process (Johnston & Jones, 2004). Namely, single productivity and multi factor productivity are the two types of measures for gauging the productivity. Where single productivity relates with a measure of output to single measure of inputs and multi factor links with bundle of inputs (all factors of production). Measuring productivity is an essential task to monitor the progress of the firm, which in turn helps to make changes where necessary to stay competitive.

Relating with the current business issue, escalating the productivity has been crucial factor for firms, especially service oriented firms (Brown & Dev, 2000) as it allows to lower the costs associated, to increase the demand in return by providing more services. Since the concept of productivity was not paid attention in the service industry, it has been given much importance over the last two decades as a result of stiff competition and challenges. In addition, maintaining productivity has been viewed as a long-term strategy, especially for hotel industry (Kilic & Okumus, 2005).

Often, productivity is perceived as a measure of output towards input, in manufacturing context, while service industry holds different dimension such as quality as the main component of productivity (Kilic & Okumus, 2005). Johnston and Johnson (2004) study supports the fact that it was given little importance for productivity in service organizations. Furthermore it is revealed that the productivity is a hypernym of utilization, efficiency, effectiveness, quality, predictability and other performance dimensions. Service productivity differs a lot from the conventional productivity perspective (Johnston & Johnson, 2004).

Simply put, productivity can be measured with the input and out variables for the manufacturing and can obtain a certain level of measure in terms of ratios or units and so on. However to gauge service productivity is a complex task since measuring and managing the inputs/outputs will be inconsistent for the simultaneous consumption of services as well as its perishability and heterogeneity. In support of this, Biege et al. (2013) states that feature of services so called IHIP (Intangibility (I), heterogeneity (H), inseparability (I) and perishability (P)) characteristics are the major constraints for measuring the productivity. In addition, when measuring the productivity of service business, customer involvement (Li & Prescott, 2009) is an inevitable part as it is evaluated on the basis of customer satisfaction (Biege et Al., 2013). Normally when measuring service productivity, some sorts of challenges are being faced and it tends to be worse when incorporating innovation and knowledge intensity of services (Biege et Al., 2013) and the heterogeneous nature of inputs and outputs makes it difficult to measure (Li & Prescott, 2009).

2.2 Benefit and Shortcomings of Robot Application in Business

When computers were invented to use for the business years ago, they have been modified for the best suit for the diverging demands. By 2030, it has been forecasted that robots will dominate all the industries irrespective of the sector as computers are today. However novel uses have been identified for industries namely automobiles, manufacturing and service industries. According to Duhamel et al. (2013), robot sales have augmented with escalated demands as a result of auto industry expansion, wage inflation and enhanced robot capabilities. In a nut shell, benefits of robots can be categorized as follows (Figure 3).

The effort of robots to improve the productivity measure starts from its efficiency in operation. By providing a consistent job nature it can deliver the jobs with more accuracy when compared to manual workers. Moreover it can increase the throughput that which directly impacts production. However, the real potential lies at working with constant speed without breaks. Built in with high strength and weight of the electric motors inside, agile and consistent tasks can be performed by robots along with greater quality. When looking into manufacturing sector, robotic automation brought drastic changes by its efficiency in operation and hence raises the productivity. For instance, Drake trailers, largest manufacturers for trailers, based in Australia were successful to up their productivity as a result of robotic automation (Lamb, 2013). Productivity was one of the major benefits of robots (Orr, 1996).

According to Wei (1995) robots robotic automation in the industry can achieve higher productivity at optimum conditions with its efficiency in production operations. Moreover robotic operations are ample for handling tedious, filthy, and complicated jobs in unsafe environments (e.g. exposure to toxic fumes or extreme temperatures) which directly impacts the pace of production process. Safety operations by robots can increase the savings since there are no concerns for health care and insurance concerns for the employers.

"It is very difficult, labor-intensive work, often in intense heat or cold, which can make it a limiting factor to the overall efficiency of the warehouse. These working conditions, coupled with handling heavy product can also lead to injuries among warehouse personnel" says Kevin Ambrose, CEO of Wynright,

US based logistics engineering company. Indeed the robots manufactured by Wynright Corporation have transfigured the logistics and distribution centre jobs by using the Robotic Truck Unloader (RTU), for unloading trailers or containers from a dock. Moreover the robotic loader could reduce the number of shipments by increased utilization of trailers (Green, 2013).

Unlike financial savings, time saving is a major advantage of robots since it can perform the work diligently. More over the accuracy in operational movements also reduce the material waste which in turn is another advantage for business. Wei (1995) categorized the benefits of robots into social and economic based on its characteristics (Figure 2).The major attributes of robots provide economic benefits which is indirectly relating with some sort of social benefits. Flexibility in operation can raise the level of volume produced and therefore the productivity can be escalated. When manual workers are being utilized for the operation, there is certain limit they can mould the things within a specific amount of time. But when they are aided with the robotic assistance they can outperform he work with better outcomes. Besides that the utilization can be made at its best with high pace. The number of output production is greatly impacted by the precision and flexibility of the robots in operation. Moreover, since everything

Figure 2. The benefits of robots Source: Taiwan Industrial Robots (1995)

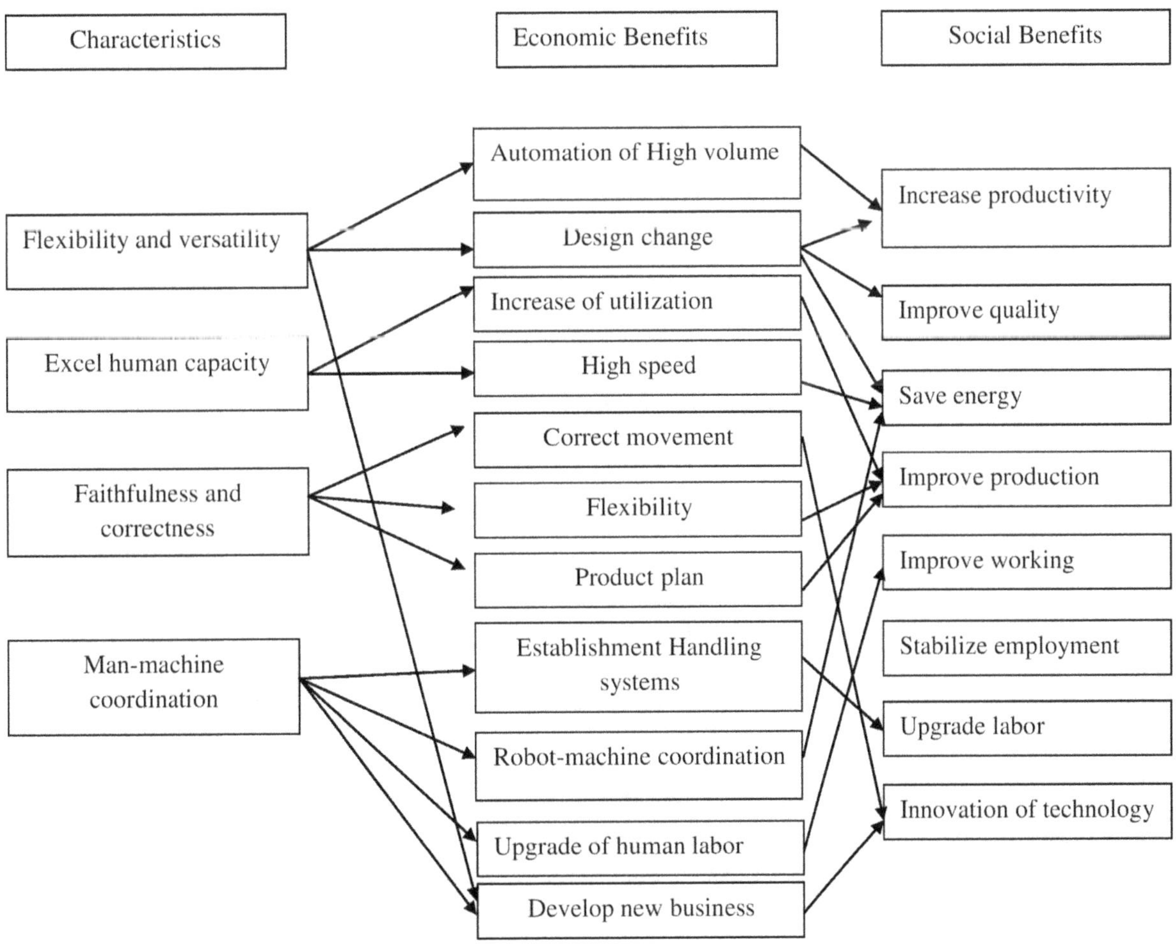

Figure 3. Benefits and Shortcomings of robots

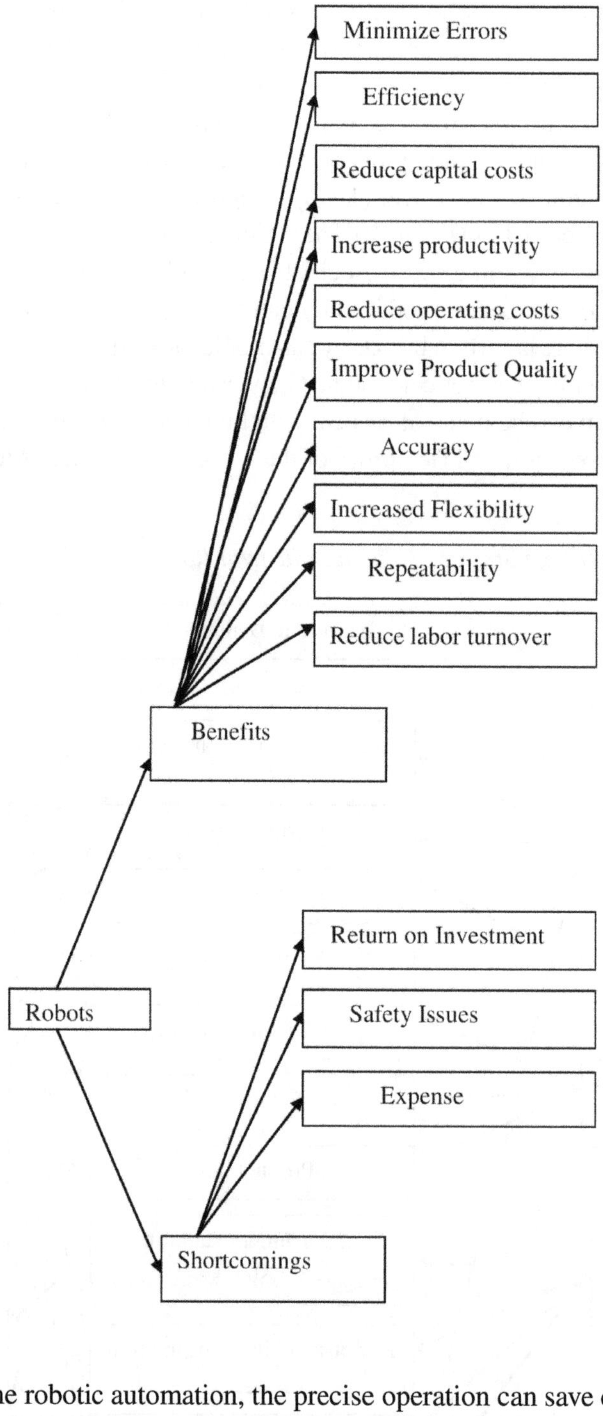

is well designed under the robotic automation, the precise operation can save energy. It has been noted that quality attribute is one of the major advantage of robots. While operations are repeatable with high precision, it has a great impact on producing quality products.

According to Petrina (2010) the following are the drivers of the usage growth and production of industrial robots

- Provides an economic exposure for business against the rising labor costs
- Unavailability of skilled labor
- Nature of monotonous, heavy and intense work
- Safety attributes for workers with reduced effects for production and environmental protection
- Increased technological accuracy and quality improvement
- Ability to use machines 365 days a year
- Improvement in flow line production and elimination of human interface for repetitive tasks

2.2.1 Benefits of Robots in Food Industry

The reliance on robots became a necessity for the food industry when the robotic automation fulfilled the requirements for the industry. According to Mahalik and Nambiar (2009), food industry includes food production, food processing and food service; all of them incorporate robotic operations for different purposes (Erzincanli & Sharp, 1997). Remarkable changes have been taken place in food industry globally. Stringent work conditions, extreme temperature, speedy process to produce low cost goods were the main requisites for the industry. Moreover the perishability of food products has urged the manufacturers to maintain efficiency and reduce lead times (Mahalik & Nambiar, 2009). Reduced waste, elevated throughput, speedy changeover times, flexibility, and product quality are the reaping benefits by robots in the industry (Rene et al. 2010). In particular, robots provide a greater contribution in handling product with new designs and shapes without any need for extra configuration or reprogramming (Spreckly, 2011)

Due to the growing trend for health consciousness, food products need delicate handling and storage (Erzincanli & Sharp, 1997) for the reason that food products are agile and get spoiled easily. In addition food safety have been given much significance since it creates a positive impact on the quality of food products and it can be achieved by delicate processes with ensured safety measures (Fernando et al, 2014). The success of the food industry is always dependent on product quality, consistency and high return on investment, and robots are meant to achieve all these necessities. Therefore robots have a rising demands for the food and beverages industry. In addition, rising labor costs was the main justification for installing robots in food industry.

Basically, maintaining hygiene is the escalating concern for the food industry as it will affect the total sales and productivity of the sector. In an article by Green (2013), states that one in six persons get sick as a result of consuming contaminated food in United States. So it has been forecasted that the food packaging and processing can maintain hygiene when robots are incorporated. Moreover higher throughput and reduced expenditure and labor saving s are the main benefits of robots in food industry (Hoffman, 2012). Unlike the industrial manufacturing robots, are flexible, as they can be mounted in ceilings and can work form any direction from top to bottom there by reducing space.

According to Green (2013) robots in food material handling increase the pace of production with increased safety measures for the personnel. It has been discovered that, robots are fit for preventing food contamination since U.S department of Agriculture approved robots for its sanitary standards. Moreover the demand for robots is increasing the food industry when it was approved by Food safety Modernization Act 2011(Hoffman, 2012). In addition some of the features like reliability, high process control, speed, flexibility and dexterity provide much benefit for the food industries to maintain productivity, safety, and hygiene performance etc.

As noted earlier, robots are able to perform repetitive tasks to yield huge benefits. Notably, robotic automation helps to reduce repetitive strain injury (RSI) Purnell (1998) since chances are more for creating injuries for manual workers.

According to Purnell (1998) employees in meat industry have to work continuously in freezing temperature and causes fatigue and injuries. As far as food manufacturers are concerned, top priority is at producing with higher safety levels and better quality (Purnell, 1988, Erzincanli & Sharp, 1997). Nevertheless, contamination of food can be eliminated by the usage of robots when it is being used for packaging and palletizing. In addition, it has been discovered that the flexibility of robotics in food manufacturing is a proactive solution to eradicate repetitive motion disorders. Presently, novel uses in palletizing such as vision assisted robotic application for ensuring product quality and solutions to end of line packaging are greatly influenced by the robotics operation (Fanuc Robotics America, 2013). Moreover the robotic trend for packaging has been increased substantially in the recent years (Figure 4).

Robots have some attributes that are particularly well suited to the food industry," "they are very flexible and can be programmed to switch from one production task to another. And they do not breathe pathogens into the workplace, thus lengthening product shelf life" says Buxton, CEO, Processing and Packaging machinery Association (PPMA) in UK. According to Spreckly (2011), food manufacturing industry in UK has embraced the use of robotics technology to a greater extent and resulted in the growth of the sector by 300%. However, by understanding the significance of robots, manufacturers are amending the technologies that will dominate the industries by providing novel contributions.

Increased labor costs and reduced technology costs will drives the usage of robots in food industry and it is anticipated that robotics hold great potential for the industry in the future. It is clear from the

Figure 4. Association of packaging and processing technologies (PMMI) survey (2008)

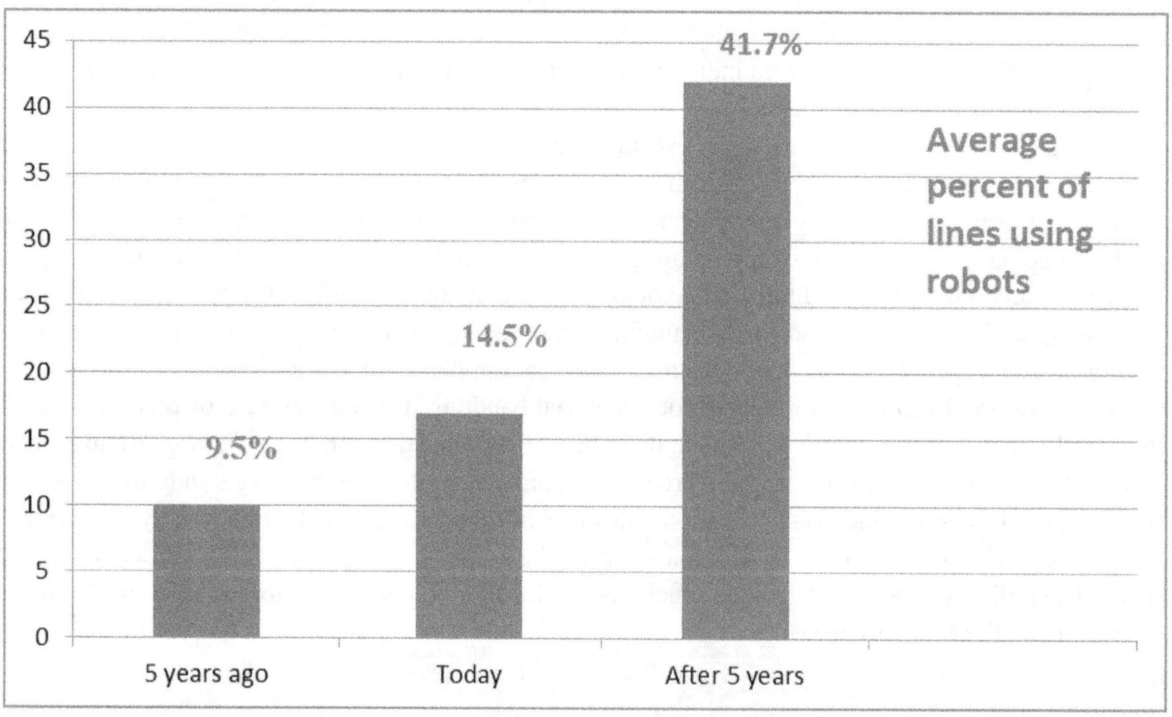

statistics that usage level of robots has been escalating since the market growth of robots are witnessing a huge growth (expected to rise by 8.5 billion in 2015), according to New York-based technology market research firm ABI. According to BARA (British Automation and Robot Association), robots usage has been doubled (Marshall, 2013), however, robots paved the way for growth of companies by increasing the productivity by saving the costs associated with, it is essential to realize its potential and takes initiative for further investment for the long term viability.

2.2.2 Robots in Fine Dining Restaurant

Despite the wide use of robots in manufacturing and service sectors, the utilization in food service industry is a novel application. The motive behind implementing the robotic system by the restaurant operators was to strive against labor issues such as labor shortage and labor turn over. While narrowing down the application of robots in food service industry, it can be categorized under kitchen operation and front staff operation. A Chinese restaurant in Pasadena, California was the first to use robots for serving the customers. They use two robots, Tanbo R1 and Tanbo R -2 (costs around$20,000) was a novel introduction for the restaurant business (Davis, 2012). Currently, several restaurants throughout the world incorporated the use of robotic arms and robotic machine for preparing and serving dishes. Mostly, use of robots in restaurants is concentrated towards technologically advanced countries such as Japan and China. For instance, FuA-Men restaurant in Japan use the automated robots for preparing dishes. The autonomous robots are using the robotic arms to do the tasks such as boil water, pour soup into bowls, and place toppings on the dish. The two mechanical robot arms are working mutually to carry out the noodle preparation. While looking into the process design, robotic arms are used in an organized manner with its precise movements in adding toppings and ingredients, consistency in taste and temperature, and accuracy of timing for food preparation. Consistently, the robots in FuA-men are successful in delivering the dishes in 1 minute 40 seconds. Hence the patrons are ensured the delivery of the noodle within a preset time. The robot arms are capable of preparing 80 bowls of noodles per day which costs around seven dollars each. The sophisticated operation of the robotic arms in FuA-Men restaurant delivers the efficient operation in the food preparation without any intervention of human chef. However there is little intervention of manual chef such as taking the order and serving the customers.

Next, a cooking robot called AIKE, was developed by Shanghai Qi Ding Food Development Co., Ltd. in China (Yingying, 2011). This high tech robot can prepare 24 traditional dishes and can store 751 recipes and can prepare the food in three minutes with the authentic tastes. Moreover it can create a green and low-carbon cooking environment by the standardized preparation techniques, cold-chain distribution and non-polluting disposal of cooking fumes. The robot is designed like an ATM machine and the operation design involves simple steps that can be handled by any worker. The process of cooking initiates with the mixing of ingredients in the robots, which resembles like a refrigerator. The robot after reading the barcode will pour all the ingredients into the wok followed by stir frying the dish. Employees only need to press buttons on the robot to choose a dish and it will display the name of the ingredients and their quantities. The ingredients are prepared by the manual workers according to the customers' requirements.

The 'I Robot' restaurant, Nanning, China is an example of another robot restaurant. It uses two robots from a Shenzen-based technology company. The robots chefs are able to make hundreds of Chinese dishes within three minutes. The robotic chef is peculiar for its efficient operation and hygiene. The ingredients are stored in the databases of the computers which monitors the robot movements. Unlike the other robot restaurant manual labor is necessary for adding the toppings and ingredients. The raw materials are prepared by the manual workers for assisting the robot operated cooking.

In addition, robot manufactured by the Japanese company MIK, designed for automatic cooking of dishes like fried rice and stir-fried vegetables is another exapmle. The wok is designed in a flat shape with an automatic stirrer, and the ingredients are mixed by the mixer, making the role of manual workers superfluous. The robot wok includes a well fixed stir fryer that provides the consistency of the operation. Precise and continuous movement of the robots avoids burning, charring, or distraction and the dishes are served with acceptable taste and flavors. The cost of the robot is US$ 10,900. Hence the owners are able to save the labor costs by the usage of the automatic wok. Moreover, the labor costs associated with the training can be another cost saving, since this automatic machine doesn't require any training to manipulate. On the other hand, Dalu robot restaurant in China and Hajime restaurant in Thailand use robots for serving the dishes. Preparing burgers round the clock without any human intervention is one of the other novel current applications of robots. Smarter restaurants, San Francisco based company has introduced robots for making burgers, with a capacity of preparing burgers with a high level of consistency and sanitation (Durden, 2014). Incorporating more automation will reduce the costs and increase profitability and it will positively impacts the successful operation of the food service industry (Stargardt, 2010)*"The benefits of using robots as ramen chefs include the accuracy of timing in boiling noodles, precise movements in adding toppings and consistency in the taste and temperature of the soup,"* says Kenji Nagaya, president of local robot manufacturer Aisei. On the other .hand, ruyi, a Chinese fast food restaurant in Singapore, have been using the robotic wok for preparing certain dishes.

As mentioned, labor shortage and labor turnover has been the main issue which drive the restaurant operators to implement robotic automated machines in their operation. In Singapore, there are 6400 food establishments, which are in surge of solving the labor shortage issues (Wong, 2013). Coupled with high labor costs and the tax for hiring foreign workers have questioned to achieve the business productivity. Tung Lok, a leading restaurant group has initiated the use of robtic wok to prepare instant dishes in its two outlets in Singapore. With its efficient operation, the robotic wok (an equipment used to cook dishes) were successful in preparing dishes by making the process less labor intensive (Spring 2012). Tung Lok Group was able to reduce 20 to 14 employees from each outlet as a result of robot automation. The wages of each worker costs an average of $24,000 annually, while the machine cost only $10,000 each. Moreover, the restaurants operating under this group consolidates the operation by using the pre packed ingredients and items hence the cooking is made easier without interrupting the taste buds of the customers.

Restaurants using the automated machines such as robots are successful in delivering dishes with consistent taste with reduced waiting time. Hence, customers who seek instant foods are always attracted to dine in such restaurants. Moreover, the machines are configured to prepare certain dishes or either help the employees to improve the pace of operation. However, food preparation using robotic machines always ensure safety and are less prone to create quality issues.

2.2.3 Shortcomings of Robots

Although robotics provides a better interface to raise productivity and decrease the cost, initial investment is the major drawback of robots, irrespective of the sectors. For instance, robots in earlier days were not confined to all industries since the initial investment was high (Figure 5). Eventually industries realized the usage benefit and invested in acquiring robotic machines for their operation. Recently, robots are being employed in restaurants for several purposes as the owners realized the use for a better service quality and productivity. However, small scale restaurants are often reluctant or unable to invest in robotics as a result of high investment.

In a survey conducted in Australia about the usage of robots, the major shortcoming of robots (Orr, 1996) was the initial investment. The survey compares the attributes of industrial robots between the year 1990 and 1993 and it was discovered the initial investment was the major drawback.

Industry needs low cost robot. For instance, consider a medium or small scale business, the cost of acquiring a new robot is expensive according to their budget, and hence they will be confronted with budget constraints in case if they bought a new one. Moreover, even after the purchase of robots, maintenance will be an expense for the business owners. Studies shows that the usage of industrial robots has an increasing trend throughout the world, but the real usage is mainly confined to Japan, European countries and North America. The situation for the rest of the economies, where there is a greater need for robots, will be facing with the price dilemma. Out of the total 75,000 units in world, Japan acquired 389,000 with 198,000 in Europe and 90,000 in North America (Sujan & Meggiolaro, 2010)

Revolution of robotization in the future is demanding that most of the jobs will be handled by robots. By 2025, half of the jobs will be handled by robots in United States. If the majority of jobs are handled by these machines it will create a profound impact on human resources globally, irrespective of the sectors

Figure 5. Shortcomings of robots Source: A longitudinal survey of robot usage in Australia (1996)

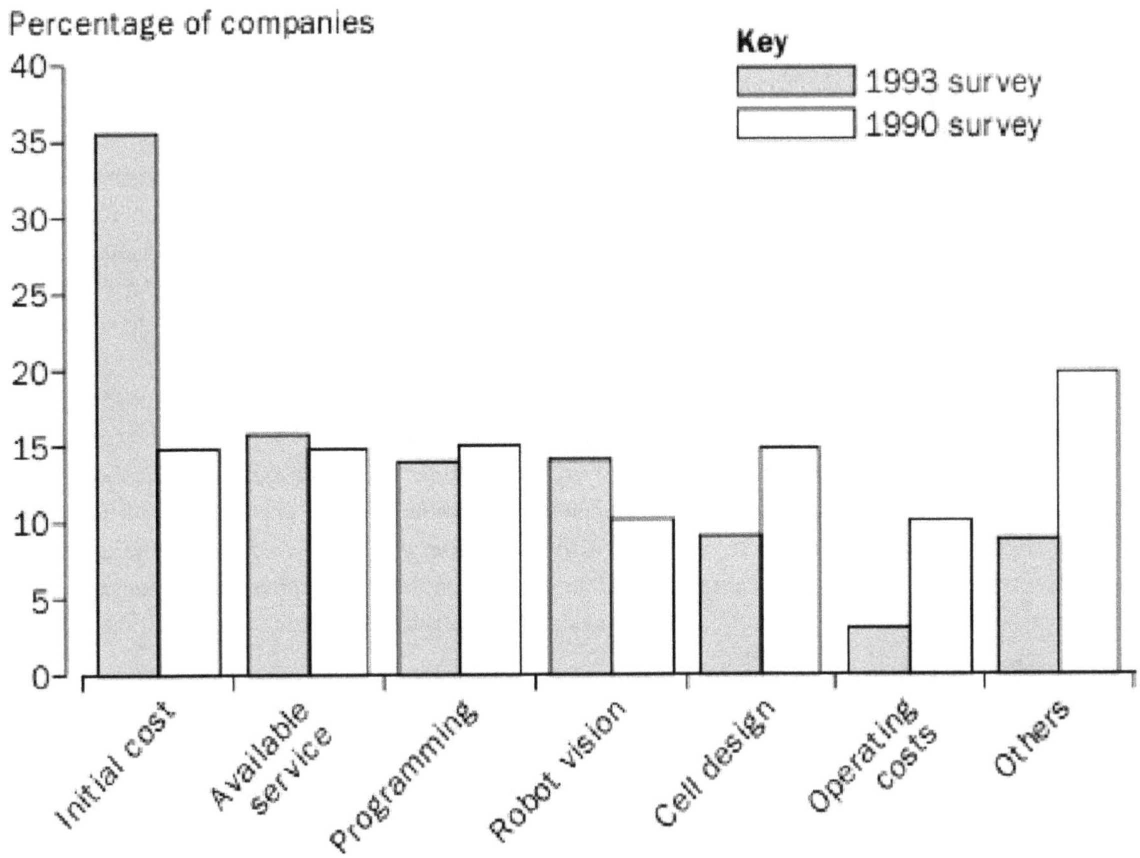

or industries. When global automotive and manufacturing industries increase the rate of robotics, job loss will became a global issue for the workers around. For instance, Google, Amazon and FedEx have started using more robotics technology for their business operation. On the flip side, it will indirectly affect the employment opportunities of prospective and experienced workers. In a nut shell, robotics is forming as an obstacle towards the income of the workers around the industries to an extent.

Moreover, when robots are introduced in to the business for novel applications, knowledge is necessary to operate it, Hence training will be required for the employees, emerging as an additional cost for the company. The production will be disrupted when robots are operated in absence of skilled employees for its programming. Besides that, innovation or creativity is not possible by this human manipulated machine.

3. CHALLENGES

3.1 Robot Application in Food Industry

It is interesting that the challenges of the food industry are common for global industries. To date, converging consumer demands, safety issues, competition and government regulations are the common challenges faced by the food and beverage industry (Thompson, 2005). Indeed the small and medium scale industries are the real fatalities as a result of these challenges. Provided with limited resources, they are reluctant or unable to compete with large scale industries. However some unique characteristics of the industry are the main drivers of the major issues. Involved with huge labor and requisite for safety and ergonomic issues (Brumson, 2011) with high turnover rate was one of the obstacles for sustained operation. To retain employees for performing mundane tasks and concern for hygiene, quality and consumer safety are the other major threats.

Moreover the above issues are solved to a great extent for the well existence of the industry. With the advancement in picking, palletizing and material handling, the demands rose for the robots for the food industry. Robots used in food industry are bit different from those in manufacturing industries. Ironically, using skilled labors was the main issue upon the introduction of robots in the industry. Although robots are used for pick and palletizing, the intelligent sensing is absent for picking variable size objects (Erzincanli & Sharp, 1997).

As such, much reliance on manual workers is needed. Robots are able to work according to the predestined function and therefore it lacks sense for decision making ability for certain circumstances. Moreover, robots can work in a limited envelope according to its design based on its number of arms or structure (Purnell, 1988). Another issue found with the robots was the unreliability of the gripper cylinders. Recently Unilever found some issues with the gripper couple of weeks after the installation (Bosch Rexroth UK, 2013). It was estimated that the cost for replacing cylinders are expensive (16000 pounds for 180 cylinders). Even though such issues can be solved to an extent, financial loss and time constraints are the effects of these issues.

3.2 Robot Application in Fine Dining Restaurant

According to Ruddick (2013), by 2025 restaurant industry will be in surge of fighting against technological challenges and efforts will be initiating to raise the profit margins, although robots will improve the efficiency. While scrutinizing the efficiency of robots in automated restaurants, it is essential to analyze whether it's apt for enhancing the service quality and product quality in long term since it ensures customer satisfaction and hence raises sales revenue (Chow et al, 2007). For instance, robots in restaurants are fit for preparing the predetermined menu, it cannot make multiple varieties of food unless it's designed so.

Since human intervention is necessary for the food preparation, the taste will depend on the worker who adds the toppings and ingredients. Hence, the consistent taste cannot be achieved and it will affect the overall rating of the food of that particular restaurant. For instance, serious issues have arisen in preparing the fried rice by robotic wok at Ruyi restaurant Singapore. Customers were complaining about the inconsistent taste of the rice prepared by robotic wok (Open Rice Singapore, 2014). The method of preparing the food by workers depends on the employees such as the timing of adding toppings and ingredients. Primarily, the delivering the quality product is the factor that drives the success of restaurants (Stargardt, 2010). Apart from the ambiance of the restaurant, quality and price are the encouraging factors, keeping the operational excellence far away for the customers to remain loyal to the business. Although opportunities are available for automating the restaurants, deep understanding to implement the automation control is necessary to flourish the business (Stargardt, 2010).

In case of any breakdown of the robotic machine, it will result in the interruption of the whole operation. Moreover, quality and taste differences will affect the customer's preference of choosing the restaurant for the dining in future. As noted earlier, unlike the industrial robots, maintenance will be an issue with the robots in automated restaurant. While hiring the employees it is essential to provide training for them to be familiar with the machine to provide standardized operation. Normally, the economies which faces labor shortage in the restaurant industry hires foreign employees, who does not possess enough skills, resulting in additional costs for training and retaining them.

While it does not seem to create any issues with restaurants which fully depend on robots, chances are more for the semi-automated restaurants to face challenges with the food preparation. Nevertheless, restaurants employing robot waiter cannot delight the customers with high quality of service, usually in high tech restaurants. Robots in restaurants create an excitement for customers and tendency for dining will be increased. Ironically, some people still believes dining as a societal activity where they still need manual employees to serve for them. Moreover, the robot waiters still need the assistance of the employees to take the order since it will be a difficult task for the old people to communicate with the robotic machines. Industries are implementing more automation for reducing the work force, human power is still essential for food service operation (Canadian restaurant and food services association, 2006).

4. SOLUTION

4.1 Alternative Solution to Solve Labor Shortage Restaurants

Unlike other industries labor shortage is one of the prime issues of food industry, since much reliance on human power is essential for its operation. Labor shortage does not mean the reduction in the numbers of labor required, it can be the shortage of skilled employees for a particular work. However impact of labor shortage does not affect the food service industry along, rather it creates adverse effects to the global economy (Canadian restaurant and food services association, 2006). In point of fact, the shortage of labors will affect the service performance of restaurant which in turn will affect the customers to stop dining. It is essential to know what really causes the shortage of labor and to sustain the business in long run. At any firm, labor shortage will occur when the number of required employees is more than the supply of available workers, who are educated enough or willing to do the job (Veneri, 1999). Moreover when the employer is not able to hire the caliber employee for a particular job, then it causes qualitative shortage of labors (Veneri, 1999). When the current business process are outsourced to the third party, less efforts are needed to strive against the labor shortage and therefore the restaurant industry will be exempted from this universal issue of labor . More over outsourcing operation will enable the industries to focus on core business activities (Fernando, 2013).

Several studies highlight the issues faced by food service industry, most notably with labor issues and productivity (Moss, 1969; Livingston, 1974). Food service industries are facing both the quantitative and qualitative shortage of labors. According to Canadian restaurant and food service association, 2006 quick service restaurant faces shortage of entry level workers, where table dining restaurant faces shortage of qualified cooks. Issues with the employees were the major concern for the industry apart from the energy, management, and production issues (Schaffer, 1984). Major issues dealt with employees were poor attitude, food service not considered as a career by many employees, high turnover, and low wages compared with other industries (Schaffer, 1984). According to Debrah (1994), rising education standard is one of the main factors since the educated youngsters are reluctant to work in the industry. Moreover, there prevails a poor image for the hotel industry to work for the new entrants, who are running behind to grab professional jobs in the advancing economy (Debrah, 1994). When firms face the labor shortages, it will be difficult for them to make the best use of the human capital advantage (Broek, 2012). Wage rate for entry level workers and promotional opportunities were the other main reasons for high turnover and labor shortage (Moss, 1969). To address these issues several suggestions has been listed as follows.

4.1.1. Training and Retention

The primary step to reduce the shortage of labor is to offer training for the employees. However sending full time employees for training creates adverse effect when they leave the job in the future (Debrah, 1994). Providing on job training will be an advantage to improve the skills of the employ for the particular job. As the food service restaurant is concerned, providing on job training for the front line workers and formal training for the chef will enhance the working capability. On job training is a widely used and effective training method in food service industry and it drives the success for the industry by the effective implementation (Gallaher, 1977). Moreover training is an effective device for affecting employee behavior, helps to reduce the anxiety of the employees, and alters the attitude of worker towards the job requirement (Gallaher, 1977). Although being cost effective method several issues such as absenteeism and labor turnover can be reduced to an extent by the training (Gallaher, 1977).

Moreover training has been paid higher attention in various segments of the food service industry. Training in food service industry comprises of levels according scale of the business. A new entry worker normally requires on job training, lasting for one or two weeks while workers in long chain operators are assigned for formal training, involves video training, seminars, short term subject based training, and computer training. A well trained employee offers advantages such as they can be assigned during the peak time or giving them extra shift will raise the efficiency of work in return for additional pay or bonuses for them.

When oriented properly, training helps to reduce the errors develops competence (Debrah, 1994), saves time, improve work flow, customer service and hence monetary savings for the business. In addition, maintaining the employees as humans rather than workers and encouraging them will promote the standard of the job being performed. When workers are treated nicely with flexible working hours with proper payment (Chong et.al; 2001), the tendency of the people to attract the industry will be high. It is a notion that the working conditions in the restaurant industry are worse and it adversely affects the tendency of the employee to stick with the job. Hence, improving the work conditions by making some changes in the work design will have a positive effect on attracting and retaining sufficient personnel.

When employers are more concerned about the skills of new hiring, making use of exit interviews and job evaluation and occupational testing (Moss, 1969) is a best alternative. As of restaurant industry, to avoid hiring of unskilled labors, giving preference to experienced employees can helps in retaining the caliber employees. In addition, retaining multi skilled employees can ensure the quality service and hence reliance on a new worker can be reduced to a great extent.

4.1.2 Reliance on Part Time Workers

The major constraint with the full time employees involves the resigning of the job at any point of their work. When well trained workers leave the job on a sudden, they can get a job anywhere at any time, whereas the management will have to hire a new employee which incurs additional cost and literally they are not able to work as the old worker unless they have enough experience. According to Moss, 1969 increased reliance on part time workers will reduce the labor shortage issues and high turnover. Hiring part time workers in the food service industry reduced costs when compared to full timers. Hiring retired people, housewives and unemployed old age people is a better option to lay off the labor shortage issues. Hiring such employees provides a win-win situation for both the parties. Moreover, Additional costs such as insurance cost can be abolished by hiring part time employees and they can be trained without paying.

4.1.3 Use of Convenience Foods

According to Dungan and Lacey (1969) convenience food were put into more use in food industry as a result of following issues.

- Increasing labor costs
- Difficulty of getting and keeping skilled or unskilled personnel
- Declining productivity of kitchen and dining room employees
- Increasing capital investment in space and kitchen equipment
- Growing patronage in commercial and institutional food service establishments magnifying the problems of manpower and overhead costs,
- Profits, narrow and difficult to maintain

Literally convenience food can be any type of foods that are precooked processed, precooked frozen foods to (Dungan & Lacey, 1969). Use of convenient foods in restaurants will reduce the time required for preparing the food and with the growing concern for health for patrons; usage of such foods will ensure safety. Moreover usage of convenient foods will ensure a consistent taste. Food items can be prepared in a bulk quantity with the mix usage of convenience foods and they can be stored for re heating hence it saves the time for preparation. However, Restaurants operating 24 hours can make the best use of convenience foods, hence increases efficiency ensuring a fast delivery of the food. Moreover fast food restaurants and other small scale restaurants in a crowded area ensure flexibility in preparing food with such foods. While intensifying the use of convenience foods, requirement of skilled labors is not essential to prepare the menu and it also ensures acceptance by the customers (Dungan & Lacey, 1969).

4.1.4 Automation and Technology

Introducing novel automation technology can resist the employee shortage in food service industry. With the advancement in the technology, introduction of automated machines can misplace the labors by ensuring the level of service and productivity. Food service industry has embraced lots of innovation and automated technologies to sustain the performance level and hence saves costs. According to Koutroumanis 2011, revolutionizing impact of technology in restaurant industry started in 1980s with the introduction of Point of Sale (POS) systems. With the POS system automation took over the job of kitchen and service staff, rather it enhanced the service quality level. Automated ordering system was one of the recent novel applications used by restaurant industry. However, technology transition in the restaurant has been increased at a steady rate hence maintained the competitiveness (Kotroumanis, 2011).

Equipments that are built in to manage the food preparation and serving will reduce the interface of the personnel to perform the tasks. Use of Vending machine, labor saving equipment, and computer assisted menu planning (Dungan & Lacey, 1969). Centralized food preparation and computer assisted menu planning, robotization of kitchen are the other suggested methods for reducing the interface manual workers in the restaurant. Furthermore, Robotic equipment such as fryer, noodle making robot, robot waiter also enhance the operation of restaurant without the involvement of much labor power. However, while the above suggestions can confronts the prevailing issues in restaurant industry, modifying the operational style can contribute to reduce the labor shortage, such as reducing the space for dining, except for the areas where crowd is high, stressing more on self-service, take away food and drive thru will minimize the required personnel for operating business

5. CONCLUSION

The objective of this chapter is to explore robotic transformation in literature and investigate its applications in food industry. Robotization in food industry aims at achieving economic benefits for the business operation along with the eradication of labor shortage issue. However the implementation of robots in restaurant industry still remains a novel application as a result of cost constraints. The level of robotic usage mainly concentrated towards developed countries with a very little consumption in the rest of the economies, where labor issues are high. Hence necessitating the fully automated restaurants can maintain the productivity and quality. In addition automated restaurants using robotic applications still confronts with various issues related with the quality of the food prepared by robots and still need reliance on manual workers. Moreover technological advancements should be made for the robotics in food industry to avoid such issues. Implementation of robotics along wouldn't be able to pioneer the business operations, while considering several alternatives will contribute for the sustained operation of industry. Labor shortage and cost constraints can be eradicated by proper training and outsourcing the robotic operations to third party, hiring part time workers and use of convenience foods. Globally, while there is an increased usage of convenience foods, it will reduce the time for preparing foods with ensured safety. When restaurants intensify the usage of convenience foods, safety concerns will be reduced and thus helps in attracting new customers by still retaining the loyal ones. Moreover, implementing fully automation will eradicate the issues related with the quality of food products and hence the labor shortage wouldn't be a barrier for the sustained operation of business.

REFERENCES

Almeida, V., Menasce, D., Peligrinelli, F., & Fonseca, R., Meira., W. Jr. (2001). Analyzing robot behavior in E- business sites. *Proceedings of the Sixth Workshop on Web Caching and Content Distribution*, Boston, Massachusetts. doi:10.1145/378420.378838

Biege, S., Lay, G., Zanker, C., & Schmall, T. (2013). Challenges of measuring service productivity in innovative, knowledge-intensive business services. *Service Industries Journal*, *33*(3/4), 378–391. doi :10.1080/02642069.2013.747514

Bogue, R. (2009). The role of robots in the food industry: A review, Industrial Robot. *International Journal (Toronto, Ont.)*, *36*(6), 531–536.

Bosch Rexroth technology eliminates gripper issues on Unilever's Twister production line (2013, Jan 25). *Bosch Rexroth*. Retrieved from http://source.theengineer.co.uk/plant-equipment/conveyors-and-line-equipment/conveyor-systems/bosch-rexroth-technology-eliminates-gripper-issues-on-unilevers-twister-production-line/2012634.article

Brown, J. R., & Dev, C. S. (2000). Improving productivity in a service business evidence from the Hotel Industry. *Journal of Service Research*, *2*(4), 339–354. doi:10.1177/109467050024003

Brumson, B. (2011, July 2). Unique robotic applications, Robotic Online, Retrieved from http://www.robotics.org/content-detail.cfm/Industrial-Robotics-Featured-Articles/Unique-Robotic-Applications/content_id/2572

Brumson, B. (2011, Sep 27 2). Scara vs. Cartesian Robots: Selecting the right type for your applications. *Robotics Online*. Retrieved from http://www.robotics.org/content- detail.cfm/Industrial-Robotics-Featured-Articles/Scara-vs-Cartesian-Robots-Selecting-the-Right-Type-for-Your-Applications/content_id/1001

Chong, P. P., Chen, Y. S., & Chen, J. C. H. (2001). IT induction in the food service industry. *Industrial Management & Data Systems*, *101*(1), 13–20. doi:10.1108/02635570110365961

Coxworth, B. (2012, Sep 18). Baxter industrial robot aims at bringing automation to smaller manufacturers. *Gizmag.com*. Retrieved from http://www.gizmag.com/rethink-robotics-baxter-industrial-robot/24183/

Davis, L. (2012). They were terrible at their job. *IO9.com*. Retrieved from http://io9.com/5904068/these-1980s-robot-waiters-were-real-but-they-were-terrible-at-their-job

Debrah, Y. A. (1994). Management of Operative Staff in a Labour-scarce Economy: The Views of Human Resource Managers in the Hotel Industry in Singapore. *Asia Pacific Journal of Human Resources*, *32*(1), 41–60. doi:10.1177/103841119403200104

Djellal, F., & Gallouj, F. (2009). *Measuring and improving productivity in services: issues, strategies and challenges*. USA: Edward Elgar Publishing.

Duhamel et al. (2013, May 20). Rethink Robotics - Finding a Market. *Stanford Case Publisher.*

Dungan, A. L., & Lacey, S. E. (1969). Convenience Foods—What Are They? How Are They Utilized? *The Cornell Hotel and Restaurant Administration Quarterly*, *10*(2), 6–12. doi:10.1177/001088046901000202

Durden, T. (2014). Meet "Smart Restaurant": The Minimum-Wage-Crushing, Burger-Flipping Robot. *Zerohedge.com.* Retrieved from http://www.zerohedge.com/news/2014-01-12/meet-smart-restaurant-minimum-wage-crushing-burger-flipping-robot?page=2

Erzincanli, F., & Sharp, J. M. (1997). A classification system for robotic food handling. *Food Control*, *8*(4), 191–197. doi:10.1016/S0956-7135(97)00048-0

Using Robots to Solve End-of-Line Issues in the Food Industry. (2014). Fanuc America Corporation. Retrieved from http://www.fanucrobotics.com/robotics-articles/Food_Robots_End_of_line_Solutions.aspx

Fernando, Y. (2012). Service Innovation along the Chain of Service Process in Airline Business. Outsourcing Management for Supply Chain Operations and Logistics Service. (pp. 185).

Fernando, Y., Ng, H. H., & Yusoff, Y. (2014). Activities, motives and external factors influencing food safety management system adoption in Malaysia. *Food Control*, *41*, 69–75. doi:10.1016/j.foodcont.2013.12.032

Gallagher, M. C. (1977). The Economics of Training Food Service Employees. *The Cornell Hotel and Restaurant Administration Quarterly*, *18*(1), 54–56. doi:10.1177/001088047701800112

Gorle, P., & Clive, A. (2011). *Positive impact of industrial robots on employment. Metra Martech/International Foundation of Robotics Report.*

Green, T. (2013). Frito-Lay and Wynright Put robots on the docks. *Robotic Business Review.* Retrieved from http://www.roboticsbusinessreview.com/article/frito_lay_and_wynright_put_robots_on_the_docks

Hoffman, M. (2012). Moving From Man to Machine. *Food Logistics.* Retrieved from http://www.foodlogistics.com/article/10630825/moving-from-man-to-machine?page=1

Ilyukhin, S. V., Haley, T. A., & Singh, R. K. (2001). A survey of automation practices in the food industry. *Food Control*, *12*(5), 285–296. doi:10.1016/S0956-7135(01)00015-9

Industrial robot statistics. (2012). *International Federation of Robotics.* Retrieved from http://www.ifr.org/industrial-robots/statistics/

Johnston, R., & Jones, P. (2004). Service productivity: Towards understanding the relationship between operational and customer productivity. *International Journal of Productivity and Performance Management*, *53*(3), 201–213. doi:10.1108/17410400410523756

Jose, S. (2010). *Global Service Robotics Market to Reach US $38.42 Billion by 2015.* Retrieved from http://www.prweb.com/releases/service_robotics/professional_robots/prweb4240924.html

Kilbane, D. (2008). *Joseph Engelberger: Robotics Move from Industry to Space To Elder Care*. Retrieved from http://electronicdesign.com/embedded/joseph-engelberger-robotics-move-industry-space-elder-care

Kilic, H., & Okumus, F. (2005). Factors influencing productivity in small island hotels: Evidence from Northern Cyprus. *International Journal of Contemporary Hospitality Management*, *17*(4), 315–331. doi:10.1108/09596110510597589

Kochan, A. (2005). BMW uses even more robots for both flexibility and quality. *International Journal (Toronto, Ont.)*, *32*(4), 318–320.

Koutroumanis, D. A. (2011). Technology's Effect on Hotels and Restaurants: Building a Strategic Competitive Advantage. *Journal of Applied Business and Economics*, *12*(1), 72–80.

Lamb, R. (2010). How have robots changed manufacturing? *Howstuffworks.com*. Retrieved from http://science.howstuffworks.com/robots-changed-manufacturing.htm

Li, X., & Prescott, D. (2009). *Measuring productivity in the service sector*. Guelph, Ontario, Canada: Canadian Tourism Human Research Council and University of Guelph.

Livingston, G. E. (1974). Changes in the food service industry. *The Cornell Hotel and Restaurant Administration Quarterly*, *15*(1), 14–19. doi:10.1177/001088047401500105

Mahalik, N. P., & Yen, M. (2009). Extending field bus standards to food processing and packaging industry: A review. *Computer Standards & Interfaces*, *31*(3), 586–598. doi:10.1016/j.csi.2008.03.027

Markoff, J. (2012). Skilled Work, Without the Worker. *New York Times*. Retrieved from http://www.nytimes.com/2012/08/19/business/new-wave-of-adept-robots-is-changing-global-industry.html?pagewanted=all&_r=1

Mickle, P. (2014), 1961: A peep into the automated future. Retrieved from http://www.capitalcentury.com/1961.html

Moran A, (2011, Aug 2), Chinese Foxconn employees could soon be replaced with robots.

Mortimer, J. (2009). Robots boost for GMM van production. Industrial Robot. *International Journal (Toronto, Ont.)*, *36*(2), 121–122.

Moss, A. L. (1969). Manpower Shortage in Food Service. *The Cornell Hotel and Restaurant Administration Quarterly*, *10*(1), 43–48. doi:10.1177/001088046901000112

Openrice Ruyi. (2014). Overview. Retrieved from http://sg.openrice.com/singapore/restaurant/ruyi/15168/

Orr, S. C. (1996). A longitudinal survey of robot usage in Australia. *Integrated Manufacturing Systems*, *7*(5), 33–46. doi:10.1108/09576069610129900

Petrina, A. M. (2011). Advances in robotics. *Automatic Documentation and Mathematical Linguistics*, *45*(2), 43–57. doi:10.3103/S000510551102004X

Purnell, G. (1998). Robotic equipment in the meat industry. *Meat Science*, *49*, S297–S307. doi:10.1016/S0309-1740(98)90056-0 PMID:22060720

Moreno Masey, R. J., Gray, J. O., Dodd, T. J., & Caldwell, D. G. (2010). Guidelines for the design of low-cost robots for the food industry. Industrial Robot. *International Journal (Toronto, Ont.)*, *37*(6), 509–517.

Restaurant, C. (2006). Help wanted: The labour shortage crisis and Canadian foodservice industry. *Canadian Restaurant and Foodservice Association*, *6*, 1–9.

PPMI. (2008). Robots on the Rise Robotics: Usage and Trends in Packaging Applications. Retrieved from http://www.pmmi.org/files/PIB/Robotics.pdf

Ruddick, P. (2013). Food service 2025: Report predicts restaurant industry's robotic future. Retrieved from http://www.bighospitality.co.uk/Trends-Reports/Foodservice-2025-Report-predicts-restaurant-industry-s-robotic-future

Schaffer, J. D. (1984). Research in the Food Service Industry: An Exploratory Study and Critique. *Journal of Hospitality & Tourism Research (Washington, D.C.)*, *9*(1), 55–71. doi:10.1177/109634808400900106

Schaller, R. R. (1997). Moore's law: past, present and future. *Spectrum, IEEE*, 34(6), 52-59.

Spreckly, A. (2011). Why robots are ideal ingredient for modern food producers. *Food Processing*. Retrieved from http://www.fponthenet.net/article/40527/Why-robots-are-ideal-ingredient-for-modern-food-producers.aspx

Automation the way forward. (2012). *Spring Singapore Business Times*. Retrieved from http://www.spring.gov.sg/NewsEvents/ITN/Pages/Automation-the-way-forward-20120619.aspx#.UzrkGaiSwj-

Stargardt, W. (2010). *Automation - The Future of Restaurant Operations Energy perspectives*. Retrieved from http://wayne-stargardt.blogspot.co.uk/2010/07/automation-future-of-restaurant.html

Sujan, V. A., & Meggiolaro, M. A. (2011). Simultaneous Localization and Mapping for Mobile Robot Teams with Visual Sensors. *Intelligent Transportation Vehicles*, 1.

Syverson, C. (2010). *What determines productivity? (No. w15712)*. National Bureau of Economic Research. doi:10.3386/w15712

Thompson, O. (2005). Key Challenges Facing Food and Beverage Manufacturers and What They Must Do To Survive. *Food & Beverage Manufacturing*. Retrieved from http://www.gilbertassociates.com/Documents/NAVFood_and_BeverageMANU_WP12.09.pdf

Tobe, F. (2012). Companies making the necessary transition from industrial to service robots, Retrieved from http://singularityhub.com/2012/06/06/companies-making-the-necessary-transition-from-industrial-to-service-robots/

van den Broek, Y. Y., & Beijer, S. S. (n. d.). Labor shortage solutions: which, when and why?.

Veneri, C. M. (1999). Can occupational labor shortages be identified using available data?. *Monthly Lab. Rev.*, *122*, 15.

Wallin, P. J. (1997). Robotics in the food industry: An update. *Trends in Food Science & Technology*, *8*(6), 193–198. doi:10.1016/S0924-2244(97)01042-X

Wang, J. (2013). 8 Companies Leading the Charge for Commercial-Use Robotics.

Wei, C. C. (1995). Taiwan's industrial robots. *Industrial Robot: An International Journal*, *22*(2), 21–23. doi:10.1108/EUM0000000004181

Wilson, M. (2010). Developments in robot applications for food manufacturing. *Industrial Robot: An International Journal*, *37*(6), 498–502. doi:10.1108/01439911011081632

Wong, T. (2013), Can food outlets up productivity without sacrificing taste? *Singapolotics*. Retrieved from http://www.singapolitics.sg/views/can-food-outlets-productivity-without-sacrificing-taste

Yingying, S. (2014), Sci-Fi chef. *China Daily Europe*. Retrieved from http://europe.chinadaily.com.cn/life/2011-03/13/content_12168517.htm

KEY TERMS AND DEFINITIONS

Disrupted: When robots are operated in absence of skilled employees for its programming.

Food Safety: Approved robots for safety, sanitary standards and permissible according to Islamic law.

Low Cost Robot: Robot which does not require high and any additional capital investment and custom software to ensure a long term endure in the industry.

Productivity: It is efficiency, effectiveness, quality, performance and it can obtain a certain level of measure in terms of ratios or units and so on.

Robot: An electro mechanical machine operated with the computer manipulating various functions to be applied for optimizes business efficiency, reduce the costs and enhance organization productivity.

Robotic Performance: Automation of robot application reached the optimum level of productivity and efficiency in operations.

Compilation of References

Automation the way forward. (2012). *Spring Singapore Business Times.* Retrieved from http://www.spring.gov.sg/NewsEvents/ITN/Pages/Automation-the-way-forward-20120619.aspx#.UzrkGaiSwj-

Bosch Rexroth technology eliminates gripper issues on Unilever's Twister production line (2013, Jan 25). *Bosch Rexroth.* Retrieved from http://source.theengineer.co.uk/plant-equipment/conveyors-and-line-equipment/conveyor-systems/bosch-rexroth-technology-eliminates-gripper-issues-on-unilevers-twister-production-line/2012634.article

Industrial robot statistics. (2012). *International Federation of Robotics.* Retrieved from http://www.ifr.org/industrial-robots/statistics/

Mobile TV Predicted To Be A Hit. (2007). *BBC News.* Retrieved from http://news.bbc.co.uk/2/hi/technology/6639249.stm

Abdulmalek, F. A., & Rajgopal, J. (2007). Analyzing the benefits of lean manufacturing and value stream mapping via simulation: A process sector case study. *International Journal of Production Economics, 107*(1), 223–236. doi:10.1016/j.ijpe.2006.09.009

Abu-Shanab, E., & Al-Tarawneh, H. (2013). Production Information Systems Usability in Jordan. In I. Management Association (Ed.), Industrial Engineering: Concepts, Methodologies, Tools, and Applications (pp. 975-989). Hershey, PA: Engineering Science Reference. doi:10.4018/978-1-4666-1945-6.ch053

Achanga, P., Shehab, E., Roy, R., & Nelder, G. (2006). Critical success factors for lean implementation within SMEs. *Journal of Manufacturing Technology Management, 17*(4), 460–471. doi:10.1108/17410380610662889

Adalsteinsson, D., & Sethian, J. (1995). A Fast Level Set method for propagating interfaces. *Journal of Computational Physics, 118*(2), 269–277. doi:10.1006/jcph.1995.1098

AdSage. (2013). *Introduction to mobisage*, Retrieved from http://mobi.adsage.cn/about

Affonso, R. C., Cheutet, V., Ayadi, M., & Haddar, M. (2013). Simulation in Product Lifecycle: Towards a better information management for design projects. *The Journal of Modern Project Management, 1*(1), 112–119.

Ait Hssain, A., & Dunod, E. D. (2000). *Optimisation des flux de production: méthodes et simulation.* Dunod.

Almeida, V., Menasce, D., Peligrinelli, F., & Fonseca, R., Meira., W. Jr. (2001). Analyzing robot behavior in E- business sites. *Proceedings of the Sixth Workshop on Web Caching and Content Distribution*, Boston, Massachusetts. doi:10.1145/378420.378838

Anderson, N. (1970). Evolutionary Significant of Virus Infection. *Nature, 227*(5265), 1346–1347. doi:10.1038/2271346a0 PMID:5455138

Arimoto, S., Miyazaki, F., & Kawamura, S. (2003). Cooperative Motion Control of Multiple Robot Arms or Fingers. *Proceeding of the IEEE International Conference on Robotics and Automation*, 4, 1407-1412. doi:10.1109/ROBOT.1987.1087820

Arimoto, S., Yun-Hui, L., & Tomohide, N. (1994). Principle of orthogonalization for hybrid control of robot arms. *Automatic Control-World Congress*; 3, 335-340.

Arnaldi, B., Fuchs, P., & Tisseau, J. (2003). *Traité de la réalité virtuelle.* Les Presses de l'Ecole des Mines de Paris.

Arteaga-Pérez, M. (1998). On the properties of a dynamic model of flexible robot manipulators. *Journal of Dynamic Systems, Measurement, and Control, 120*(1), 8–14. doi:10.1115/1.2801326

Asada, H., & Slotline, J.-J. (1986). *Robot Analysis and Control.* USA: John Wiley & Son.

Ashby, M. F. (2005). *Materials selection in mechanical design, MRS BULLETIN.* Cambridge, UK: Cambridge University Press.

Asimov, I. (1994). The Robots of Dawn, Spectra publication, Washington D.C., USA

Ates, A., & Bititci, U. (2011). Change process: A key enabler for building resilient SMEs. *International Journal of Production Research, 49*(18), 5601–5618. doi:10.1080/00207543.2011.563825

Aurenhammer, F. (1991). Voronoi Diagrams: A survey of a fundamental geometric data structure. *ACM Computing Surveys, 23*(3), 345–405. doi:10.1145/116873.116880

Aurenhammer, F., & Klein, R. (2000). In J. Sack & J. Urrutia (Eds.), *Handbook of computational geometry* (pp. 201–290). doi:10.1016/B978-044482537-7/50006-1

Aurich, J. C., Yang, X., Schröder, S., Hering-Bertram, M., Biedert, T., Hagen, H., & Hamann, B. (2012). Noise investigation in manufacturing systems: An acoustic simulation and virtual reality enhanced method. *Journal of Manufacturing Science and Technology, 5*(4), 337–347. doi:10.1016/j.cirpj.2012.09.010

Baglin, G., & Capraro, M. (1999). *L'Entreprise Lean Production ou la PME compétitive par l'action collective.* Presses Universitaires de Lyon.

Bailetti, A. J., & Litva, P. F. (1995). *Integrating Customer Requirements into Product Design, the Journal of Product Innovation Management, 12 (1), Baker, M. & Hart, S. (1999). Product Strategy and Management.* Harlow, UK: Prentice Hall.

Bal, M., & Hashemipour, M. (2009). Virtual factory approach for implementation of holonic control in industrial applications: A case study in die-casting industry. *Robotics and Computer-integrated Manufacturing, 25*(3), 570–581. doi:10.1016/j.rcim.2008.03.020

Barot, C., Lourdeaux, D., Burkhardt, J.-M., Amokrane, K., & Lenne, D.BAROT. (2013). V3S: A Virtual Environment for Risk-Management Training Based on Human-Activity Models. *Presence (Cambridge, Mass.), 22*(1), 1–19. doi:10.1162/PRES_a_00134

Beal, A. (2008). *Gestão Estratégica da informação: como transformar a informação e a tecnologia da informação em fatores de crescimento e alto desempenho nas organizações.* São Paulo: Atlas.

Belding, T. (1995). The Distributed Genetic Algorithm Revisited. In *Proceedings of the 6th International Conference on Genetic Algorithms* (pp. 114–121).

Bel, G., & Kieffer, J. P. (2002). *Pilotage assisté par la simulation discrète, Méthodes du pilotage des systèmes de production, 99 - 127.* France: Hermès Science Europe Ltd.

Berchet, C. (2000). *Modélisation pour la simulation d'un système d'aide au pilotage industriel.* France: National Polytechnic Institute of Grenoble.

Bhattacharya, P., & Gavrilova, M. L. (2008). Roadmap-based path planning-Using the Voronoi diagram for a clearance-based shortest path. *Robotics & Automation Magazine, IEEE, 15*(2), 58–66. doi:10.1109/MRA.2008.921540

Biege, S., Lay, G., Zanker, C., & Schmall, T. (2013). Challenges of measuring service productivity in innovative, knowledge-intensive business services. *Service Industries Journal, 33*(3/4), 378–391. doi:10.1080/02642069.2013.747514

Blum, H. (1967). In M. I. T. Press (Ed.), *"A transformation for extracting new descriptors of shape," in Models for Perception of Speech and Visual Form, W. W. Dunn* (pp. 153–171). Cambridge, Mass.

Bogue, R. (2009). The role of robots in the food industry: A review, Industrial Robot. *International Journal (Toronto, Ont.), 36*(6), 531–536.

Bonitz, R. G., & Hsia, T. C. (1994). Force decomposition in cooperating manipulators using the theory of metric spaces and generalized inverses. *IEEE Transactions on Robotics and Automation, 2*, 1521–1527. doi:10.1109/ROBOT.1994.351372

Bonitz, R. G., & Hsia, T. C. (1996). Internal force-based impedance control for cooperating manipulators. *IEEE Transactions on Robotics and Automation, 12*(1), 78–89. doi:10.1109/70.481752

Bortoff, S. A. (2000). Path planning for UAVs. *Proceedings of the 2000 American Control Conference*, Vol. 1. *No. 6. IEEE.*

Boudouh, T., Boxberger, J., & Gomes, S. (2013). Project Management and Lean Engineering: An Industrial Application. *The Journal of Modern Project Management, 1*(1), 50–55.

Breu, H., Gil, J., Kirkpatrick, D., & Werman, M. (1995). Linear time Euclidean Distance Transform algorithms. *IEEE Transactions on Pattern Analysis and Machine Intelligence, 17*(5), 529–533. doi:10.1109/34.391389

Brown, J. R., & Dev, C. S. (2000). Improving productivity in a service business evidence from the Hotel Industry. *Journal of Service Research, 2*(4), 339–354. doi:10.1177/109467050024003

Brown, K., Schmied, H., & Tarondeau, J. C. (2003). Success Factors in R&D: A Meta-Analysis of the Empirical Literature and Derived Implications for Design Management. *Design Management Journal, 2*, 72–87.

Brumson, B. (2011, July 2). Unique robotic applications, Robotic Online, Retrieved from http://www.robotics.org/content-detail.cfm/Industrial-Robotics-Featured-Articles/Unique-Robotic-Applications/content_id/2572

Brumson, B. (2011, Sep 27 2). Scara vs. Cartesian Robots: Selecting the right type for your applications. *Robotics Online.* Retrieved from http://www.robotics.org/content- detail.cfm/Industrial-Robotics-Featured-Articles/Scara-vs-Cartesian-Robots-Selecting-the-Right-Type-for-Your-Applications/content_id/1001

Buche, C., Bossard, C., Querrec, R., & Chevaillier, P. (2010). Pegase: A generic and adaptable intelligent system for virtual reality learning environments. *International Journal of Virtual Reality, 9*(4), 1–13.

Bui, T. N., & Moon, B. (1994). A New Genetic Approach for the Travelling Salesman Problem, *Proceeding of the 1st IEEE Conference on Evolutionary Computing*, 1, (pp: 7-12).

Burdea, G., & Coiffet, P. (1993). *La Réalité Virtuelle.* Paris: Hermès Sciences Publications.

Burnard, K., & Bhamra, R. (2011). Organisational resilience: Development of a conceptual framework for organisational responses. *International Journal of Production Research, 49*(18), 5581–5599. doi:10.1080/00207543.2011.563827

Bzymek, Z. M., Nunez, M., Li, M., & Powers, S. (2008). Simulation of a machining sequence using delmia/quest software. *Computer-Aided Design and Applications, 5*(4), 401–411. doi:10.3722/cadaps.2008.401-411

C. A. T. I. A. V5 API Documentation (2005). *Dassault Systèmes*. Retrieved from http://catiadoc.free.fr/online/CAAScdDmiTechArticles/CAADmiTocActivity.htm

Caccavale, F., Chiacchio, P., & Chiaverini, S. (1999). Stability analysis of a joint space control law for a two-manipulator system.*Proceeding of the 35th IEEE Conference on Decision Control*, 3, 85-88. doi:10.1109/9.739077

Caccavale, F., Chiacchio, P., & Chiaverini, S. (2000). Task-Space regulation of cooperative manipulators. *Automatica*, *36*(6), 879–887. doi:10.1016/S0005-1098(99)00215-0

Cad Shroer MPDS4 Factory Layout. (2013, August). Retrieved from: http://www.cadschroer.com/index.php?screen=1&ziel=ProductsMPDS&thema=mask1.php&id=351&land=com

Caggiano, A., & Teti, R. (2012). Digital Manufacturing Cell Design for Performance Increase. *Procedia CIRP*, *2*, 64–69. doi:10.1016/j.procir.2012.05.041

Camuffo, A., Furlan, A., & Rettore, E. (2007). Risk Sharing in Supplier Relations: An Agency Model for the Italian Air-conditioning Industry. *Strategic Management Journal*, *28*(12), 1257–1266. doi:10.1002/smj.635

Cantu-paz, E. (2000). Topologies, Migration Rates, and Multi-Population Parallel Genetic Algorithms. In *Proceedings of the Genetic and Evolutionary Computation Conference* (pp. 91–98).

Capi, G. (2007). Multiobjective Evolution of Neural Controllers and Task Complexity. *IEEE Transactions on Robotics*, *23*(6), 1225–1234. doi:10.1109/TRO.2007.910773

Caudell, T., & Mizell, D. (1992). Augmented reality: an application of heads-up display technology to manual manufacturing processes, *Proceedings of the Hawaii International Conference on Systems Science, HI, USA*, 659-669. doi:10.1109/HICSS.1992.183317

Chang, K. S., Holmberg, R., & Khatib, O. (1995). The augmented object model: cooperative manipulation and parallel mechanisms dynamics.*Proceedings of the 2000 IEEE International Conference on Robotics and Automation*, 1, 470-475. doi:10.1109/ROBOT.2000.844099

Chen, T. L., & Kemp, C. C. (2010). Lead me by the hand: Evaluation of a direct physical interface for nursing assistant robots. In *2010 5th ACM/IEEE International Conference on Human-Robot Interaction (HRI)* (pp. 367–374). IEEE. doi:10.1109/HRI.2010.5453162

Chen, T. L., & Kemp, C. C. (2011). A Direct Physical Interface for Navigation and Positioning of a Robotic Nursing Assistant. *Advanced Robotics*, *25*(5), 605–627. doi:10.1163/016918611X558243

Chevaillier, P., Trinh, T. H., Barange, M., Devillers, F., Soler, J., De Loor, P., & Querrec, R. (2011). Semantic modelling of virtual environments using MASCARET.*Proceedings of the Fourth Workshop on Software Engineering and Architectures for Realtime Interactive Systems, IEEE VR*, Singapore.

Chiacchio, P., Chiaverini, S., & Siciliano, B. (1996). Direct and inverse kinematics for coordinated motion tasks of a two-manipulator system. *Journal of Dynamic Systems, Measurement, and Control*, *118*(4), 691–697. doi:10.1115/1.2802344

Chiang, C. S. (1992). *The Euclidean Distance Transform.* [Ph. D. Thesis]. Dept. Comp. Sci. Purdue University.

Chiavenato, I., & Sapiro, A. (2003). *Planejamento Estratégico. Fundamentos e Aplicações.* Rio de Janeiro: Elsevier.

Choi, B. K., & Jerard, R. B. (1998). *Sculptured surface machining: theory and applications.* Dordrecht: Kluwer Academic Publishers. doi:10.1007/978-1-4615-5283-3

Choi, S., & Cheung, H. (2008). A versatile virtual prototyping system for rapid product development. *Computers in Industry*, *59*(5), 477–488. doi:10.1016/j.compind.2007.12.003

Chong, P. P., Chen, Y. S., & Chen, J. C. H. (2001). IT induction in the food service industry. *Industrial Management & Data Systems*, *101*(1), 13–20. doi:10.1108/02635570110365961

Choset, H. (1996). *Sensor based motion planning: The hierarchical generalized Voronoi graph*. Ph. D. Thesis, California Institute of Technology, Pasadena, California.

Choset, H. (2005). *Principles of Robot Motion: Theory, Algorithms, and Implementations*. The MIT Press.

Chryssolouris, G., Mavrikios, D., Pappas, M., Xanthakis, V., & Smparounis, K. (2009). In L. Wang & A. Y. C. Nee (Eds.), *A Web and Virtual Reality-based Platform for Collaborative Product Review and Customisation* (pp. 137–152). Collaborative Design and Planning for Digital Manufacturing. doi:10.1007/978-1-84882-287-0_6

Claver, J.-F., Pitt, D., & Gelinier, J. (1996). *Gestion de flux en entreprise: modélisation et simulation*. France: Hermès Science Publications.

Clay (2010). *iZigg, SMS Mobile Media 90210*. Retrieved from http://www.mobilemediaforbusiness.com/14/izigg-mobile-media-marketing-90210/

Cody, R. L., Cosmas, J., & Tsekleves, E. (2009). Open-standards Rich Media Mobile Platform & Rapid Service Creation Tool. *IEEE Global Mobile Congress*, (pp. 1-6). doi:10.1109/GMC.2009.5295896

Cordier, M. O. (2006). Comparing diagnosability in continuous and discrete-events systems.*InProceedings of the 17 the International workshop on principles of Diagnosis*. 55-60.

Correa, H. L., & Gianese, I. G. N. (1993). *Just in time, MRPII e OPT: um enfoque estratégico*. São Paulo: Atlas.

Coxworth, B. (2012, Sep 18). Baxter industrial robot aims at bringing automation to smaller manufacturers. *Gizmag. com*. Retrieved from http://www.gizmag.com/rethink-robotics-baxter-industrial-robot/24183/

Critchlow, J. (1985). *Introduction to Robotics*. Michigan, USA: Collier Macmillan.

Crute, V., Ward, Y., Brown, S., & Graves, A. (2003). Implementing Lean in aerospace-challenging the assumptions and understanding the challenges. *Technovation*, *23*(12), 917–928. doi:10.1016/S0166-4972(03)00081-6

Dangelmaier, W., Fischer, M., Gausemeier, J., Grafe, M., Matysczok, C., & Mueck, B. (2005). Virtual and augmented reality support for discrete manufacturing system simulation. *Computers in Industry*, *56*(4), 371–383. doi:10.1016/j.compind.2005.01.007

Dariush, B., Gienger, M., Arumbakkam, A., Zhu, Y., Jian, B., Fujimura, K., & Goerick, C. (2009). Online transfer of human motion to humanoids. *International Journal of Humanoid Robotics*, *6*(02), 265–289. doi:10.1142/S021984360900170X

Dauchez, P., & Zapata, R. (1985). Co-ordinated control of two cooperative manipulators: the use of a kinematic model. *Proc. 15th Int. Symp. Ind. Robots*. 641–648.

Davenport, T. (1998). Putting the enterprise into the enterprise system. *Harvard Business Review*, *76*(4), 121–131. PMID:10181586

Davenport, T. (2000). *Mission Critical*. Boston, MA: Harvard Business School Press.

David, R., & Alla, H. (1999). *Du Grafcet aux réseaux de Petri*. France: Hermès Science Publications.

Davila, T. (2000). An Empirical Study on the Drivers of Management Control Systems, Design in New Product Development. *Accounting, Organizations and Society*, *25*(4–5), 383–409. doi:10.1016/S0361-3682(99)00034-3

Davis, L. (2012). They were terrible at their job. *IO9.com*. Retrieved from http://io9.com/5904068/these-1980s-robot-waiters-were-real-but-they-were-terrible-at-their-job

Davis, J. L. (1988). *Wave Propagation in Solids and Fluids*. Springer. doi:10.1007/978-1-4612-3886-7 de Berg, M., van Krefeld, M., Overmars, M., & Schwarzkopf, O. (2000). *Computational geometry: Algorithms and applications, (2nd Ed)*. Springer. doi:10.1007/978-3-662-04245-8

Debrah, Y. A. (1994). Management of Operative Staff in a Labour-scarce Economy: The Views of Human Resource Managers in the Hotel Industry in Singapore. *Asia Pacific Journal of Human Resources*, *32*(1), 41–60. doi:10.1177/103841119403200104

DeJong, K. A. (1975). *An Analysis of the Behavior of a Class of Genetic Adaptive Systems*. PhD Dissertation, Department of Computer and Communication Sciences, University of Michigan, Ann Arbor, MI.

Dellinger, W. F., & Anderson, J. N. (1992). Interactive force dynamics of two robotic manipulators grasping a non-rigid object.*Proceeding of IEEE International Conference on Robotics and Automation*, 3, 2205-2210. doi:10.1109/ROBOT.1992.219930

Delmia/Quest User Manual. (2006). Delmia/Quest.

Demeter, K., & Matyusz, Z. (2010). (in press). The impact of lean practices on inventory turnover.[Corrected Proof.]. *International Journal of Production Economics*.

Deterding, S., Sicart, M., Nacke, L., O'Hara, K., & Dixon, D. (2011). *Gamification, Using Game-design Elements in Non-gaming Contexts. Proceedings of the 2011 annual conference extended abstracts on human factors in computing systems*, (pp. 2425-2428). doi:10.1145/1979742.1979575

Dhillon, B. S. (2007). *Applied Reliability and Quality*. London: Springer-Verlag.

Dias, A. H. F., & de Vasconcelos, J. (2002). Multiobjective genetic algorithms applied to solve optimization problems. *IEEE Transactions on Magnetics*, *38*(2), 1133–1136. doi:10.1109/20.996290

Díaz Baca, C. S. (2007). *Manipulación cooperativa robot-robot y humano-robot. Aplicación a sistemas flexibles de desensamblado automático*. (Doctoral dissertation), Universidad de Alicante.

Dijkstra, E. (1959). A note on two problems in conexion with graphs. *Numerische Mathematik*, *1*(1), 269–271. doi:10.1007/BF01386390

Djellal, F., & Gallouj, F. (2009). *Measuring and improving productivity in services: issues, strategies and challenges*. USA: Edward Elgar Publishing.

Doil, F., & Schreiber, W., A. T. P. C. (2003). Augmented Reality for Manufacturing Planning.*Proceeding of the Workshop on Virtual Environment*, 71-76.

Dong, S., & Yun-Hui, L. (2001). Position and Force Tracking of a Two-Manipulator System Manipulating a Flexible Beam Payload.*Proceedings 2001 ICRA. IEEE International Conference on Robotics and Automation*, 4, 3484-3488. doi:10.1109/ROBOT.2001.933157

Duddy, W. J., Nissink, J. W., Allen, F. H., & Milner-White, E. J. (2004). Mimicry by asx- and ST-turns of the four main types of {beta}-turn in proteins. *Protein Science*, *13*(11), 3051–3055. doi:10.1110/ps.04920904 PMID:15459339

Duffy, V. G., & Salvendy, G. (2000). Concurrent engineering and virtual reality for human resource planning. *Computers in Industry*, *42*(2-3), 109–125. doi:10.1016/S0166-3615(99)00065-2

Duhamel et al. (2013, May 20). Rethink Robotics - Finding a Market. *Stanford Case Publisher*.

Dungan, A. L., & Lacey, S. E. (1969). Convenience Foods—What Are They? How Are They Utilized? *The Cornell Hotel and Restaurant Administration Quarterly*, *10*(2), 6–12. doi:10.1177/001088046901000202

Durden, T. (2014). Meet "Smart Restaurant": The Minimum-Wage-Crushing, Burger-Flipping Robot. *Zerohedge.com.* Retrieved from http://www.zerohedge.com/news/2014-01-12/meet-smart-restaurant-minimum-wage-crushing-burger-flipping-robot?page=2

Eddy, J., & Lewis, K. (2002). Visualization of multidimensional design and optimization data using cloud visualization, *ASME Conference Proceedings*, 899-908. doi:10.1115/DETC2002/DAC-34130

Edward, L., Lourdeaux, D., Lenne, D., Barthes, J. P., & Burkhardt, J. M. (2008). Modelling autonomous virtual agent behaviours in a virtual environment for risk. *International Journal of Virtual Reality*, *7*(3), 13–22.

Eiben, A. E., & Schippers, C. A. (1998). On evolutionary exploration and exploitation. *Fundamenta Informaticae*, *35*(1-4), 35–50.

Ernst, H. (2002). Success Factors of New Product Development: A Review of the Empirical Literature. *International Journal of Management Reviews*, *4*(1), 1–40. doi:10.1111/1468-2370.00075

Erzincanli, F., & Sharp, J. M. (1997). A classification system for robotic food handling. *Food Control*, *8*(4), 191–197. doi:10.1016/S0956-7135(97)00048-0

Fernando, Y. (2012). Service Innovation along the Chain of Service Process in Airline Business. Outsourcing Management for Supply Chain Operations and Logistics Service. (pp. 185).

Fernando, Y., Ng, H. H., & Yusoff, Y. (2014). Activities, motives and external factors influencing food safety management system adoption in Malaysia. *Food Control*, *41*, 69–75. doi:10.1016/j.foodcont.2013.12.032

Fillatreau, P., Fourquet, J.-Y., Bolloc'h, R. L., Cailhol, S., Datas, A., & Puel, B. (2013). Using virtual reality and 3D industrial numerical models for immersive interactive checklists. *Computers in Industry*, *64*(9), 1253–1262. doi:10.1016/j.compind.2013.03.018

Flash, T., & Hogans, N. (1985). The Coordination of Arm Movements: Mathematical Model'. *The Journal of Neuroscience*, *5*(7), 1688–1703. PMID:4020415

Fogel, D. B. (1995). Phenotype, Genotype and Operators in Evolutionary Computation, *Proceedings of the IEEE International Conference on Evolutionary Computation*, ICEC '95, (pp:875-880), Perth, Australia, IEEE press, Piscataway, NJ. doi:10.1109/ICEC.1995.489143

Fonseca Ferreira, N. M., & Tenreiro Machado, J. A. (2003). Fractional Order Position/Force Control of Two Cooperating Manipulators.*Proceedings of the ICCC 2003 IEEE International Conference on Computational Cybernetics*, 8(29), 29-31. doi:10.1155/2010/375858

Fontanili, F. (1999). Intégration d'outils de simulation et d'optimisation pour le pilotage d'une ligne d'assemblage multiproduit à transfert asynchrone. University of Paris XIII, France.

Fountas, N. A., Kimpenis, A. A., & Vaxevanidis, N. M. (2012). Computational Techniques in Statistical Analysis and Exploitation of CNC Machining Experimental Data. In J.P. Davim (Ed.), Computational Methods for Optimizing Manufacturing Technology: Models and Techniques, (pp. 111-143). IGI-Global. doi:10.4018/978-1-4666-0128-4.ch005

Fountas, N.A., Vaxevanidis, N.M., Stergiou, C.I., Benhadj-Djilali. R. (n. d.). Evaluation of 3- and 5-Axis Sculptured Surface Machining in CAM Environment through Design of Experiments (in press). *International Journal of Computer Integrated Manufacturing, 28*(3), 278-296.

Fountas, N. A., Kimpenis, A. A., Vaxevanidis, N. M., & Davim, J. P. (2012). Single and multi-objective optimization methodologies in CNC machining. In J. P. Davim (Ed.), *Statistical and Computational Methods in Manufacturing* (pp. 187–218). Berlin, Heidelberg: Springer-Verlag. doi:10.1007/978-3-642-25859-6_5

Fründ, J., Gausemeier, J., Matysczok, C., & Radkowski, R. (2005). Using Augmented Reality Technology to Support Automobile Development. *Lecture Notes in Computer Science*, *3168*, 289–298.

Fujii, S., & Kurono, S. (1975). Coordinated computer control of a pair of manipulators.*Proceeding 4th IFToMM World Congress,*411-417.

Gagvani, N., & Silver, D. (1997). *Parameter controlled skeletonization of three dimensional objects (Tech. Rep.). CAIP-TR-216.* Rutgers State University of New Jersey.

Gallagher, M. C. (1977). The Economics of Training Food Service Employees. *The Cornell Hotel and Restaurant Administration Quarterly*, *18*(1), 54–56. doi:10.1177/001088047701800112

Garrido, S., Moreno, L., Abderrahim, M., & Martin, F. (2006). Path planning for mobile robot navigation using Voronoi Diagram and Fast Marching. In *Proc of IROS'06. Beijing. China.* (pp. 2376-2381). doi:10.1109/IROS.2006.282649

Garrido, S., Moreno, L., & Blanco, D. (2006). Voronoi diagram and Fast Marching applied to path planning. In *Proc. of ICRA* (pp. 3049-3054). doi:10.1109/ROBOT.2006.1642165

Garrido, S., Moreno, L., Blanco, D., & Munoz, M. L. (2007). Sensor-based global planning for mobile robot navigation. *Robotica*, *25*(02), 189–199. doi:10.1017/S0263574707003384

Gausemier, J., & Frund, J. M. C. (2002). AR-Planning Tool Design Flexible Manufacturing Systems with Augmented Reality.*Proceedings of 8th Euro graphics Workshop on Virtual Environment*, 19-25.

Gerbaud, S., Mollet, N., Ganier, F., Arnaldi, B., & Tisseau, J. (2008). GVT: a platform to create virtual environments for procedural training.*Proceedings of IEEE VR Conference*, 225-232. doi:10.1109/VR.2008.4480778

Gesellschaft, F. R. I. (2006). Carl Adam Petri und die Petrinetze. *Informatik-Spektrum*, *29*(5), 369–381. doi:10.1007/s00287-006-0107-7

Giua, A., & Seatzu, C. (2007). A system theory view of Petri nets. *Advances in Control Theory and Applications*, *353*, 99–127.

Gogg, T. J., & Mott, J. R. A. (1993). Introduction to simulation.*InProceedings of the 25th conference on winter simulation*, 9-17. doi:10.1145/256563.256571

Goldberg, D. E. (1989). *Genetic Algorithms in Search, Optimization, and Machine Learning.* Reading, Massachusetts: Addison-Wesley.

Gorle, P., & Clive, A. (2011). *Positive impact of industrial robots on employment. Metra Martech/International Foundation of Robotics Report.*

Goulding, J., Nadim, W., Petridis, P., & Alshawi, M. (2012). Construction industry offsite production: A virtual reality interactive training environment prototype. *Advanced Engineering Informatics*, *26*(1), 103–116. doi:10.1016/j.aei.2011.09.004

Green, T. (2013). Frito-Lay and Wynright Put robots on the docks. *Robotic Business Review.* Retrieved from http://www.roboticsbusinessreview.com/article/frito_lay_and_wynright_put_robots_on_the_docks

Grienitz, V., & Hausicke, M. (2013). Strategic Optimization of Future Manufacturing Process with Grafem, Technology Roadmaps and Scenarios Technique.*Proceedings of the 22th International Conference on Production Research,Foz do Iguaçu, Brazil.*

Gudiño-Lau, J., Arteaga, M. A., Muñoz, L. A., & Parra-Vega, V. (2004). On the Control of Cooperative Robots without Velocity Measurements. *IEEE Transactions on Control Systems Technology*, *12*(4), 600–608. doi:10.1109/TCST.2004.824965

Gueaieb, W., Karray, F., Al-Sharhan, S., & Basir, O. (2002). A hybrid adaptive fuzzy approach for control of cooperative manipulators. *Proceedings. ICRA '02.IEEE International Conference on Robotics and Automation*, 2, 2153-2158. doi:10.1109/ROBOT.2002.1014858

Haghirian, P., & Madlberger, M. (2007). Consumer Attitude towards Advertising via Mobile Devices – An Empirical Investigation among Austrian Users. *International Journal of Mobile Communications*, 48–67. doi:10.1504/IJMC.2007.011489

Hao, X., & Varshney, A. (2006) Geometry guided computation of 3D electrostatics for large biomolecules. Computer *Aided Geometric Design,* 545–557.

Harris, E. N., & Morgenthaler, G. W. (2004). Planning, implementation and optimization of future space missions using an immersive visualization environment (IVE) machine. *Acta Astronautica, 55*(1), 69–78. doi:10.1016/j.actaastro.2003.11.002 PMID:15786593

Hayati, S. (1988). Hybrid Position/force control of multi-arm cooperating robots.*Proceedings of the IEEE International Conference on Robotics and Automation*, 3, 82-89. doi:10.1109/ROBOT.1986.1087650

Hayward, S., & Milner-White, E. J. (2008). The geometry of alpha-sheet: Implications for its possible function as amyloid precursor in proteins. *Proteins-Structure Function and Bioinformatics, 71*(1), 415–425. doi:10.1002/prot.21717 PMID:17957773

Ho, C.-F. (1996). Information technology implementation strategies for manufacturing organizations: A strategic alignment approach. *International Journal of Operations & Production Management, 16*(7), 77–100. doi:10.1108/01443579610119171

Hoffman, M. (2012). Moving From Man to Machine. *Food Logistics.* Retrieved from http://www.foodlogistics.com/article/10630825/moving-from-man-to-machine?page=1

Holland, J. (1992). *Adaptation in Natural and Artificial Systems* (2nd Ed.). Massachusetts: MIT Press.

Holland, J. H. (1975). *Adaptation in Natural and Artificial Systems. University of Michigan Press*. Ann Arbor, Michigan: MIT Press.

Holling, C. S. (1973). Resilience and stability of ecological systems. *Annual Review of Ecology and Systematics, 4*(1), 1–23. doi:10.1146/annurev.es.04.110173.000245

Holling, C. S. (2001). Understanding the complexity of economic, ecological, and social systems. *Ecosystems, 4*(5), 390–405. doi:10.1007/s10021-001-0101-5

Holz, D., Nieuwenhuisen, M., & Droeschel, D. (2013). Active Recognition and Manipulation for Mobile Robot Bin Picking.

Hsu, P. (1986). Coordinated Control of Multiple Manipulators Systems. *IEEE Transactions on Robotics and Automation, 9*(4), 400–410. doi:10.1109/70.246051

Hsu, P. (1989). Control of Multiple Manipulator system-trajectory tracking load distribution, internal force control and decentralized architecture.*Proceeding of the IEEE International Conference on Robotics and Automation*, 2, 1234-1239. doi:10.1109/ROBOT.1989.100149

Hundal, M. S. (1991). Use of Functional Variants in Product Development, *ASME Design Theory and Methodology Conference*, Miami, Florida, USA.

Hu, Y. R., Goldenberg, A. A., & Zhou, C. (1995). Motion and force control of coordinated robots during constrained motion tasks. *The International Journal of Robotics Research, 14*(4), 351–365. doi:10.1177/027836499501400404

Ifinedo, P. (2007). An empirical study of ERP success evaluations by business and IT managers. *Information Management & Computer Security, 15*(4), 270–282. doi:10.1108/09685220710817798

Ilyukhin, S. V., Haley, T. A., & Singh, R. K. (2001). A survey of automation practices in the food industry. *Food Control*, *12*(5), 285–296. doi:10.1016/S0956-7135(01)00015-9

Jayaram, S., Jayaram, U., Kim, Y., DeChenne, C., Lyons, K., Palmer, C., & Mitsui, T. (2007). Industry case studies in the use of immersive virtual assembly. *Journal of Virtual Reality*, *11*(4), 217–228. doi:10.1007/s10055-007-0070-x

Jazar, R. N. (2007). *Theory of Applied Robotics : Kinematics, Dynamics and Control*. New York: Springer. doi:10.1007/978-0-387-68964-7

Jeong, D. H., Ziemkiewicz, C., Ribarsky, W., & Chang, R. (2009). *Understanding principal component analysis using a visual analytics tool. Charlotte Visualization Center*. UNC Charlotte; doi:10.1109/SBR-LARS.2012.59

Johnston, R., & Jones, P. (2004). Service productivity: Towards understanding the relationship between operational and customer productivity. *International Journal of Productivity and Performance Management*, *53*(3), 201–213. doi:10.1108/17410400410523756

Jose, S. (2010). *Global Service Robotics Market to Reach US $38.42 Billion by 2015*. Retrieved from http://www.prweb.com/releases/service_robotics/professional_robots/prweb4240924.html

Kalpakjian, S., & Schmid, S. R. (2008). Manufacturing Processes for Engineering Materials 5th Ed., Pearson; Prentice Hall.

Kantonen, T., Woodward, C., & Katz, N. (2010). Mixed reality in virtual world teleconferencing, *Proceedings of the IEEE Conference on Virtual Reality*, 179-182.

Karagiannis, S., Iakovakis, V., Kechagias, J. D., Fountas, N. A., & Vaxevanidis, N. M. (2013, September 13-16). Prediction of Surface Texture Characteristics in Turning of FRPs Using ANN, *Proceedings of the 14th EANN Conference Engineering Applications of Neural Networks, Part I, CCIS 383*, EANN 2013, (pp: 144–153), Halkidiki, Greece. doi:10.1007/978-3-642-41013-0_15

Kaufmann, H., & Schmalstieg, D. (2003). Mathematics and Geometry Education with Collaborative Augmented Reality. *Computers & Graphics*, *27*(3), 339–345. doi:10.1016/S0097-8493(03)00028-1

Kawato, M., Maeda, Y., Uno, Y., & Suzuki, R. (1990). Trajectory Formation of Arm Movement by Cascade Neural Network Model Based on Minimum Torque Change. *Biological Cybernetics*, *62*(4), 275–288. doi:10.1007/BF00201442 PMID:2310782

Keerthi, S., Huang, C., & Gilbert, E. (1999). Equidistance diagram- a new roadmap method for path planning. In *Proc. ieee int. conf. on Robotics and Automation* (pp. 682-687).

Kelby, S. (2015) *Adobe Photoshop Layer basics*, Retrieved from http://helpx.adobe.com/en/photoshop/using/layer-basics.html

Kelly, R., & Santibáñez, V. (2003). *Control de Movimiento de Robots Manipuladores*. Madrid: Pearson Educación.

Kersting, P., & Zabel, A. (2009). Optimizing NC-tool paths for simultaneous five-axis milling based on multi-population multi-objective evolutionary algorithms. *Advances in Engineering Software*, *40*(6), 452–463. doi:10.1016/j.advengsoft.2008.04.013

Kilbane, D. (2008). *Joseph Engelberger: Robotics Move from Industry to Space To Elder Care*. Retrieved from http://electronicdesign.com/embedded/joseph-engelberger-robotics-move-industry-space-elder-care

Kilic, H., & Okumus, F. (2005). Factors influencing productivity in small island hotels: Evidence from Northern Cyprus. *International Journal of Contemporary Hospitality Management*, *17*(4), 315–331. doi:10.1108/09596110510597589

Kilpatrick J. (2003). Lean principles. *Utah manufacturing Extension Partnership*.

Klingstam, P., & Gullander, P. (1999). Overview of simulation tools for computer aided production engineering. *Computers in Industry*, *38*(2), 173–186. doi:10.1016/S0166-3615(98)00117-1

Klouwenberg, M.K., KOO, W.J.D., & van Schaik, J. A. M. (. (1995). Establishing business strategy with information technology. *Information Management & Computer Security*, *3*(5), 8–20. doi:10.1108/09685229510104945

Kochan, A. (2005). BMW uses even more robots for both flexibility and quality. *International Journal (Toronto, Ont.)*, *32*(4), 318–320.

Kofman, J., Wu, X., Luu, T. J., & Verma, S. (2005). Teleoperation of a robot manipulator using a vision-based human-robot interface. *Industrial Electronics. IEEE Transactions*, *52*(5), 1206–1219.

Kousuge, K., Kogan, M., Furuta, K., & Nosaki, K. (1989). Coordinated Motion Control of Robot Arm Based on Virtual Internal Model.*Proceeding of IEEE International Conference on Robotics and Automation*, 2, 1097-1102. doi:10.1109/ROBOT.1989.100127

Koutroumanis, D. A. (2011). Technology's Effect on Hotels and Restaurants: Building a Strategic Competitive Advantage. *Journal of Applied Business and Economics*, *12*(1), 72–80.

Krafcik, J. F. (1988). Triumph of the lean production system. *Sloan Management Review*, *30*(1), 41–52.

Krimpenis, A. A., Fountas, N. A., Noeas, G. D., Iordanidis, D. M., & Vaxevanidis, N. M. (2013, June 3-7). On the 3D Parametric Modeling of Manufacturing Systems.*7th International Working Conference "Total Quality Management – Advanced and Intelligent Approaches", 473-479, Belgrade, Serbia*

Kubota, N., Shimojima, K., & Fukuda, T. (1996). Virus-evolutionary genetic algorithm for a self-organizing manufacturing system. *Computers & Industrial Engineering*, *30*(4), 1015–1026. doi:10.1016/0360-8352(96)00049-6

Ladeveze, N., Fourquet, J.-Y., & Puel, B. (2010). Interactive path planning for haptic assistance in assembly tasks. *Computers & Graphics*, *34*(1), 17–25. doi:10.1016/j.cag.2009.10.007

Lafortune, C. S. (2008). *Introduction to Discrete Event Systems* (2nd Ed.). Boston: Springer - Verlag.

Lamb, R. (2010). How have robots changed manufacturing? *Howstuffworks.com.* Retrieved from http://science.howstuffworks.com/robots-changed-manufacturing.htm

Laudon, K. C., & Laudon, J. P. (2007). *Sistemas de informação gerenciais. 7.* São Paulo: Prentice-Hall.

Law, F. L., Kasirun, Z. M., & Gan, C. K. (2011). Gamification towards Sustainable Mobile Application. *Proceedings of IEEE Software Engineering 2011 5th Malaysian Conference*, (pp.349-353). doi:10.1109/MySEC.2011.6140696

Lee, J., Han, S., & Yang, J. (2011). Construction of a computer-simulated mixed reality environment for virtual factory layout planning. *Computers in Industry*, *62*(1), 86–98. doi:10.1016/j.compind.2010.07.001

Leu, M. C., ElMaraghy, H. A., Nee, A. Y., Ong, S. K., Lanzetta, M., & Putz, M. et al. (2013). CAD model based virtual assembly simulation, planning and training. *Manufacturing Technology*, *62*(2), 799–822.

Li, H., Chan, G., & Skitmore, M. (2012). Visualizing safety assessment by integrating the use of game technology. *Automation in Construction*, *22*(0), 498–505. doi:10.1016/j.autcon.2011.11.009

Lindskog, E., Berglund, J., Vallhagen, J., & Johansson, B. Visualization Support for Virtual Redesign of Manufacturing Systems. In: Procedia 7, Nr. 0, S. 419 – 424, 2013. doi:10.1016/j.procir.2013.06.009

Lindskog, E., Berglund, J., Vallhagen, J., & Johansson, B. (2013). Visualization Support for Virtual Redesign of Manufacturing Systems. *Journal of Procedia*, *7*(0), 419–424.

Liu, Y., Arimoto, S., Parra-Vega, V., & Kitagaki, K. (1997). Decentralized adaptive control of multiple manipulators in cooperation. *International Journal of Control*, *67*(5), 649–674. doi:10.1080/002071797223938

Liu, Y., Jiang, Y., & Li, L. (2011). Multi-objective performance optimization of redundant robots using differential evolution. *Proceedings of 2011 6th International Forum on Strategic Technology*, (2), 410–414. doi:10.1109/IFOST.2011.6021052

Livingston, G. E. (1974). Changes in the food service industry. *The Cornell Hotel and Restaurant Administration Quarterly*, *15*(1), 14–19. doi:10.1177/001088047401500105

Li, X., & Prescott, D. (2009). *Measuring productivity in the service sector*. Guelph, Ontario, Canada: Canadian Tourism Human Research Council and University of Guelph.

Lopez de Lacalle, L. N., Lamikiz, A., Sanchez, J. A., & Salgado, M. A. (2007). Tool path selection based on the minimum deflection cutting forces in the programming of complex sculptured surfaces milling. *International Journal of Machine Tools & Manufacture*, *47*(2), 388–400. doi:10.1016/j.ijmachtools.2006.03.010

Lowensohn, J. (2013). Apple's App Store downloads hit 50 billion. *CNET*. Retrieved from http://www.cnet.com/news/apples-app-store-downloads-hit-50-billion/

Luecke, G. R., & Lai, K. W. (1997). A joint error-feedback approach to internal force regulation in cooperating manipulator systems. *Journal of Robotic Systems*, *14*(9), 631–648. doi:10.1002/(SICI)1097-4563(199709)14:9<631::AID-ROB1>3.0.CO;2-M

Mahalik, N. P., & Yen, M. (2009). Extending field bus standards to food processing and packaging industry: A review. *Computer Standards & Interfaces*, *31*(3), 586–598. doi:10.1016/j.csi.2008.03.027

Mahdjoub, M., Monticolo, D., Gomes, S., & Sagot, J.-C. (2010). A Collaborative Design for Usability approach supported by Virtual Reality and a Multi-Agent System embedded in a PLM environment. *Computer Aided Design*, *42*(5), 402–413. doi:10.1016/j.cad.2009.02.009

Mahkovic, R., & Slivnik, T. (1998). Generalized Local Voronoi Diagram of visible region. In *Inproc. IEEE Int. Conf. on Robotics and Automation, (pp.*349-355*),Leuven, Belgium.* doi:10.1109/ROBOT.1998.676424

Manca, D., Brambilla, S., & Colombo, S. (2013). Bridging between Virtual Reality and accident simulation for training of process-industry operators. *Advances in Engineering Software*, *55*, 1–9. doi:10.1016/j.advengsoft.2012.09.002

Marc, J., Belkacem, N., & Marsot, J. (2007). Virtual reality: A design tool for enhanced consideration of usability "validation elements". *Safety Science*, *45*(5), 589–601. doi:10.1016/j.ssci.2007.01.004

Marinho, M. M., Geraldes, A. A., Bó, A. P., & Borges, G. A. (2012, October). Manipulator control based on the dual quaternion framework for intuitive teleoperation using Kinect. In *Robotics Symposium and Latin American Robotics Symposium (SBR-LARS), 2012 Brazilian,* (pp. 319-324). IEEE.

Marion, N., Querrec, R., & Chevaillier, P. (2009). Integrating knowledge from virtual reality environments to learning scenario models. A meta-modeling approach. *Proceedings of the International conference of Computer Supported Education*, 254-259, Lisbon, Portugal.

Marion, N., Septseault, C., Boudinot, A. & Querrec, R. (2007). GASPAR: Aviation management on aircraft carrier using virtual reality, *Proceedings of Cyberworlds*, Hanover, Germany.

Markoff, J. (2012). Skilled Work, Without the Worker. *New York Times*. Retrieved from http://www.nytimes.com/2012/08/19/business/new-wave-of-adept-robots-is-changing-global-industry.html?pagewanted=all&_r=1

Markt, P. L. & Mayer, M. H., (1997). WITNESS simulation software: a flexible suite of simulation tools, 7(10), 711-717.

Marnewick, C., & Labuschagne, L. (2005). A conceptual model for enterprise resource planning (ERP). *Information Management & Computer Security, 13*(2), 144–155. doi:10.1108/09685220510589325

Marri, H. B., Gunasekaran, A., Irani, Z., & Putnik, G. (2008). Collaborative Alliance for the Implementation of Computer Integrated Manufacturing in Small and Medium Enterprises. In G. Putnik & M. Cruz-Cunha (Eds.), *Encyclopedia of Networked and Virtual Organizations* (pp. 216–224). Hershey, PA: Information Science Reference; doi:10.4018/978-1-59904-885-7.ch029

MathWorks. (2012). Bioinformatics Toolbox: User's Guide (R2012a). Retrieved from www.mathworks.com/help/pdf_doc/bioinfo/bioinfo_ug.pdf

Matthiassen, B. (1997). *Design for robustness and reliability – improving the quality consciousness in engineering design.* Lyngby: Department of Control and Engineering Design, Technical University of Denmark.

Mauch, S. (2003). *Efficient algorithms for solving static Hamilton-Jacobi equations.* Doctoral dissertation, California Inst. of Technology.

Michieletto, S., Chessa, N., & Menegatti, E. (2013, November). Learning how to approach industrial robot tasks from natural demonstrations. In *Advanced Robotics and its Social Impacts (ARSO), 2013 IEEE Workshop on* (pp. 255-260). IEEE. doi:10.1109/ARSO.2013.6705538

Michieletto, S., Tosello, E., Romanelli, F., Ferrara, V., & Menegatti, E. (2014). ROS-I Interface for COMAU Robots. In Simulation, Modeling, and Programming for Autonomous Robots (pp. 243-254). Springer International Publishing. doi:10.1007/978-3-319-11900-7_21

Mickle, P. (2014), 1961: A peep into the automated future. Retrieved from http://www.capitalcentury.com/1961.html

Miller, T., & Monaghan, C. (2012). Apple Press Release Library, Apple's App Store Downloads, Top 25 Billion. *Apple.com.* Retrieved from http://www.apple.com/pr/library/2012/03/05Apples-App-Store-Downloads-Top-25-Billion.html

Milner-White, E. J., Nissink, J. W., Allen, F. H., & Duddy, W. J. (2004). Recurring main-chain anion-binding motifs in short polypeptides: Nests. *Acta Crystallographica. Section D, Biological Crystallography, 60*(11), 1935–1942. doi:10.1107/S0907444904021390 PMID:15502299

Milner-White, E. J., & Russell, M. (2008). Predicting the conformations of peptides and proteins in early evolution. *Biology Direct, 3*(3), 1–9. PMID:18226248

Milner-White, E. J., & Russell, M. J. (2005). Sites for phosphates and iron-sulfur thiolates in the first membranes: 3 to 6 residue anion-binding motifs (nests). *Origins of Life and Evolution of the Biosphere, 35*(1), 19–27. doi:10.1007/s11084-005-4582-7 PMID:15889648

Milner-White, E. J., Watson, J. D., Qi, G., & Hayward, S. (2006). Amyloid formation may involve alpha- to beta sheet interconversion via peptide plane flipping. *Structure (London, England), 14*(9), 1369–1376. doi:10.1016/j.str.2006.06.016 PMID:16962968

Mobile Media Consumption . (2014). *Limelight.com.* Retrieved from http://blog.limelight.com/2011/06/a-look-at-the-numbers-mobile-media-consumption-2010-2011/

Mohamed, Z., & Capi, G. (2012). Development of a New Mobile Humanoid Robot for Assisting Elderly People. *Procedia Engineering, 41*(Iris), 345–351. doi:10.1016/j.proeng.2012.07.183

Mohamed, Z., Mano, M., Kitani, M., & Capi, G. (2013). Adaptive Humanoid Robot Arm Motion Generation by Evolved Neural Controllers. *International Journal of Control Automation and System*, 1–6.

Mohamed, N., & Singh, J. K. G. (2012). A conceptual framework for information technology governance effectiveness in private organizations. *Information Management & Computer Security*, *20*(2), 88–106. doi:10.1108/09685221211235616

Mohamed, Z., Kitani, M., & Capi, G. (2013a). Adaptive Arm Motion Generation of Humanoid Robot Operating in Dynamic Environments. *Industrial Robot. International Journal (Toronto, Ont.)*.

Mohamed, Z., Kitani, M., & Capi, G. (2013b). Optimization of Robot Arm Motion in Human Environment. *International Journal of Enhanced Research Publication*, *2*(10), 1–7.

Moran A, (2011, Aug 2), Chinese Foxconn employees could soon be replaced with robots.

Moreno Masey, R. J., Gray, J. O., Dodd, T. J., & Caldwell, D. G. (2010). Guidelines for the design of low-cost robots for the food industry. Industrial Robot. *International Journal (Toronto, Ont.)*, *37*(6), 509–517.

Mortimer, J. (2009). Robots boost for GMM van production. Industrial Robot. *International Journal (Toronto, Ont.)*, *36*(2), 121–122.

Moss, A. L. (1969). Manpower Shortage in Food Service. *The Cornell Hotel and Restaurant Administration Quarterly*, *10*(1), 43–48. doi:10.1177/001088046901000112

Motwani, J. R., & Ullman, J. (2007). *Introduction to Automata Theory, Languages, and Computation*. USA: Prentice Hall.

Mujber, T., Szecsi, T., & Hashmi, M. (2004). Virtual reality applications in manufacturing process simulation. *Journal of Materials Processing Technology*, *155*, 1834–1838. doi:10.1016/j.jmatprotec.2004.04.401

Munaro, M., Antonello, M., Moro, M., Ferrari, C., Clemente, G., Pagello, E., & Menegatti, E. (2014). FibreMap: Automatic Mapping of Fibre Orientation for Draping of Carbon Fibre Parts. In *ROS-Industrial in European Research Projects, IAS-13 Workshop on* (pp. 272–275).

Murata, T. (1989). Petri nets: Properties, analysis and applications. *Proceedings of the IEEE*, *77*(4), 541–580. doi:10.1109/5.24143

Nagatani, K., Choset, H., & Thrun, S. (1998). Towards exact localization without explicit localization with the generalized Voronoi graph. In *Proc. IEEE int. conf. On Robotics and Automation* (pp. 342-348). Leuven, Belgium. doi:10.1109/ROBOT.1998.676421

Nakahara, H., & Sagawa, T. (1989). *Virus Theory of Evolution*. Tokyo, Japan: Tairyusha.

Nakamura, Y., Naga, K., & Yoshikawa, T. (1989). Mechanics of coordinative manipulation by multiple robotics mechanism.*Proceeding of the IEEE International Conference on Robotics and Automation*, 4, 991-998. doi:10.1109/ROBOT.1987.1087941

Nakano, E., Imamizu, H., Osu, R., Uno, Y., Gomi, H., Yoshioka, T., & Kawato, M. (1999). Quantitative Examinations of Internal Representations for Arm Trajectory Planning : Minimum Commanded Torque Change Model Quantitative Examinations of Internal Representations for Arm Trajectory Planning : Minimum Commanded Torque Change Model. *Journal of Neurophysiology*, *81*, 2140–2155. PMID:10322055

Nakano, E., Ozaki, S., Ishida, T., & Kato, I. (1974). Cooperational control of the anthropomorphous manipulator MELAR.*Proceeding 4th International Symposium on Industrial Robots*. 251-260.

Naniwa, T., Arimoto, S., & Wada, K. (1997). Learning and adaptive controls for coordination of multiple manipulators without knowing physical parameters of an object.*Proceedings of IEEE International Conference on Robotics and Automation*, 2, 1496-1502. doi:10.1109/ROBOT.1997.614350

Nash, J. F. (1951). Equilibrium Points in n-person games. *The Annals of Mathematics*, *54*, 289.

Ng, L. X., Ong, S. K., & Nee, A. Y. C. (2010). ARCADE: A Simple and Fast Augmented Reality Computer-Aided Design Environment Using Everyday Objects.*Proceedings of IADIS Interfaces and Human Computer Interaction 2010 Conference*, 227–234.

Ng, L. X., Oon, S. W., Ong, S. K., & Nee, A. Y. C. (2011). GARDE: A Gesture-based Augmented Reality Design Evaluation System. *International Journal of Interactive Design and Manufacturing*, *5*(2), 85–94. doi:10.1007/s12008-011-0117-9

O'Brien, J. A. (2004). *Sistemas de Informação e as decisões gerenciais na era da Internet*. São Paulo: Saraiva.

Ochoa, R., Ochoa, J., & Muskus, C. (2011) Automation of bioinformatic tools to detect gene fusion events in the Leishmania braziliensis and Leishmania major genomes*Computing Congress (CCC),* 6th. doi:10.1109/COLOMCC.2011.5936324

Ogniewicz, R., & Kubler, O. (1995). Hierarchic Voronoi Skeletons. *Pattern Recognition*, *28*(3), 343–359. doi:10.1016/0031-3203(94)00105-U

Okabe, A., Boots, B., & Sugihara, K. (1992). Spatial Tessellations: Concepts and Aplications of Voronoi Diagrams"., 1992. (J. Wiley & Sons, Eds.). Chichester, UK.

Oliveira, D. P. R. (2011). *Sistemas de Informações Gerenciais*. São Paulo: Atlas.

Openrice Ruyi. (2014). Overview. Retrieved from http://sg.openrice.com/singapore/restaurant/ruyi/15168/

Orr, S. C. (1996). A longitudinal survey of robot usage in Australia. *Integrated Manufacturing Systems*, *7*(5), 33–46. doi:10.1108/09576069610129900

Osborne, M. J. (2003). *An Introduction to GameTheory*. Oxford, UK: Oxford University Press.

Ozturk, E., & Budak, E. (2007). Modeling of 5-axis milling processes. *Machining Science and Technology. International Journal (Toronto, Ont.)*, *11*(3), 287–311.

Pahl, G., & Beitz, W. (1993). In K. Wallace, (Ed.), *Engineering Design: A Systematic Approach*. New York: Springer-Verlag.

Parra-Vega, V., Rodríguez-Ángeles, A., Arimoto, S., & Hirzinger, G. (2001). High precision constrained grasping with cooperative adaptive control. *Journal of Intelligent & Robotic Systems*, *32*(3), 235–254. doi:10.1023/A:1013987209547

Pegden, C. D., Sadowski, R. P., & Shannon, R. E. (1995). Introduction to Simulation Using SIMAN. New York, NY, USA: McGraw-Hill, Inc.

Perdereau, V., & Drouin, M. (1996). Hybrid external control for two robot coordinated motion. *Robotica*, *14*(2), 141–153. doi:10.1017/S0263574700019056

Petrina, A. M. (2011). Advances in robotics. *Automatic Documentation and Mathematical Linguistics*, *45*(2), 43–57. doi:10.3103/S000510551102004X

Pfeffer, L. E., & Cannon, R. H. Jr. (1993). Experiments with a Dual-Armed, Cooperative, Flexible Drivetrain Robot System. *Proceedings of International Conference on Robotics and Automation*, 3, 601-608. doi:10.1109/ROBOT.1993.291839

Pires, E. J. S., de Moura Oliveira, P. B., & Machado, J. T. (2007). Manipulator trajectory planning using a MOEA. *Applied Soft Computing*, *7*(3), 659–667. doi:10.1016/j.asoc.2005.06.009

Pollack, M. E., Engberg, S., Matthews, J. T., Dunbar-jacob, J., Mccarthy, C. E., & Thrun, S. (2002). *Pearl : A Mobile Robotic Assistant for the Elderly*.

Pope, S., & Muller, T. (2011). Apple Press Release Library, Apple's App Store Downloads, Top 15 Billion. *Apple.com*. Retrieved from https://www.apple.com/pr/library/2011/07/07Apples-App-Store-Downloads-Top-15-Billion.html

Pouliquen, M., Bernard, A., Marsot, J., & Chodorge, L. (2007). Virtual hands and virtual reality multimodal platform to design safer industrial systems. *Computers in Industry*, *58*(1), 46–56. doi:10.1016/j.compind.2006.04.001

PPMI. (2008). Robots on the Rise Robotics: Usage and Trends in Packaging Applications. Retrieved from http://www.pmmi.org/files/PIB/Robotics.pdf

Purnell, G. (1998). Robotic equipment in the meat industry. *Meat Science*, *49*, S297–S307. doi:10.1016/S0309-1740(98)90056-0 PMID:22060720

Qi, Y. (2011). Risk of Social Disruption and Integration Function Development of News Products of Traditional Media in the Digital Age. *Proceedings of 6th International Conference*, (pp.70-74). IEEE.

Qianlima (2014). *Mobile Market Reaches 5 Billion*, Retrieved from http://digi.163.com/14/0521/20/9SQ0GDET001618JV.html

Querrec, R., Vallejo, P., & Buche, C. (2013). MASCARET: create virtual learning environments from system modelling. Engineering Reality of Virtual Reality, *SPIE Electronic Imaging Conference*, San Francisco, CA, USA.

Quinsat, Y., & Sabourin, L. (2007). Optimal selection of machining direction for three-axis milling of sculptured parts. *International Journal of Advanced Manufacturing Technology*, *33*(7-8), 684–692. doi:10.1007/s00170-006-0515-5

Ramabalan, S., Saravanan, R., & Balamurugan, C. (2008). Multi-objective dynamic optimal trajectory planning of robot manipulators in the presence of obstacles. *International Journal of Advanced Manufacturing Technology*, *41*(5-6), 580–594. doi:10.1007/s00170-008-1506-5

Ramadorai, A., Tarn, T., & Bejczy, A. (1992). Task definition, decoupling and redundancy resolution by nonlinear feedback in multi-robot object handling.*Proceeding of the IEEE International Conference on Robotics and Automation*, 1, 467-474. doi:10.1109/ROBOT.1992.220296

Raymond, C. W., James, M., & Graham, R. (2009). Automated design process modelling and analysis using immersive virtual reality. *Computer Aided Design*, *42*(12), 1082–1094.

René. Chevance., J. (2001). Méthodologie en matière de performance des systèmes. Techniques de l'ingénieur. Techniques de l'ingénieur, Paris.

Restaurant, C. (2006). Help wanted: The labour shortage crisis and Canadian foodservice industry. *Canadian Restaurant and Foodservice Association*, *6*, 1–9.

Richir, S. & Fuchs, P. (2003). Réalité virtuelle et conception Méthodes. *Techniques de l'ingénieur.*

Rivera Dueñas, J. C. (2012). *Control de posición/fuerza-torque para manipuladores con compensación de fricción.* (Doctoral dissertation), Available from UNAM Theses database. (tes.TES01000681507)

Romanelli, F., & Ferrara, V. (2014). *C5GOpen: The latest generation of the Industrial Robots Open Control System for University and SMEs.* Comau.

Rosenbaum, D. A., Loukopoulos, L. D., Meulenbroek, R. G.J., Vaughan, J., & Engelbrecht, S. E. (1995). Planning reached by evaluation stored posture.pdf. *Psychological Review*, *102*(1), 28–67. doi:10.1037/0033-295X.102.1.28 PMID:7878161

Roterman, I. (1995). The geometric analysis of peptide backbone structure and its local deformations. *Biochimie*, *77*(3), 204–216. doi:10.1016/0300-9084(96)88126-0 PMID:7647113

Ruddick, P. (2013). Food service 2025: Report predicts restaurant industry's robotic future. Retrieved from http://www.bighospitality.co.uk/Trends-Reports/Foodservice-2025-Report-predicts-restaurant-industry-s-robotic-future

Sagredo Hernández, L. R. (2006). *Control de robots cooperativos con un observador*. (Unpublished master dissertation), Available from UNAM Theses database. (tes.TES01000650102)

Sahar, G., & Hollerbach, J. M. (1986). Planning of Minimum - Time Trajectories for Robot Arms. *The International Journal of Robotics Research*, *5*(3), 90–100. doi:10.1177/027836498600500305

Schaffer, J. D. (1984). Research in the Food Service Industry: An Exploratory Study and Critique. *Journal of Hospitality & Tourism Research (Washington, D.C.)*, *9*(1), 55–71. doi:10.1177/109634808400900106

Schaller, R. R. (1997). Moore's law: past, present and future. *Spectrum, IEEE*, 34(6), 52-59.

Scheider, S. A., & Cannon, S. H. (1992). Object impedance control for cooperative manipulation: Theory and experimental result. *IEEE Transactions on Robotics and Automation*, *8*(3), 383–394. doi:10.1109/70.143355

Sciavicco, L., & Siciliano, B. (2001). *Modelling and Control of Robot Manipulators*. London: Springer.

Sethian, J. (1996). *Level Set Methods*. Cambridge University Press.

Shah, R., & Ward, P. T. (2003). Lean manufacturing: Context, practice bundles, and performance. *Journal of Operations Management*, *21*(2), 129–149. doi:10.1016/S0272-6963(02)00108-0

Shirley, P. & Marschner, S. (2009). *Fundamentals of Computer Graphics*, Wellesley, Massachusetts: A K Peters, Ltd.

Shore, H., & Arad, R. (2003/2004). Product robust design and process robust design: Are they the same? *Quality Engineering*, *16*(2), 193–207. doi:10.1081/QEN-120024007

Shuang, X. (2011). Consumer Adoption of Mobile Advertising: An Empirical Investigation among China Users. *Proceedings of IEEE Product Innovation Management (ICPIM), 2011 6th International Conference*, (pp.461-464). doi:10.1109/ICPIM.2011.5983702

Siciliano, B., & Villani, L. (1999). *Robot Force Control*. Klumer Academic Publishers. doi:10.1007/978-1-4615-4431-9

SIMCORE & CLARTE. (2013, August). Ergo-wide: Analyse ergonomique en réalité virtuelle. Retrieved from: http://www.simcore.fr/Ergonomie.asp

Smith, R. W. (1987). Computer processing of line images: A survey. *Pattern Recognition*, *20*(1), 7–15. doi:10.1016/0031-3203(87)90013-6

Spong, M., Lewis, F., & Abdallah, C. (1993). *Robot Control : Dynamics, Motion Planning, and Analysys*. New York: IEEE Press.

Spreckly, A. (2011). Why robots are ideal ingredient for modern food producers. *Food Processing*. Retrieved from http://www.fponthenet.net/article/40527/Why-robots-are-ideal-ingredient-for-modern-food-producers.aspx

Sreng, J., Lécuyer, A., Mégard, C., & Andriot, C. (2006). Using visual cues of contact to improve interactive manipulation of virtual objects in industrial assembly/maintenance simulations. *Journal of Visualization and Computer Graphics*, *12*(5), 1013–1020. doi:10.1109/TVCG.2006.189 PMID:17080829

Stargardt, W. (2010). *Automation - The Future of Restaurant Operations Energy perspectives*. Retrieved from http://wayne-stargardt.blogspot.co.uk/2010/07/automation-future-of-restaurant.html

Stark, R., Israel, J. H., & Wöhler, T. (2010). Towards Hybrid Modeling Environments-Merging Desktop-CAD and Virtual Reality-technologies. *Annals of CIRP*, *59*(1), 179–182. doi:10.1016/j.cirp.2010.03.102

Strougo, R., & Wenderlich, R. (2011). *Learning Cocos2D: A Hands-On Guide to Building iOS Games with Cocos2D, Box2D, and Chipmunk*. Boston, Massachusetts: Addison-Wesley.

Stuckler, J., Grave, K., Klas, J., Muszynski, S., Schreiber, M., & Tischler, O., Behnke, S. (2009). Dynamaid : Towards a Personal Robot that Helps with Household Chores. In *Proceeding of RSS Workshop on Mobile Manipulation in Human Environments*.

Sturges, R. H., O'Shaughnessy, K., & Kilani, M. I. (1990). Representation of Aircraft Design Data for Supportability, Operability, and Producibility Evaluations, *Carnegie Mellon University Engineering Design Research Center Report,* 01-30-90.

Sturges, R. H., O'Shaughnessy, K., & Reed, R. G. (1993). Systematic Approach to Conceptual Design. *Concurrent Engineering*, *1*(2), 93–105. doi:10.1177/1063293X9300100202

Sudha, N., Nandi, S., & Sridharan, K. (1999). A parallel algorithm to construct Voronoi diagram and its vlsi architecture. In *Proc. IEEE Int. Conf. on Robotics and Automation, (pp.*1683-1688*)*. doi:10.1109/ROBOT.1999.770351

Suh, N. P. (1988). The Principles of Design. New York: Oxford University Press.

Sujan, V. A., & Meggiolaro, M. A. (2011). Simultaneous Localization and Mapping for Mobile Robot Teams with Visual Sensors. *Intelligent Transportation Vehicles,* 1.

System, M. R. a product design support system of CANON, (2012, May). Retrieved from: http://www.actinnovation.com/innovation-technologie/canon-mr-system-realite-augmentee-4831.html

Syverson, C. (2010). *What determines productivity? (No. w15712)*. National Bureau of Economic Research. doi:10.3386/w15712

Szewczyk, J., Morel, G., Bidaud, P., & Basañez Villaluenga, L. (1997). Distributed Impedance Control of Multiple Robot Systems, *Proceeding of the IEEE International Conference on Robotics and Automation*. USA. 2, 1801-1806. doi:10.1109/ROBOT.1997.614414

Taguchi, G. (1986). *Introduction to quality engineering – designing quality into products and processes*. Tokyo, Japan: Asian Productivity Organization.

Taisbak, C. M. (1980). An Archimedean proof of Heron's formula for the area of a triangle; reconstructed. *Centaurus*, *24*(1), 110–116. doi:10.1111/j.1600-0498.1980.tb00368.x

Takase, K., Inoue, H., Sato, K., & Hagiwara, S. (1974). The design of an articulated manipulator with torque control ability.*Proceeding 4th. International Symposium on Industrial Robots*. 261-270.

Takeda, N., Kakudate, S., Nakahira, M., Shibanuma, K., & Tesini, A. (2008). Development of a virtual reality simulator for the ITER blanket remote handling system. *Fusion Engineering and Design*, *83*(10-12), 1837–1840. doi:10.1016/j.fusengdes.2008.05.042

Takus, D. A., Profozich, & David, M. (1997). ARENA SOFTWARE TUTORIAL, *Winter Simulation Conference*.

Tamaki, H., Kita, H., Shimizu, N., Maekawa, K., & Nishikawa, Y. (1994). A Comparison Study of Genetic Codings for the Travelling Salesman Problem, *The 1st IEEE Conference on Evolutionary Computing*, 1, (pp: 1-6), Florida.

Tao Jian, M., & Luh, J. Y. (1991). Position and Force Control for Two Coordinating Robots.*Proceeding of the IEEE International Conference on Robotics and Automation*, 1, 176-181. doi:10.1109/ROBOT.1991.131575

Tarault, A.; Bourdot, P.; Vézien, J. & Sacari. (2005). An immersive remote driving interface for autonomous vehicles, *Computational Science – ICCS*, 3-15.

Tarn, T. J., Bejczy, A. K., & Yun, X. (1987). Design of dynamics control of two cooperating robot arm.*Proceeding IEEE International Conference on Robotics and Automation*, 4, 7-13. doi:10.1109/ROBOT.1987.1088028

Tarn, T. J., Bejczy, A. K., & Yun, X. (1988). New nonlinear control algorithms for multiple robot arms. *IEEE Transactions on Aerospace and Electronic Systems, 24*(5), 571–583. doi:10.1109/7.9685

Teamcenter Manufacturing Plant Simulation. (2013, August). Retrieved from: http://www.robertirobotics.com/ugs/efactory/emplant/_files/fs_tecnomatix_em_plant.pdf

Tecnomatix Factory Layout Simulation. (2013, August). Retrieved from: http://www.plm.automation.-siemens.com/zh_cn/Images/7656_tcm78-64291.pdf

Tennant, G. (2002). *Design for Six Sigma: launching new products and services without failure*. Hampshire, UK: Gower Publishing Limited.

Teo, W. L., & Tan, K. S. (2013). Adoption of Information Technology Governance in the Electronics Manufacturing Sector in Malaysia. In I. Management Association (Ed.), Industrial Engineering: Concepts, Methodologies, Tools, and Applications (pp. 887-906). Hershey, PA: Engineering Science Reference. doi:10.4018/978-1-4666-1945-6.ch049

Tesic, R., & Banerjee, P. (1999). Exact collision detection using virtual objects in virtual reality modeling of a manufacturing process. *Journal of Manufacturing Systems, 18*(5), 367–376. doi:10.1016/S0278-6125(00)87639-6

The PyMOL Molecular Graphics System (2010) Version 1.5.0.4 Schrödinger, LLC.

Thompson, O. (2005). Key Challenges Facing Food and Beverage Manufacturers and What They Must Do To Survive. *Food & Beverage Manufacturing*. Retrieved from http://www.gilbertassociates.com/Documents/NAVFood_and_BeverageMANU_WP12.09.pdf

Tinós, R., Terra, M., & Bergrman, M. (2002). *Dynamic Load-Carrying Capacity of Cooperative Manipulators with Passive Joints*. Congreso Brasileiro de Automática.

Tobe, F. (2012). Companies making the necessary transition from industrial to service robots, Retrieved from http://singularityhub.com/2012/06/06/companies-making-the-necessary-transition-from-industrial-to-service-robots/

Tolio, T., Sacco, M., Terkaj, W., & Urgo, M. (2013). Virtual Factory: An Integrated Framework for Manufacturing Systems Design and Analysis. *Procedia, 7*, 25–30.

Tomiyama, T. (2006). A Classification of Design Theories and Methodologies. *Proceedings of the 2006 ASME IDETC*, Paper No. DETC2006-99444, ASME. (CD-ROM) doi:10.1115/DETC2006-99444

Tomiyama, T. (2005). Meijer BR, Directions of Next Generation Product Development. In H. A. ElMaraghy & W. H. ElMaraghy (Eds.), *Advances in Design* (pp. 27–35). London: Springer.

Tomiyama, T., Gu, P., Jin, Y., Lutters, D., Kind, C., & Kimura, F. (2009). Design methodologies: Industrial and educational applications. *CIRP Annals-Manufacturing Technology, 58*(2), 543–565. doi:10.1016/j.cirp.2009.09.003

Torrance, G. M., Leader, D. P., Gilbert, D. R., & Milner-White, E. J. (2009). A novel main chain motif in proteins bridged by cationic groups: The niche. *Journal of Molecular Biology, 385*(4), 1076–1086. doi:10.1016/j.jmb.2008.11.007 PMID:19038265

Tosello, E., Michieletto, S., Bisson, A., Pagello, E., & Menegatti, E. (2014, June). A learning from demonstration framework for manipulation tasks. *Proceedings of ISR/Robotik 2014; 41st International Symposium on Robotics,* (pp. 1–7). VDE.

Tsitsiklis, J. N. (1995). Efficient algorithms for globally optimal trajectories. *IEEE Transactions on Automatic Control, 40*(9), 1528–1538. doi:10.1109/9.412624

Uchiyama, M., & Dauchez, P. (1988). A symmetric hybrid position/force control scheme for the coordination of two robots.*Proc. 1988 IEEE Int. Conf. on Robotics and Automation*, 1, 350-356. doi:10.1109/ROBOT.1988.12073

Uchiyama, M., & Dauchez, P. (1993). Symmetric kinematic formulation and non-master/slave coordinated control of two-arm robots. *Advanced Robotics*, *7*(4), 361–383. doi:10.1163/156855393X00221

Uchiyama, M., Iwasawa, N., & Hakomori, K. (1987). Hybrid position/force control for coordination of a two-arm robot. *Proc. 1987 IEEE Int. Conf. on Robotics and Automation*, 4, 1242-1247. doi:10.1109/ROBOT.1987.1087766

Ulrich, K. T., & Eppinger, S. D. (2004). *Product Design and Development* (3rd Ed.). McGraw Hill, Irwin.

Uno, Y., Kawato, M., & Suzuki, R. (1989). Formation and Control of Optimal Trajectory in Human Multijoint Arm Movement. *Biological Cybernetics*, *101*, 89–101. PMID:2742921

Ur-Rehman, R., Caro, S., Chablat, D., & Wenger, P. (2010). Multi-objective path placement optimization of parallel kinematics machines based on energy consumption, shaking forces and maximum actuator torques: Application to the Orthoglide. *Mechanism and Machine Theory*, *45*(8), 1125–1141. doi:10.1016/j.mechmachtheory.2010.03.008

Using Robots to Solve End-of-Line Issues in the Food Industry. (2014). Fanuc America Corporation. Retrieved from http://www.fanucrobotics.com/robotics-articles/Food_Robots_End_of_line_Solutions.aspx

Vahrenkamp, N., Scheurer, C., Asfour, T., Kuffner, J., & Str, H. (2008). Adaptive Motion Planning for Humanoid Robots. Proceedings of *2008 IEEE/RSJ International Conference on Intelligent Robots and Systems* (pp. 22–26).

Vahrenkamp, N., Kaiser, P., Asfour, T., & Dillmann, R. (2011). RDT+: A parameter-free algorithm for exact motion planning.*2011 IEEE International Conference on Robotics and Automation*, 715–722. doi:10.1109/ICRA.2011.5979777

van den Broek, Y. Y., & Beijer, S. S. (n. d.). Labor shortage solutions: which, when and why?.

Vaxevanidis, N. M., Kechagias, J. D., Fountas, N. A., & Manolakos, D. E. (2013, July 3-5). Three Component Cutting Force System Modeling and Optimization in Turning of AISI D6 Tool Steel Using Design of Experiments and Neural Networks, *Proceedings of the World Congress on Engineering* WCE 2013, Vol I, (pp: 1-5), London, U.K.

Veneri, C. M. (1999). Can occupational labor shortages be identified using available data?. *Monthly Lab. Rev.*, *122*, 15.

VirtualiTeach. Assistance de Conception, CLEMI, CLARTE & CEA-List, (2013, August). Retrieved from: http://www.ac-creteil.fr/retrouvezlactualite-avril2013-pedagogicconcrctc.html

VR4D: Virtual Reality for Design, (2013, August). Retrieved from: http://www.lecoledesign.com/fr/actualites/bdd/actualite/1247

Wada, Y., Kaneko, Y., Nakano, E., Osu, R., & Kawato, M. (2001). Quantitative examinations for multi joint arm trajectory planning--using a robust calculation algorithm of the minimum commanded torque change trajectory. *Neural Networks : The Official Journal of the International Neural Network Society, 14*(4-5), 381–93. Retrieved from http://www.ncbi.nlm.nih.gov/pubmed/11411627

Walker, B., Carpenter, S., Anderies, J., Abel, N., Cumming, G. S., & Janssen, M. et al. (2002). Resilience management in social-ecological systems: A working hypothesis for a participatory approach. *Conservation Ecology*, *6*(1), 14–35.

Walker, B., Holling, C. S., Carpenter, S., & Kinziq, A. (2004). Resilience, adaptability and transformability in social – ecological systems. *Ecology and Society*, *9*(2), 5. Retrieved from http://www.ecologyandsociety.org/vol9/iss2/art5

Walker, I. D., Freeman, R., & Marcus, S. (1991). Analysis of motion and internal grasped by multiple cooperating manipulators. *The International Journal of Robotics Research*, *10*(4), 396–409. doi:10.1177/027836499101000408

Walker, M., Kim, D., & Dionise, J. (1989). Adaptive Coordinated motion control of two manipulators arms.*Proceeding of the IEEE International Conference on Robotics and Automation*, 2, 1084-1090. doi:10.1109/ROBOT.1989.100125

Wallin, P. J. (1997). Robotics in the food industry: An update. *Trends in Food Science & Technology*, *8*(6), 193–198. doi:10.1016/S0924-2244(97)01042-X

Wang, J. (2013). 8 Companies Leading the Charge for Commercial-Use Robotics.

Wang, J., Jensen, G., & Red, W. (2006). Global Finish Machining Using CM^2. *Computer-Aided Design and Applications*, *3*(1-4), 405–415. doi:10.1080/16864360.2006.10738486

Ward, J., & Peppard, J. (1996). Reconciling the IT/business relationship: A troubled marriage in need of guidance. *The Journal of Strategic Information Systems*, *5*(1), 37–65. doi:10.1016/S0963-8687(96)80022-9

Warkentin, A., Ismail, F., & Bedi, S. (2000). Multi-point tool positioning strategy for 5-axis machining of sculptured surfaces. *Computer Aided Geometric Design*, *17*(1), 83–100. doi:10.1016/S0167-8396(99)00040-0

Watson, J. D., & Milner-White, E. J. (2002). A novel main-chain anion-binding site in proteins: The Nest. A particular combination of phi,psi values in successive residues gives rise to anion-binding sites that occur commonly and are found often at functionally important regions. *Journal of Molecular Biology*, *315*(2), 171–182. doi:10.1006/jmbi.2001.5227 PMID:11779237

Watson, J. D., & Milner-White, E. J. (2002). The conformations of Polypeptide Chains where the main-chain parts of successive residues are enantiomeric. Their occurrence in Cation and Anion-binding regions of proteins. *Journal of Molecular Biology*, *315*(2), 183–191. doi:10.1006/jmbi.2001.5228 PMID:11779238

Webb, P., Pollard, C., & Ridley, G. (2006). Attempting to define IT governance: wisdom or folly?*Proceedings of the 39th Hawaii International Conference on System Sciences, Kauai, HI*, 1-10. doi:10.1109/HICSS.2006.68

Weber-Jahnke, J. H., & Stier, J. (2009). Virtual prototyping of automated manufacturing systems with Geometry-driven Petri nets. *Computer Aided Design*, *41*(12), 942–951. doi:10.1016/j.cad.2009.06.012

Wei, C. C. (1995). Taiwan's industrial robots. *Industrial Robot: An International Journal*, *22*(2), 21–23. doi:10.1108/EUM0000000004181

Wen, J. T., & Kreutz-Delgado, K. (1992). Motion and force control of multiple robotic manipulators. *Automatica*, *28*(4), 729–743. doi:10.1016/0005-1098(92)90033-C

Whitehead, T. N. (1954). *The design and use of instruments and accurate mechanism*. New York, USA: Dover Publications.

Wiese, E., Israel, J. H., Zöllner, C., Pohlmeyer, A. E., & Stark, R. (2009). The Potential of Immersive 3D-sketching Environments for Design Problem-solving.*Proceedings of 13th International Conference on Human–Computer Interaction*, 485–489.

Williams, D., & Khatib, O. (1993). The virtual linkage: a model for internal forces in multi-grasp manipulation.*Proc. 1993 IEEE Int. Conf. on Robotics and Automation*. 1025-1030. doi:10.1109/ROBOT.1993.292110

Wilmarth, S. A., Amato, N. M., & Stiller, P. F. (1999). MAPRM: A probabilistic roadmap planner with sampling on the medial axis of the free space. In *Proc. IEEE Int. Conf. on Robotics and Automation* (p. 1024-1031). doi:10.1109/ROBOT.1999.772448

Wilson, M. (2010). Developments in robot applications for food manufacturing. *Industrial Robot: An International Journal*, *37*(6), 498–502. doi:10.1108/01439911011081632

Womack, J., & Jones, D. (2005). System Lean: Penser l'entreprise au plus juste. Village mondial, 2ème édition, Paris.

Womack, J. P., Jones, D. T., & Ross, D. (1990). The Machine that Changed the World. *The International Journal of Human Factors in Manufacturing*, *4*(3), 1522–7111.

Wong, T. (2013), Can food outlets up productivity without sacrificing taste? *Singapolotics*. Retrieved from http://www.singapolitics.sg/views/can-food-outlets-productivity-without-sacrificing-taste

Woong, K., & Beom, H. L. (1999). General Redundancy Optimization Method for Cooperating Manipulators Using Quadratic Inequality Constraints. *IEEE Transactions on Systems, Man, and Cybernetics. Part A, Systems and Humans*, *29*(1), 41–51. doi:10.1109/3468.736359

Xinhua, L., Youhui, L., & Qi, L. (2011). Virtual Assembly and Simulation of Vibration Sieve Based on a Human-Interface Environment. *Procedia Engineering*, *15*(0), 2988–2992. doi:10.1016/j.proeng.2011.08.562

Yatziv, L., Bartesaghi, A., & Sapiro, G. (2005). A fast O(n) implementation of the Fast Marching algorithm. *Journal of Computational Physics*, *212*(2), 393–399. doi:10.1016/j.jcp.2005.08.005

Yen, D. C., Chou, D. C., & Chang, J. (2001). A synergic analysis for web-based enterprise resources planning systems. *Computer Standards & Interfaces*, *24*(4), 337–346. doi:10.1016/S0920-5489(01)00105-2

Yingying, S. (2014), Sci-Fi chef. *China Daily Europe*. Retrieved from http://europe.chinadaily.com.cn/life/2011-03/13/content_12168517.htm

Yin, R. (2005). *Case Study Research: Design and Methods*. London: Sage.

Yoshida, K., Kurazame, R., & Umentani, Y. (1991). Dual-arm coordination in space free-flying robot.*Proceeding of the IEEE International Conference on Robotics and Automation*, 3, 2516-2521. doi:10.1109/ROBOT.1991.132004

Yoshikawa, T., & Zheng, X. Z. (1990). Coordinated dynamic hybrid position/force control for multiple robot manipulators handling one constrained object.*Proceedings of 1990 IEEE International Conference on Robotics and Automation*, 2, 1178-1183. doi:10.1109/ROBOT.1990.126156

Yun, X., & Kumar, V. (1991). An approach to simultaneous control of trajectory and interaction forces in dual-arm configuration. *IEEE Transactions on Robotics and Automation*, *5*(5), 618–625. doi:10.1109/70.97873

Zeroudi, N., Fontaine, M., & Necib, K. (2012). Prediction of cutting forces in 3-axes milling of sculptured surfaces directly from CAM tool path. *Journal of Intelligent Manufacturing*, *23*(5), 1573–1587. doi:10.1007/s10845-010-0460-x

Zhang, X., & Xiong, K. (2012). A Conceptual Model of User Adoption of Mobile Advertising, *Proceedings of IEEE Computer Science and Electronics Engineering2012 International Conference*, (pp.124-127). doi:10.1109/ICCSEE.2012.454

Zhang, X., Cheung, W. K., Luo, Z., & Tong, F. (2010). A Game Theoretic Approach for Sensitive Information Sharing in Supply Chain. *International Journal of Applied Logistics*, *1*(4), 1–12. doi:10.4018/jal.2010100101

Ziaee, M., Fathian, M., & Sadjadi, S. J. (2006). A modular approach to ERP system selection A case study. *Information Management & Computer Security*, *14*(5), 485–495. doi:10.1108/09685220610717772

About the Contributors

Zongwei Luo is an associate prof1essor in South University of Science and Technology of China. Previously he was a senior researcher at the E-business Technology Institute, The University of Hong Kong (China). He is the founding Editor-in-Chief of the International Journal of Applied Logistics and serves as an associate editor and/or editorial advisory board member in many international journals. Dr. Luo's recent interests include applied research and development in the area of big data and network computing; service science and computing; innovation management and sustainable development; technology adoption and risk management; and e-business model and practices, especially for logistics and supply chain management.

***

Mohamed-Amine Abidi received his engineering degree in computer science in September, 2012 from the University of Manouba, Tunisia. He is currently pursuing a Ph.D. degree in the virtual reality engineering (virtual reality and simulation) at the National Engineering School of Saint-Etienne (ENISE), France. From 2012 to 2015, he works as a lecturer at ENISE. He is the author of two papers and co-author of one book.

Marco Antonio Arteaga-Pérez was born in Huejutla de Reyes, Hidalgo, Mexico, on September 9, 1967. He received the B.S. degree in Computer Engineering and the M.S. degree in Electrical Engineering from the Universidad Nacional Autónoma de México (UNAM), in 1991, and 1993, respectively. In 1998, he received the Dr. Ing. degree from the Gerhard-Mercator University, in Duisburg, Germany. Since 1998 he has been a Professor in the Department of Electrical Engineering, of the School of Engineering at the National University of Mexico. His main research interests are robotics and control.

Patrick Baert received his engineering degree in Mechanical engineering from the National Engineering School of Saint-Etienne (ENISE) in June 2004. He received his master degree with specialization in control in June 1996. He received the researcher engineer degree in 2007. He worked mainly on aspects of powder metallurgy as part of implementation of thick coatings and additive manufacturing modeling. Since 2010, he has been in charge of the development of virtual reality activities in the LTDS lab and more specifically the engineering research group, Perception. He is the co-author of about five research articles and one book.

Saikat Kumar Basu did his Masters in Botany (Specialization: Microbiology) from the University of Calcutta (India) and in Agricultural Studies from the University of Lethbridge (Canada). Currently he is finishing his doctoral studies in Biomolecular Sciences at the Department of Biological Sciences, University of Lethbridge. Saikat received the Graduate Citizenship Award (2009), the Alberta Graduate Student Award (2009), the CSA Best Article Award (2010), the SAEA Volunteer Award (2011), the AFC Celebrate Agriculture in Alberta Photo Award (2012) and the ASPB Peggy Thompson Publication Award (2014). His practice areas include Bioengineering and Cell Biology. Saikat enjoys traveling, soccer, cricket and photography in his leisure time.

Redha Benhadj-Djilali is a Chartered Engineer, Member of the Institution of Mechanical Engineers and Member of the British Computer Society. He specialised in CAD/CAM/CAE, engineering design and manufacturing automation and has been at Kingston University since 1994. He worked in the manufacturing industry before starting his academic career at Kingston University where he first held a research fellowship, then a position as a lecturer. His research project was based on the development of a pneumatic array of air jet proximity to tactile sensing device for manufacturing parts identification. He also held a research fellowship for more than two years, during that time he was involved in promoting research in manufacturing automation and assembly. Dr Redha Benhadj-Djilali also carried out further research in the application of the tactile sensing device to monitor slip and gripping force to assist robot part handling and assembly. He is on the editorial Boards of three international Journal, Sensor Review, Industrial Robots and Assembly Automation and review other international journals/conferences.

Andrea Bisson was born in 1989 in Noventa Vicentina, Vicenza, Italy. He obtained his Bachelor's and Master's degrees in Computer Engineering in 2012 and 2014, respectively, from the University of Padua. He has developed his master thesis work at the Autonomous Intelligent System Laboratory under the supervision of Prof. Emanuele Menegatti. The manuscript is entitled: "Development of an Interface for Intuitive Teleoperation of Comau Manipulator Robots using RGB-D Sensors". He has contributed as co-author to a number of scientific papers regarding robotics and teleoperation. Now, he is currently working as Software Engineer at Comau S.p.A - Robotics since November 2014.

Genci Capi received the B.E. degree in mechanical engineering from the Polytechnic University of Tirana, Tirana, Albania, in 1993, and the PhD degree in information systems engineering from Yamagata University, Yamagata, Japan, in 2002. From 2002 to 2004, he was a Researcher in the Department of Computational Neurobiology, Advanced Telecommunications Research Institute International (ATR), Kyoto, Japan. In 2004, he joined the Department of System Management, Fukuoka Institute of Technology, as an Assistant Professor, and in 2006, he was promoted to Associate Professor. He is currently a Professor in the Department of Electrical and Electronic Systems Engineering, University of Toyama, Toyama, Japan. His current research interests include intelligent robots, multi-robot systems, humanoid robots, learning, and evolution.

Pierre Chevaillier is a researcher in computer science at the Lab-STICC, UMR CNRS 6285. He is also a professor in computer science at the Department of Software Engineering at ENIB (Ecole Nationale d'ingénieurs de Brest). Since 2008 to 2011, he was the director of the CERV (European Center for Virtual Reality). Since 2014, he is the director of research at ENIB. He is the author or co-author of many research articles and some books.

Sumukh Deshpande completed his Bachelor in Bioinformatics from VIT, Vellore, India in 2008 and Masters of research in Bioinformatics from The University of Glasgow, UK in 2009. Thereafter he worked as a Research Assistant in University of Calgary, Canada. Thereafter he joined Jaivik Data Consulting as a Bioinformatics Research Associate. He also worked as Bioinformatics Scientist in Premier Biosoft. Currently he is working as Lecturer and Bioinformatician in Molecular Diagnostics and the Personalised Therapeutic Unit in the University of Hail. His areas of research interests include Next-generation sequencing data analysis and statistical analysis of whole genome and exome data using bioinformatics tools and databases.

Yudi Fernando is a Senior Lecturer in Operations Management at the Graduate School of Business, Universiti Sains Malaysia (USM). He received his MBA and PhD from the School of Management at USM also holds a Bachelor's degree in Business and Management from UNAND, Indonesia. Dr. Yudi also worked for several years in the electronics industry. Thus far, he has published in several journals such International Journal of Services and Operations Management, International Journal of Applied Logistics, Food Control, International Journal of Information Management, International Journal Logistics Systems and Management, International Journal of Operations and Supply Chain Management, International Journal of Productivity and Quality Management, International Journal of Value Chain Management, Benchmarking: An International Journal, and so on.

Valentina Ferrara born in 1983 in Turin (Italy) studied Electronic and Computer Science Engineer at Politecnico di Torino and Université de Technologie de Compiègne obtaining a Double-Degree in 2011 with a Final Work made at Commissariat à l'énergie atomique et aux énergies alternatives de Fontenay-Aux-Roses (France) on "Development of a Real-Time Application for the collection and the localization of radiation data for the remediation of contaminated soils". After several years of experience in software and firmware development in automotive field, Ferrara obtained a Second level Master's Degree in Industrial Automation from the Politecnico di Torino in 2013. Ferrara has been working as software developer for Comau Robotics since 2012.

Nikolaos A. Fountas received his Mechanical Engineering Diploma from1 the School of Pedagogical & Technological Education (ASPETE) of Athens, Greece in 2004 and his MSc degree from Kingston University in 2008. Since 2004, he has worked as a design and manufacturing engineer for several industries, mainly in the fields of CNC Programming, Machine Tools / Manufacturing and Product Design. Currently, he is a Ph.D. student at Kingston University-UK (engineering faculty) and he works as Laboratory Assistant under contract, teaching CNC Machine Tools - CNC Programming, Materials Processing Technology and CAD/CAM/CAE Systems in the Department of Mechanical Engineering Technology Educators, School of Pedagogical & Technological Education (ASPETE), Athens, Greece. His main research interests fall into the areas of CNC Programming, CAD/CAM/CAE Systems and Software Automation, Sculptured Surface Machining and Optimization through Artificial Intelligence / Soft Computing Techniques.

Santiago Garrido received the B.S. degree in mathematics, a B.S. degree in physics science, and a Master's degree in computer science, robotics and automation, and a Ph.D. degree from the Universidad Carlos III de Madrid (UC3M), Madrid, in 2000. In 1997, he joined the Department of Systems Engineering and Automation, UC3M, where he works as Associated Professor. He has been involved in several mobile robotics projects. His research interests include mobile robotics, mobile manipulators, environment modeling, sensor based robot motion, path planning and mobile robot global localization problems.

Barbara Lyonnet is an associate professor at the University of Nantes in LEMNA Laboratory, France. She has a Ph.D. thesis on industrial systems. Her research interests focuses on lean manufacturing, production processes and logistics. She has published in the field of engineering applications, risks, critical resources and Lean manufacturing. She is the author or co-author of many research articles and some books.

Anas Mathath is a student at the Graduate School of Business, Universiti Sains, Malaysia (USM), pursuing a Master's of Business Administration (International Business). He holds a Bachelor's in Commerce (Information Technology) from Bharathiar University, India.

Emanuele Menegatti is Associate of University of Padua. His research interests are in the field of Robot Vision. In particular, he has been working on omnidirectional vision, on distributed vision systems, on 2D and 3D industrial robot vision, and on RGB-D vision algorithms for mobile robots. Currently, Menegatti is teaching "3D Data Processing" and "Computer Architecture" and "Educational Robotics for teachers". Menegatti was Coordinator of the European project Thermobot and Local investigator for the European Projects 3DComplete and FibreMap, and Focus. He has been Principal Investigator of several other research grants. In 2005, Menegatti was a founding partners of two spin-off of the University of Padua: IT+Robotics srl and EXiMotion srl. Menegatti published more than 30 journal papers, over 120 conference papers, 2 books, 4 book chapters, and two industrial patents.

Stefano Michieletto is a post-doc fellow at the Intelligent Autonomous Systems Laboratory (IAS-Lab) of the University of Padua, Italy. He received his Ph.D. in Science and Technology Information in 2014 from the Department of Information Engineering (DEI) of the University of Padua under the supervision of Prof. Emanuele Menegatti. He obtained his Bachelor and Master degrees in Computer Engineering in 2005 and 2008, respectively, from the University of Padua. He won a research grant regarding the study of probabilistic models for convey physiological signals to robot motions. His research interests are also in the field of machine learning, robot programming by demonstration, manipulator robots, and human-robot interaction using RGB-D sensors.

Zulkifli Mohamed received a Diploma and a Bachelors in Engineering degree from Universiti Teknologi MARA, Malaysia, in 2000 and 2003, Master in Engineering from Universiti Teknologi Malaysia, in 2006, and PhD in Engineering from University of Toyama Japan. He has been working as Maintenance Engineer for 2 years in Ameraal Beltech Sdn Bhd and currently, he worked as a Senior Lecturer in Faculty of Mechanical Engineering, Universiti Teknologi MARA, Malaysia His research interests include robot development, vision system, mechatronic design, mobile humanoid robots and intelligent robot. He is a graduate member of IEEE, IEEE-RAS, The Institution of Engineers Malaysia and Board of Engineers Malaysia.

Luis Moreno received the B.S. degree in automation and electronics engineering and the Ph.D. degree from the Universidad Politécnica de Madrid, Spain, in 1984 and 1988, respectively. From 1988 to 1994, he was an Associate Professor with the Universidad Politécnica de Madrid. In 1994, he joined the Department of Systems Engineering and Automation, Universidad Carlos III de Madrid (UC3M), Madrid, where he is Full Pofessor. He has been involved in many mobile robotics projects. His research interests include mobile robotics, mobile manipulators, environment modeling, path planning, and mobile robot global localization problems.

Pooja Purohit completed her Bachelors in Bioinformatics from Amity University, Noida, India in 2010 and Masters of Technology in Biotechnology from RGPV, Bhopal, India in 2013. After completing her B. Tech she worked as Research Assistant in Maulana Azad National Institute of Technology (MANIT) from 2010 to 2011. After her postgraduation, she also worked as Lecturer in Jawaharlal Nehru P.G. College, Bhopal. Currently she is working as Lecturer in Almas International, Hail. Her areas of interest are in protein structure analysis and docking.

Zhuang Qixing is a graduate of the University of Hong Kong in the Department of Computer Science. He has a B.Eng. in Software Engineering, from Beijing Normal University, Zhuhai, 2011. He has written total lines of code exceeding 70,000. While in Beijing Normal University, Zhuhai, he received numerous awards, including National Scholarship (0.1%, 2010), the Top Ten Outstanding Graduates of University (0.2%, 2011), the Special Grade Scholarship of University (2010), the Outstanding League Cadre of University (2010), and the First Prize in ACM-ICPC of University (2010). Mr. Zhuang has diversified working and project experiences. He was the Client Development Engineer of MMORPG (intern) in LineKong Entertainment Technology Co., Ltd. (Zhuhai), from 2010.11-2011.2. He was responsible for Network Game Development - Tank War, 2010.7-2010.10, where he was focused on design and implementation of class diagram and used the TCP/IP and IO completion Port with C++, Winsock and Lua configuration. He worked on Game Software Development - Capture the Goldfish, 2009.8-2009.10, where he designed class diagram and reconstructed the last game. He added the communication between computer and control device. While working on Game Software Development - Whack-a-mole, 2009.7-2009.8, he designed class diagram and implemented it with C++, DirectX9 and XML configuration.

Fabrizio Romanelli earned his Master Degree with full marks in 2005 as an Information-Automation Engineer, from the University of Rome – Tor Vergata. He joined Comau in 2007 as software engineer, working in the software motion-control research and development group of the Robotics Business Unit. He has experience in robot-machine interaction, obstacle avoidance, space management and robot-machine haptic interfaces, in the context of SMErobot, SMErobotics, Robofoot, X-Act and other European research projects. His experience in the system software, robotics and automation has been developed in the context of several internal projects. His technical experience ranges from real-time operating systems (VxWorks, Linux RT) on embedded systems to industrial robotics (motion and trajectory planning, external sensors integration, open control systems).

Arsalan Safari is an Assistant Professor at Qatar University, and a Research Affiliate at Massachusetts Institute of Technology. His research interests include worldwide innovation, patenting, entrepreneurship and operations management. He has completed his postdoctoral program at MIT, Sociotechnical Systems Research Center in the USA where he focused on innovation, enterprise architecture and transformation. He received his Ph.D. in management sciences from the University of Waterloo, Canada and his master of engineering in design and manufacturing at the University of Western Ontario (Western University), Canada. He holds a bachelor of science in industrial engineering, and he is a Certified Six Sigma Black Belt from American Society for Quality (ASQ). Dr. Safari worked for nearly six years in the banking industry in senior levels in Canada.

Pablo Sanchez-Sanchez was born in Puebla, Mexico, on January 2, 1976. He received the B. S. degree in Electronic Engineering from Technological Institute of Puebla in 1999 and the M. S. degree in Automation and Control from Autonomous University of Puebla in 2005. He is currently pursuing the Dr. degree in Electrical Engineering at National University of Mexico. Since 2006, he is a Professor at the Faculty of Electronics at the Autonomous University of Puebla. His main research interests are robotics, control and design.

Vijay Sangam developed professional expertise in the areas of creating and implementing lean and agile Supply Chain in manufacturing and trading environments. In order to align Supply Chain objectives with Organisational Goals, Vijay developed skills in creating a Core Competency Strategy to identify the core functions to saving Costs and enhancing Productivity and at the same time researching the outsourcing of non-core functions from Victoria University. In the globalised economy, maximization of Business Growth (market share), EVA and the maximization of Operating Profit after Capital Charge are key priorities. Vijay acquired the unique skills of the mapping of business processes to develop a gap analysis and to find solutions to maximize sustainable benefits in the areas of Cost and Velocity. Further, he also worked as a Lecturer at Massey University, New Zealand, teaching Logistics and Supply Chain Management to the post graduate students and supervising research at PG level. Vijay's expertise lies in identifying opportunities to save costs; optimize resources; reduce lead time to market (B2C & B2B); align demand with supply and reengineering logistics and supply chain business processes. Vijay worked in IT, Electronics, 3PL, Consulting, Teaching, Retail, FMCG, Automotive, and Resources Industries.

Himanshu Shee is currently a Senior Lecturer specialising in supply chain & logistics management and SAP-APO module. He has had a successful teaching experience across Australia, China, Singapore, Kuala Lumpur, Fiji and India. Dr Shee has a good track record of research, with publications and higher degree research supervision with publication in many peer-reviewed journals. He has presented papers in POMS, EUROMA, Academy of Management, ANZAM-OM and IS-DSI India. He authored a book on Firm Competitiveness: Case of software firms. He supervised many Ph. D. & DBA research scholars in the supply chain area. With expertise in SPSS and SEM, his broad area of supervision includes: sustainable supply chains; the S&OP process; logistics outsourcing; the buyer-supplier relationship; and Australian automotive supply chain management. Dr Shee has been appointed as an Associate editor and reviewer for the International Journal of Global Business and Competitiveness (IJGBC), and as a reviewer for supply chain management, an international journal, journal of business review and conference papers. Recently he was nominated to the International Advisory Board as a member in ISDSI-India for 2015. He has chaired many technical sessions at ISDSI-India and IIM-Bangalore conferences.

Constantinos Stergiou received his Mechanical Engineering degree from the National Technical University Athens and his Ph.D. from the Technische Universität Darmstadt. He is a Professor at Piraeus University of Applied Sciences and is the Head of the Mechanical Engineering Department. Since May 2013, he has been appointed as an Honorary Professor at Kingston University London - Faculty of Science, Engineering & Computing. In 2002, he organized and is the Academic Director of the MSc in Advanced Industrial and Manufacturing Systems, a collaboration with the Faculty of Science, Engineering & Computing of Kingston University London. He has professional experience at technical bureaus and research projects. He lectures in Engineering Design and Computer Aided Design, has written 8 books, most of them are the official Handbooks given out to engineering students at various Universities and TEI across Greece. He has organized training programs for manufacturing plants in countries both in and outside Europe. He has over 65 published articles in international Journals and Conferences, is supervising 4 Ph.D. students and has supervised 65 MSc theses, 25 BSc theses and 50 industrial experience semester students.

Jiang Tao is a graduate from the University of Hong Kong. He received his Bachelor of Science in Opto-Electronic Information Science and Technology from Harbin Institute of Technology (HIT). He is familiar with the Electric Circuit, the Integrated Electronic Technique, the Basic Electronic Technique, the Technology of Computer Organization, the Principle and Application of Single-Chip Computer, the Digital Signal Processing (Bilingual Teaching), the Principle of Automatic Control, Applied Optics, Physical Optics, Information Optics, Optoelectronic Instrument Design, Design and Equipment of Optoelectronic Mechatronic, Optical Fiber Communication Technology and Digital Image Processing. He has worked on various assignments of railway-related technologies, including installation and debugging, organization and analysis of the Safety Information Transmission System in the Baoji-Chengdu Line Post-Disaster Reconstruction Program. After completion of the project, he moved to the research and development department, and is responsible for programing Test Software using C language, serial debugging and CAN bus data transfer.

Rosario Toscano received his engineering degree from the Conservatoire National des Arts et Metiers (CNAM) in March 1994. He received his master degree with specialization in control in June 1996. He received the Ph.D. from the Ecole Centrale de Lyon in November 2000. He received the HDR degree from University of Jean Monnet of Saint-Etienne in December 2007. He is currently a professor at the Ecole Nationale d'Ingenieurs de Saint-Etienne (ENISE). He is the author or co-author of about sixty research articles and five books.

Nikolaos M. Vaxevanidis is Professor of Manufacturing Technology in the Department of Mechanical Engineering Technology Educators, School of Pedagogical & Technological Education (ASPETE), Athens, Greece and Director of the Laboratory of Manufacturing Processes & Machine Tools (LMProMaT) in the aforementioned Department. He received Diploma (1985) and Ph.D. (1996) in Mechanical Engineering from National Technical University of Athens, Greece. He served as Lecturer in the Hellenic Air Force Academy (2000-2006) and he was also assoc. Professor under contract (2003-2010) in the Department of Mechanical Engineering/University of Thessaly, Greece. He is a member of ASME, ASM and ASQ. His scientific interests include manufacturing technology, surface engineering, tribology and quality management. He has published more than 100 papers on these topics in International Journals and International Conference Proceedings. He is member of the Editorial Board of the Int. J. of Machining and Machinability of Materials and the Int. J. of Manufacturing, Materials, and Mechanical Engineering.

Liu Yang is a graduate of the University of Hong Kong (HKU) in the Department of Computer Science. In HKU, he participated in a project of a cloud based service platform for media integration and ads distribution applying gamification in a mobile device. The platform supports flexible interactions and collaboration among different participants to distribute media and ads, applying gamification largely to enhance user experience and engagement. With the assistances of Cloud, Augmented Reality and Layering, this platform can save computation and storage of mobile device as well as game resource construction. His thesis gave examples to demonstrate the practical value of this service platform (i.e., digital city touring). It can also be extended to other areas, such as e-learning. The future work of Mr. Liu is to investigate the areas where he can excellently apply this service model to achieve both system and business requirements such as income revenue.

Feng Yi is a graduate of the University of Hong Kong in the Department of Computer Science. He received his Bachelors of Science in Applied Computer Science and Computer Science from Northern Arizona University and Lanzhou University of Technology. He is proficient in Microsoft Office (Word, Excel, Access), has experience of User Interface design of Web-based applications, and is familiar with Database Management and design. He possesses an exceptional understanding of application architecture design and understands techniques of working communication. Mr. Feng is proficient in communication with clients. He received awards of Outstanding Student Leader from 2004 to 2007—5 times.

Index

K

L

M

N

O

P

R

S

T

V

W

www.ingramcontent.com/pod-product-compliance
Lightning Source LLC
LaVergne TN
LVHW060418300926
849162LV00006B/1461